Smart Water Technology for Sustainable Management in Modern Cities

Jorge A. Ruiz-Vanoye
Universidad Politécnica de Pachuca, Mexico

Ocotlán Díaz-Parra
Universidad Politécnica de Pachuca, Mexico

IGI Global
Scientific Publishing
Publishing Tomorrow's Research Today

Vice President of Editorial	Melissa Wagner
Managing Editor of Acquisitions	Mikaela Felty
Managing Editor of Book Development	Jocelynn Hessler
Production Manager	Mike Brehm
Cover Design	Phillip Shickler

Published in the United States of America by
IGI Global Scientific Publishing
701 East Chocolate Avenue
Hershey, PA, 17033, USA
Tel: 717-533-8845
Fax: 717-533-8661
Website: https://www.igi-global.com E-mail: cust@igi-global.com

Library of Congress Cataloging-in-Publication Data

Names: Ruiz Vanoye, Jorge, 1975- editor. | Diaz-Parra, Ocotlan, 1969- editor.
Title: Smart water technology for sustainable management in modern cities / edited by Jorge Ruiz-Vanoye, Ocotlan Diaz-Parra.
Description: Hershey, PA : IGI Global Scientific Publishing, [2025] | Includes bibliographical references and index. | Summary: "This book delves into the dynamic intersection of Artificial Intelligence (AI) and Smart Water technologies within the framework of smart cities"-- Provided by publisher.
Identifiers: LCCN 2024053573 (print) | LCCN 2024053574 (ebook) | ISBN 9798369380741 (hardcover) | ISBN 9798369380758 (paperback) | ISBN 9798369380765 (ebook)
Subjects: LCSH: Municipal water supply. | Water quality management. | Water-supply engineering. | Artificial intelligence. | Green technology.
Classification: LCC TD345 .S63 2025 (print) | LCC TD345 (ebook) | DDC 307.1/416--dc23/eng/20250213
LC record available at https://lccn.loc.gov/2024053573
LC ebook record available at https://lccn.loc.gov/2024053574

British Cataloguing in Publication Data
A Cataloguing in Publication record for this book is available from the British Library.

Table of Contents

Detailed Table of Contents

seek to reduce water scarcity, optimize efficiency and sustainability in water use, improve the quality of water for human consumption, maintain real-time monitoring and management, among other aspects relevant to everyday life. This chapter aims to identify opportunities and challenges in water management through the ethical and appropriate use of Artificial Intelligence to achieve SDG 6 and propose viable alternative solutions.

Ashay Devidas Shende, K.D.K. College of Engineering, India
Shrikant A. Tekade, Government College of Engineering, Nagpur, India
Arpan Arunrao Deshmukh, G.H. Raisoni College of Engineering and Management, India
Sandeep Prabhudas Tembhurkar, SSVPS Bapusaheb Shivajirao Deore College of Engineering, Dhule, India
P. Selvakumar, Department of Science and Humanities, Nehru Institute of Technology, India

Water quality monitoring is a crucial process that involves the assessment and analysis of water characteristics to determine its suitability for various uses, from drinking and agriculture to recreation and aquatic life support. This monitoring plays a pivotal role in environmental stewardship and public health protection by identifying contaminants, tracking pollution sources, and guiding regulatory decisions. However, the practice is fraught with challenges stemming from technological limitations, resource constraints, and the dynamic nature of water systems. This article explores the significance of water quality monitoring, the challenges faced in implementing effective monitoring programs, and the advancements and strategies employed to overcome these hurdles. In agriculture, water quality directly affects crop productivity and soil health, especially in irrigation-dependent regions. Poor water quality, characterized by high salinity or heavy metal contamination, can impair soil fertility and reduce crop yields, thereby jeopardizing food security and economic stability.

This chapter discusses the significance of real-time water quality monitoring utilizing AI sensors and algorithms, emphasizing water as a renewable yet finite resource. With freshwater representing only 2.5% of Earth's water supply, its uneven distribution among continents necessitates efficient management. The increasing global population exacerbates pressure on these limited resources, particularly through leaks that contribute significantly to water loss during transportation. Advanced monitoring systems employing various sensor technologies—such as differential pressure, ultrasonic, and electromagnetic sensors—facilitate continuous tracking of essential water quality parameters. Algorithms, particularly neural networks and decision trees, enhance predictive modeling, enabling timely detection of contaminants. These innovations not only optimize resource management but also promote sustainability, ultimately protecting public health and the environment by facilitating informed decision-making in water treatment and agricultural practices.

Clean water is essential for human health and environmental sustainability. This paper explores the intricate link between water quality, public health, and ecosystem stability, highlighting the dangers posed by contaminants such as heavy metals, pathogens, and chemicals. It examines the health risks associated with polluted water and emphasizes the urgent need for sustainable solutions to mitigate these threats. By addressing water contamination through advanced treatment technologies, policy reforms, and community-driven initiatives, this work underscores the critical role of clean water in fostering a healthier, more sustainable future for all.

This article presents a combination of tools to determine if a city could be considered Smart Water City. Smart Water cities are those that use advanced technologies and innovative solutions to improve water management and ensure the long-term sustainability of water supplies. This includes optimizing water use, reducing water losses, treating, and reusing wastewater, and conserving water resources. This paper proposes the use of the evaluation of the water sensitive city index applied in the house of quality, to identify the fulfillment of the seven goals that the water sensitive city index manages within the scheme of the house of quality and thus precisely observe the areas of opportunity for decision making.

This chapter explores the use of artificial intelligence (AI) models for forecasting water demand and managing water supply systems efficiently. It highlights the need for accurate prediction models in urban areas, where fluctuating water demand poses significant challenges. Various AI techniques, including Artificial Neural Networks (ANN), Convolutional Neural Networks (CNN), and Long Short-Term Memory (LSTM) networks, are discussed for their ability to handle complex consumption data. The methodology for building these models is also addressed, covering data collection, preparation, and model training and validation. Optimization algorithms, such as Particle Swarm Optimization (PSO) and Genetic Algorithms (GA), are emphasized to enhance accuracy and robustness. Finally, a practical example illustrates the application of these AI methods in water management, highlighting AI's potential to improve the sustainability of water supply systems and identifying areas for future research to refine these models across various urban settings.

This document explores the application of artificial intelligence (AI) in optimizing wastewater treatment processes, focusing on key areas such as chemical dosing, aeration control, and sedimentation. The integration of advanced AI techniques, including neural networks, genetic algorithms, and optimization methods like Particle Swarm Optimization (PSO) and Bayesian Optimization, has significantly enhanced the efficiency and reliability of these processes. Through the implementation of machine learning models and real-time control systems, AI has enabled more precise adjustments, leading to reduced operational costs, improved resource management, and better compliance with environmental regulations. The study also highlights the potential for AI-driven innovations to further enhance sustainability in wastewater management, presenting opportunities for future research to expand on these advancements and address emerging challenges in the sector.

Water quality deterioration has become increasingly severe recently due to various factors. Pollution, climate change, and population growth all contribute to this issue. Contaminants such as microorganisms, heavy metals, and excessive nitrogen and phosphorous disrupt water pH levels, posing significant health risks. Despite advancements in IoT for maintaining pH balance through chlorine and fluoride addition after disinfection, security threats like distributed denial of service attacks, data manipulation, and session hijacking affect water treatment plant operations. This compromises water safety, leading to health problems and reduced life expectancy. To address these challenges, we propose integrating AI and blockchain technology into water treatment plant management. Our method utilizes a standard dataset on water quality parameters such as pH and total hardness to classify water as potable or non-potable.

The increase in the urban population and climate change have driven the development of Smart Water systems, which integrate artificial intelligence to improve drinking water management. AI optimizes water distribution, monitors water quality in real-time, detects leaks, and manages demand efficiently, thus addressing current challenges in water resource management. The aim of this research is to analyse the applications of AI in drinking water management within Smart Water systems. The method used includes a literature review, case studies and an analysis of the data obtained. The results show that AI improves the management of drinking water through continuous monitoring of its quality, accurate leak detection, optimization of the distribution and efficient use of water resources, demand prediction and management, and predictive maintenance. In addition, it reduces energy consumption in water treatment and distribution. However, there are technical, economic, and regulatory challenges that need to be addressed in order to achieve effective and sustainable implementation.

This introduction explores the transformative potential of AI technologies in addressing complex challenges facing drinking water systems, while also examining the ethical considerations and technical hurdles that must be navigated for responsible and effective deployment. Drinking water is a fundamental resource essential for human health, economic prosperity, and ecosystem integrity. However, managing water quality and distribution systems presents significant challenges, exacerbated by population growth, urbanization, climate change impacts, aging infrastructure, and emerging contaminants. Traditional methods of water quality monitoring and management rely on periodic sampling, laboratory analysis, and manual intervention, which are often time-consuming, resource-intensive, and may not provide real-time insights needed to prevent waterborne diseases or respond swiftly to contamination

Preface

The management of water resources has become a defining challenge of the 21st century. With increasing pressures from urbanisation, climate change, and population growth, traditional water management strategies are struggling to meet the complex demands of modern societies. This book, Smart Water Technology for Sustainable Management in Modern Cities, provides a comprehensive exploration of how cutting-edge technologies, particularly Artificial Intelligence (AI) and Internet of Things (IoT)-enabled systems, are revolutionising water management practices to create resilient and sustainable cities.

At the heart of this book lies the concept of Smart Water Systems, which integrate advanced sensors, real-time data analytics, AI-driven predictive models, and sustainable design principles to optimise water use and management. These systems address critical issues such as water scarcity, quality monitoring, leakage detection, and demand forecasting, offering a pathway to more efficient, adaptive, and environmentally responsible water governance.

The convergence of AI, IoT, and data science has opened unprecedented opportunities to transform water management practices, enabling cities to tackle challenges such as pollution control, resource allocation, and climate resilience. This book presents these innovations in a structured manner, balancing technical depth with practical applicability.

Water scarcity is no longer confined to arid regions; it is a global crisis exacerbated by rising populations, industrialisation, and climate unpredictability. Cities, which are home to more than half of the world's population, face the dual challenge of providing equitable access to clean water while safeguarding their ecosystems. Smart Water Technology for Sustainable Management in Modern Cities positions itself as a critical resource for understanding and addressing these challenges through technology and innovation.

The urgency of this topic aligns with global sustainability agendas, particularly the United Nations' Sustainable Development Goals (SDG), such as SDG 6, which aims to ensure availability and sustainable management of water and sanitation for all. This book is not only timely but essential in equipping policymakers, urban planners, engineers, and researchers with the tools and insights needed to build water-resilient cities.

This book is designed for a diverse audience, including:

- Academics and researchers exploring advancements in water management and smart technologies.
- Urban planners and policymakers seeking sustainable solutions for urban water challenges.
- Environmental and civil engineers focusing on the implementation of AI and IoT in water infrastructure.
- Graduate students and educators looking for comprehensive educational resources.
- Technology developers aiming to innovate in the field of water management systems.

By bridging theory and practice, the book provides actionable insights that cater to both academic and professional audiences.

CHAPTER OVERVIEW

Chapter 1: Foundations of Smart Water and Artificial Intelligence Technologies. This chapter introduces the foundational concepts of Smart Water technologies and Artificial Intelligence (AI), offering a comprehensive framework for understanding their pivotal role in addressing global water challenges. It details the components, benefits, and barriers to adoption, providing a roadmap for the subsequent chapters of the book.

Chapter 2: Challenges and Opportunities in Implementing AI for Smart Water. This chapter delves into the ethical and practical challenges associated with the integration of AI in water management, with a particular focus on achieving Sustainable Development Goal (SDG) 6. It identifies avenues for innovation and proposes strategies to overcome existing barriers.

Chapter 3: Water Quality Monitoring and Control with AI. This chapter examines the critical importance of water quality monitoring in protecting public health and ensuring environmental sustainability. It explores the transformative potential of AI in tackling pressing issues such as pollution, health risks, and agricultural productivity. By leveraging advanced analytical tools and predictive models, this chapter highlights how AI-driven approaches enhance the precision, efficiency, and reliability of water quality monitoring systems.

Chapter 4: Real-Time Water Quality Monitoring. This chapter emphasises the significance of real-time water quality monitoring facilitated by AI algorithms. It discusses the application of these technologies in detecting contaminants, minimising water loss, and optimising resource allocation to support sustainable water management.

Chapter 5: The Critical Role of Water Quality. This chapter explores the intricate relationship between water quality, public health, and environmental sustainability. It underscores the necessity of advanced treatment technologies and regulatory frameworks to address water contamination and protect ecological and human health.

Chapter 6: Applying the Water Sensitive City Index. Using the Water Sensitive City Index within a House of Quality framework, this chapter evaluates strategies for urban water governance. It identifies opportunities for improvement in Smart Water city initiatives and highlights areas requiring focused intervention to enhance sustainability.

Chapter 7: Water Demand Forecast and Efficient Supply Management. This chapter introduces AI-driven forecasting models and optimisation algorithms as indispensable tools for effectively managing urban water supply systems. It showcases their potential to provide precise demand predictions and improve supply chain efficiency.

Chapter 8: Optimising Water Use Through Smart Models. This chapter discusses the integration of advanced AI models, Internet of Things (IoT) technologies, and ensemble learning techniques to optimise water distribution, detect leaks, and predict water quality. These innovations enhance operational efficiency and sustainability.

Chapter 9: Precision Agriculture and AI-driven Resource Optimisation. Focusing on the agricultural sector, this chapter examines how precision irrigation and AI technologies contribute to water conservation, resource efficiency, and environmental sustainability. It highlights the economic and ecological benefits of adopting AI-driven practices in agriculture.

Chapter 10: Innovations and Efficiency in Wastewater Treatment. This chapter evaluates the application of AI in optimising wastewater treatment processes. It discusses advancements in chemical dosing, aeration, and sedimentation control, demonstrating how AI technologies enhance operational efficiency and ensure compliance with environmental standards.

Chapter 11: An Optimal Method of Assessing Water Quality. This chapter explores the integration of AI and blockchain technology to improve water quality monitoring and security. It addresses critical challenges in water treatment plant operations, offering innovative solutions for enhancing safety and reliability.

Chapter 12: Artificial Intelligence and Applications in Drinking Water Management. This chapter evaluates the role of AI in improving the management of drinking water systems. It highlights AI's contributions to real-time quality monitoring, leak detection, distribution optimisation, and predictive maintenance, ensuring resource sustainability and operational efficiency.

Chapter 13: AI Applications in Drinking Water Management. Building on the themes of Chapter 12, this chapter explores the ethical, technical, and practical considerations of deploying AI in drinking water systems. It provides actionable recommendations for achieving effective and sustainable implementation.

Chapter 14: Integrating AI in Green and Blue Infrastructure. The final chapter envisions a future where AI is seamlessly integrated into urban infrastructure to support sustainable development. It emphasises the importance of balancing ecological conservation with urbanisation and highlights the transformative potential of AI in creating sustainable smart cities.

Smart Water Technology for Sustainable Management in Modern Cities is a significant contribution to the evolving field of water management. By bridging technological innovation with practical implementation, it empowers cities to address water-related challenges effectively. This book not only highlights the transformative potential of AI and Smart Water technologies but also underscores their necessity in achieving sustainable development goals. Together, these insights pave the way for a future where water is managed as a resilient and sustainable resource, ensuring the health of ecosystems and communities alike.

Jorge A. Ruiz-Vanoye
Universidad Politécnica de Pachuca, Mexico

Ocotlán Díaz-Parra
Universidad Politécnica de Pachuca, Mexico

Chapter 1
Foundations of Smart Water and Artificial Intelligence Technologies

Jorge A. Ruiz-Vanoye
https://orcid.org/0000-0003-4928-5716
Universidad Politécnica de Pachuca, Mexico

Ocotlán Diaz-Parra
Universidad Politécnica de Pachuca, Mexico

Francisco Marroquín-Gutiérrez
Universidad Politécnica de Pachuca, Mexico

Julio C. Salgado-Ramírez
https://orcid.org/0000-0003-1666-9924
Universidad Politécnica de Pachuca, Mexico

Julio Cesar Ramos-Fernández
https://orcid.org/0000-0002-9997-6550
Universidad Politécnica de Pachuca, Mexico

Juan M. Xicotencatl-Pérez
https://orcid.org/0009-0008-8554-4192
Universidad Politécnica de Pachuca, Mexico

Luis Arturo Ortiz-Suarez
Universidad Politécnica de Pachuca, Mexico

ABSTRACT

The increasing global demand for water, compounded by the challenges posed by climate change, urbanisation, and population growth, necessitates the adoption of innovative solutions for water management. Smart Water technologies, which encompass the integration of advanced sensors, data analysis, and automated systems, offer a promising approach to optimising water use and enhancing sustainability. While challenges remain, the benefits of adopting these technologies are substantial, warranting further investment and research. As global water challenges intensify,

DOI: 10.4018/979-8-3693-8074-1.ch001

the role of Smart Water systems will become increasingly critical in ensuring the sustainable management of this vital resource. This chapter explores the components, benefits, and challenges of Smart Water systems, providing a comprehensive overview of their role in modern water management.

1. INTRODUCTION

Water management in cities faces unprecedented challenges due to scarce water resources, climate change, rapid urbanisation, and increasing demand. These interrelated factors have created an urgent need for smart solutions that optimise water use and distribution in urban environments.

The improvement in quality of life and economic development has also increased the demand for water in sectors such as agriculture, industry, and domestic use (Bertel et al., 2014). Competition for water between these sectors can lead to conflicts and tensions, making integrated water resources management more crucial than ever (Amorós et al., 2014). Smart solutions, including the use of Information and Communication Technologies (ICTs), can facilitate more efficient and equitable water management, helping to balance the needs of different users (Carvajal & Mazo, 2019; Maestre-Góngora, 2015).

The intersection of water scarcity, climate change, urbanisation, and growing water demands underscores the urgent need for Smart Water Management solutions. These solutions not only optimise the use of water resources but also promote sustainability and resilience in cities in the face of future challenges.

Smart Water Management refers to the integration of advanced technologies, particularly ICTs, into urban water systems to enhance efficiency, sustainability, and resilience. This concept has emerged as a critical response to the challenges posed by urbanisation, climate change, and water scarcity. Smart Water Systems utilise real-time data collection, analytics, and automated controls to optimise water supply, distribution, and treatment processes, thereby improving overall water management (Aivazidou et al., 2021; Lassiter & Leonard, 2022). The International Water Resources Association defines Smart Water Management as a means to achieve sustainable urban water systems through the application of innovative technologies (Aivazidou et al., 2021).

The role of smart water in urban water management is multifaceted. It encompasses the management of various water sources, including drinking water, stormwater, and wastewater, through a holistic approach that integrates both natural and engineered systems (Popartan et al., 2022). By employing smart technologies such as sensors, machine learning algorithms, and IoT devices, urban water managers can monitor water quality, predict demand, and respond to leaks or contamination events in real time (Lassiter & Leonard, 2022; Dada, 2024). This proactive management not only enhances operational efficiency but also supports climate adaptation and mitigation efforts by enabling cities to better manage their water resources in the face of changing environmental conditions (Lassiter & Leonard, 2022; Fu, 2023).

Smart Water Cities (SWCs) represent an innovative approach to urban water management, integrating advanced technologies to enhance the efficiency, sustainability, and resilience of water systems. The concept of SWCs encompasses urban municipalities that leverage ICTs and the Internet of Things (IoT) to optimise various aspects of the urban water cycle, including supply, distribution, and quality management (Dasallas, 2024; Okoli, 2024). This integration facilitates real-time monitoring and data-driven decision-making, which are essential for addressing the challenges posed by urbanisation, climate change, and population growth (Okoli, 2023; Nardo, 2021). One of the key technologies underpinning Smart Water Management is the Smart Water Metering System (SWMS), which provides accurate, real-time data on water usage. This capability not only enables better leak detection and repair but also promotes water conservation through behavioural changes among consumers (Okoli, 2023). Furthermore, the implementation of IoT solutions allows for decentralised water management, which can reduce energy consumption compared to traditional centralised systems (Aldegheishem et al., 2022). As cities increasingly adopt these technologies, they can expect improvements in operational efficiency, reduced water wastage, and enhanced service delivery (Gade, 2021).

2. SMART WATER AND ARTIFICIAL INTELLIGENCE TECHNOLOGIES

Smart Water refers to the application of advanced technologies, such as sensors, data analysis and automated systems, to manage, monitor and optimise water use more efficiently and sustainably. The Fig.1 contains the components of Smart Water:

- Continuous and real-time monitoring. Smart Sensors capture real-time data on parameters such as water quality (pH, temperature, dissolved oxygen), flow, pressure, and water levels.
- Data transmission and remote control. The Internet of Water Things (IoWT) represents a significant advancement in the management and monitoring of water resources through the integration of IoT technologies. By utilizing sensors, data analytics, and automated systems, IoWT facilitates real-time monitoring of water quality and consumption, thereby promoting sustainable water management practices.

- Prediction and optimisation. Big Data refers to the vast volumes of structured and unstructured data generated from various sources, including sensors and IoT devices. In the context of Smart Water Management, Big Data plays a crucial role in several areas: Data Collection and Integration, Enhanced Decision-Making, Predictive Analytics and Risk Management.
- Automatic control and adjustment. Systems that automatically adjust water flow, distribution or treatment based on data received in real time.
- Reduction of energy consumption. Optimisation of energy use in water systems, such as pumps and treatment plants, based on data to minimise consumption.
- Sustainable water use. Efficient use of water resources, minimising losses, leaks and guaranteeing water quality, in line with environmental sustainability objectives.
- Security and data protection. Protection of Smart Water Systems against cyber threats that can compromise the security of data and infrastructure.
- Compliance with rules and regulations. Policies and regulations play a crucial role in ensuring that Smart Water infrastructure meets environmental, safety and efficiency standards. Laws and regulations also encourage the use of smart technologies for water management.
- Training and public awareness. Education is essential for both the public and water system operators to understand the value and operation of Smart Water. This also includes training programmes for infrastructure operators.
- Positive social and environmental impact. Smart Water's impact includes not only improving water management, but also creating a more sustainable, equitable and healthy environment for communities. The effects of access to clean water and resource conservation have implications for public health, agriculture and sustainability.

Figure 1. Components of smart water

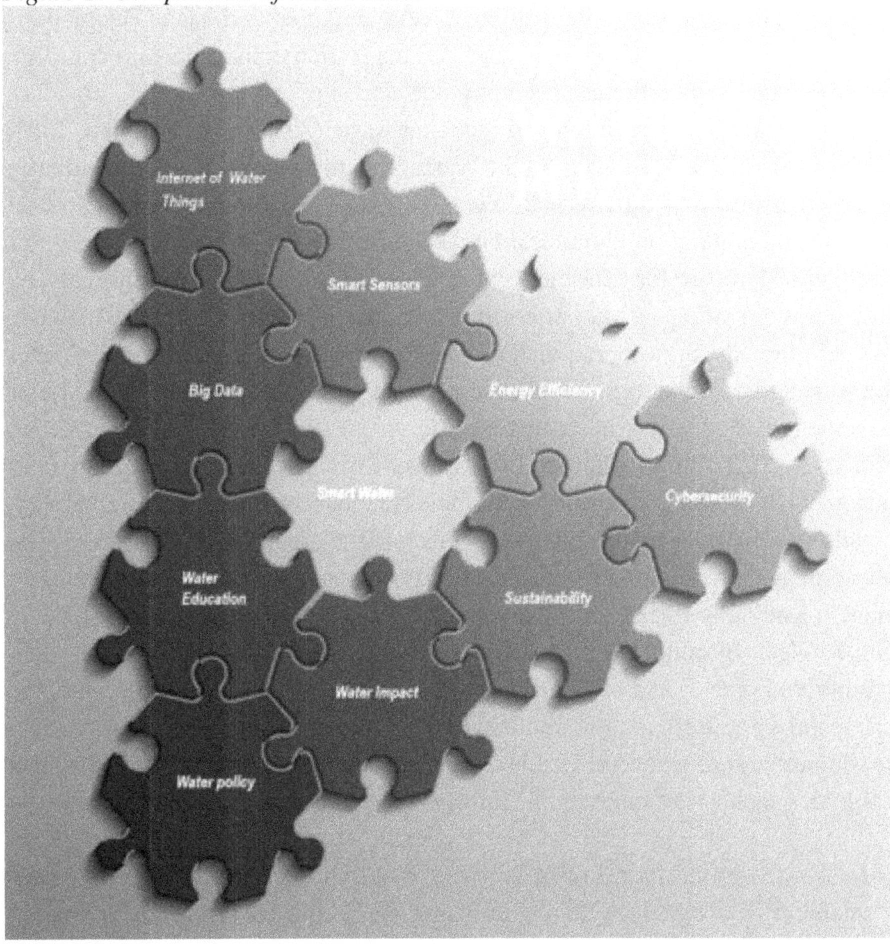

Smart Water refers to the integration of advanced technologies, particularly ICTs and the IoT, into urban water systems to enhance the efficiency, sustainability, and resilience of water management practices. This concept encompasses a range of innovative solutions designed to optimize the monitoring, distribution, and quality of water resources in urban environments. Smart Water Systems leverage real-time data collection, analytics, and automated controls to improve decision-making processes and operational efficiency, thereby addressing the challenges posed by urbanization, climate change, and water scarcity (Popartan et al., 2022; Dada, 2024). The role of smart water in urban water management is critical, as it facilitates a more integrated and holistic approach to managing the urban water cycle. This includes not only the supply and distribution of drinking water but also the management of wastewater, stormwater, and other water resources (Fu, 2023; Somayaji et al., 2022). By employing smart technologies such as smart meters, sensors, and data analytics, urban water managers can monitor water quality, detect leaks, and predict demand fluctuations in real time (Dada, 2024; Drogkoula, 2023). This proactive management approach not only enhances operational efficiency but also supports sustainable practices by promoting water conservation and reducing waste (Alshehri & Rahman, 2023; Chang & Guo, 2020). Moreover, Smart Water Systems contribute to the resilience of urban water infrastructure by enabling cities to adapt to changing environmental conditions and population dynamics. For instance, the use of predictive analytics can help cities prepare for extreme weather events, such as floods or droughts, by optimizing water storage and distribution strategies (Chang et al., 2023; Adewusi, 2024). Additionally, the integration of citizen engagement tools within Smart Water Systems fosters participatory governance, allowing residents to contribute to water management decisions and promoting a culture of sustainability (Alghadeer, 2023; Garzón et al., 2022). Smart water represents a transformative approach to urban water management, leveraging advanced technologies to enhance efficiency, sustainability, and resilience. By integrating smart solutions into the urban water cycle, cities can better address the complex challenges of water management in the 21st century, ultimately leading to improved water security and environmental sustainability.

Smart Water Systems are primarily aimed at optimising water resources, sustainability and operational efficiency in water management. These objectives are fundamental to address today's challenges of water scarcity, climate change and increasing urbanisation, and are achieved through the implementation of advanced technologies such as the Internet of Things (IoT) and artificial intelligence (AI):

- Resource optimisation in Smart Water Systems refers to the efficient management of available water, minimising waste and ensuring that it is used as effectively as possible. SWMS facilitate water distribution and control, enabling critical decisions based on real-time data (Aivazidou et al., 2021). This includes early leak detection, consumption monitoring and the implementation of dynamic tariffs that incentivise responsible water use (Lassiter & Leonard, 2022). By integrating smart metering technologies, systems can provide accurate information on water use, enabling water managers to make informed decisions and improve long-term planning (Popartan et al., 2022).
- Sustainability is a central pillar of Smart Water Systems, as they seek not only to meet current water needs, but also to preserve resources for future generations. This is achieved by implementing water management practices that consider the entire water cycle, integrating green and blue infrastructure solutions (Dada, 2024). For example, sustainable water management in urban settings can include rainwater harvesting, wastewater recycling and aquatic ecosystem restoration (Fu, 2023). In addition, Smart Water Systems promote

resilience to climate change by adapting to variations in water availability and mitigating the impacts of extreme weather events (Somayaji et al., 2022).

- Operational efficiency in Smart Water Systems refers to the improvement of water management processes through automation and data analysis. The implementation of IoT technologies enables continuous monitoring of water infrastructure, making it easier to identify problems before they become crises (Drogkoula, 2023). For example, the use of AI-based predictive models can help anticipate changes in water quality and demand, thereby optimising water distribution and treatment (Alshehri & Rahman, 2023). In addition, collaboration between different actors, such as local authorities, utilities and citizens, is essential to improve operational efficiency and foster a more integrated approach to water management (Chang & Guo, 2020).

Smart Water Systems are central to addressing contemporary challenges in water management. Through resource optimization, sustainability and operational efficiency, these systems not only improve water quality and availability, but also promote a more responsible and conscious use of this vital resource.

One of the primary advantages of incorporating AI into Smart Water Systems is enhanced efficiency in water resource management. AI technologies, particularly machine learning algorithms, can analyze vast datasets generated by sensors and IoT devices to optimize water distribution and usage. For instance, AI can predict water demand patterns, enabling utilities to allocate resources more effectively and reduce waste (Shaikh, 2024; Mohammed, 2024). Moreover, AI-driven predictive analytics can improve the management of groundwater resources by forecasting potential shortages and facilitating timely interventions (Shaikh, 2024; Shaikh, 2024). Additionally, AI can significantly enhance water quality monitoring. By employing advanced sensor networks and real-time data analytics, AI systems can detect contaminants and anomalies in water quality, ensuring compliance with health standards (Geetha & Gouthami, 2016). This capability not only safeguards public health but also aids in the rapid response to contamination events, thereby minimizing potential risks to communities (Geetha & Gouthami, 2016). Furthermore, the integration of AI in Smart Water Systems fosters sustainability. AI applications can optimize irrigation practices in agriculture, leading to more efficient water use and reduced environmental impact (Mohammed, 2024). This is particularly relevant in regions facing water scarcity, where precision irrigation technologies powered by AI can ensure that crops receive the exact amount of water needed, thus conserving this vital resource (Mohammed, 2024).

Despite the promising benefits, the integration of AI into Smart Water Systems is fraught with challenges. One significant hurdle is the ethical and privacy concerns associated with data collection and usage. The deployment of AI systems often requires extensive data on water usage patterns and quality, raising issues regarding data privacy and security. Ensuring that data is collected and utilized responsibly is crucial to gaining public trust and acceptance of AI technologies in water management. Another challenge lies in the technical complexities of implementing AI solutions. The integration of AI into existing water management systems requires significant investment in infrastructure and technology, as well as the upskilling of personnel to manage and operate these advanced systems. Many organizations face difficulties transitioning from pilot projects to full-scale implementations, often getting stuck in pilot purgatory, where initial trials do not lead to broader adoption or tangible benefits (Kuğuoğlu et al., 2021). Moreover, the variability in data quality and availability can hinder the effectiveness of AI models. Inconsistent or incomplete data can lead to inaccurate predictions and suboptimal decision-making, ultimately undermining the potential benefits of AI in Smart Water Systems (Nishant et al., 2020; Mr, 2021). Addressing these data-related challenges is essential for the successful deployment of AI technologies in this domain. While the integration of AI in Smart Water Systems offers substantial benefits in terms of efficiency, sustainability, and water quality management, it also presents significant challenges related to ethics, implementation, and data quality. A balanced approach that addresses these challenges while leveraging the advantages of AI will be crucial for the future of Smart Water Management .

3. FUNDAMENTALS OF ARTIFICIAL INTELLIGENCE IN WATER MANAGEMENT

The application of artificial intelligence (AI) in water management is revolutionising the way distribution networks are optimised, leakage predicted and monitored in real time. This approach not only improves operational efficiency, but also contributes to sustainability and cost reduction in water resource management.

Optimising water distribution networks is one of the main benefits of AI integration. AI systems can analyse large volumes of data from sensors and IoT devices to identify consumption patterns and optimise water distribution. For example, in Saudi Arabia, projects using AI have been implemented to improve water distribution management, which is crucial in a context of increasing water scarcity and population growth (Alghadeer, 2023). In addition, AI enables the implementation of predictive models that help utilities anticipate demand and adjust supply accordingly, reducing waste and improving efficiency (Alghadeer, 2023).

Leak detection and location is another field where AI has proven to be invaluable. Leaks in water distribution networks can result in significant losses of resources and high costs. Research has shown that the use of smart sensors, together with machine learning algorithms, can improve the accuracy of leak detection. For example, a recent study proposed a two-stage model that uses data analysis to predict leakage incidents and their specific locations in distribution networks (Tyagi et al., 2023). Furthermore, techniques based on neural networks and deep learning methods have been developed that have shown accuracy rates of over 90% in leak detection (Şahin & Yüce, 2023; Tsai et al., 2022). In addition, technologies such as acoustic and pressure sensors are being used to detect leaks more effectively. These technologies allow water utilities to identify potential leaks before they become major problems, which not only saves resources but also minimises environmental impact (Liang et al., 2020; Nikoloska et al., 2020). For example, the use of fibre optic sensors has proven to be effective in locating leaks in piping systems, providing an innovative solution to a persistent problem (Inalegwu, 2024).

Real-time monitoring is essential for efficient water management. AI enables the implementation of monitoring systems that provide real-time data on the quality and quantity of water in distribution networks. This is essential to ensure that the water supplied meets quality standards and to proactively detect problems. For example, systems have been developed that use wireless sensor networks to collect data on pipe condition and water quality, allowing companies to respond quickly to any anomalies (Liu et al., 2019). The ability to perform real-time analysis also facilitates informed decision-making and optimisation of daily operations (Liang et al., 2020; Liu et al., 2019).

The application of AI in water management offers significant opportunities to optimise networks, predict leakage and perform real-time monitoring. However, it is crucial to address the challenges associated with implementing these technologies, such as data quality and the need for adequate infrastructure. With a balanced approach, AI can transform water management, making it more efficient and sustainable.

The integration of machine learning (ML), neural networks, optimization algorithms, and real-time data processing is revolutionizing Smart Water Management systems. This chapter synthesizes recent research on these technologies, highlighting their applications, benefits, and challenges in enhancing water resource management. By leveraging advanced data analytics, these systems can optimize water usage, improve efficiency, and ensure sustainability.

- Machine learning is a subfield of artificial intelligence that focuses on developing algorithms that allow machines to improve their performance on a specific task by analysing data and experience, without being explicitly programmed to perform the task (Mitchell, 1997). These algorithms learn patterns from data and can make predictions or decisions based on new information. Machine learning can be applied to water management to forecast water demand, detect leaks and anomalies in water distribution networks, and optimise the use of water resources through predictive analysis of large volumes of data (Adamowski & Caruso, 2012).

- Artificial neural networks are computational models inspired by the functioning of the human brain. They are composed of layers of nodes (artificial neurons) that process information in a distributed and parallel manner. Each node is connected to other nodes through weights that are adjusted during the learning process to optimise the model's performance (McCulloch & Pitts, 1943). Neural networks are used in water management to model complex phenomena such as river flow prediction, water quality, and pollutant detection. They are particularly useful when systems cannot be effectively modelled by traditional mathematical equations (Zhang et al., 1998).
- Optimisation algorithms are techniques designed to find the best possible solution to a given mathematical problem, subject to constraints. They are used to minimise or maximise an objective function by adjusting system variables within a given search space (Kelley, 1960). In water management, optimisation algorithms can be applied to design more efficient distribution networks, reduce energy consumption in water treatment, and minimise leakage losses. They also optimise the scheduling of water use in smart irrigation systems, maximising water efficiency (Labadie, 2004).
- Real-time data processing refers to the ability of a system to process and analyse data immediately after it is generated, with minimal latencies. These systems are characterised by rapid reaction to events, allowing instantaneous decision making (Codd, 1970). In water management, real-time data processing is used to monitor distribution networks, assess water quality and manage reservoir levels or river flows. Data collected from smart sensors are analysed in real time, allowing rapid reaction to problems such as leakage or pollution (Abrahart & See, 2000).

Table 1 contains the most common artificial intelligence algorithms applied to water management.

Table 1. Artificial intelligence algorithms applied to water management.

Research	Algorithms	Contribution
Gao et al. (2020)	Deep neural networks	Accurate real-time water demand forecasting
Huang et al. (2018)	Support Vector Machines (SVM)	Detection of water leaks in urban pipelines using smart sensors
Li et al. (2017)	k-means algorithm	Classification of water quality in water bodies using sample analyses

continued on following page

Table 1. Continued

Research	Algorithms	Contribution
Mounce et al. (2016)	Genetic algorithms	Optimisation of the design of water distribution networks to minimise losses.
Wu et al. (2019)	Decision trees	Predicting pollution levels in industrial wastewater
Zhang et al. (2021)	Artificial Neural Networks (ANN)	Water flow modelling in smart irrigation systems
Riahi et al. (2020)	Multi-objective optimisation algorithms	Optimising the efficiency of wastewater treatment plants
Wang et al. (2015)	Evolutionary algorithms	Optimising water use in IoT-based agricultural systems
Li et al. (2019)	Deep learning	Predictive analysis of drought and water availability in arid regions
Zhang et al. (2018)	Convolutional Neural Networks (CNN)	Detection of pollutants in water using satellite imagery
Al-Juaidi et al. (2017)	Clustering algorithms	Real-time water use management in smart cities
Zhou et al. (2020)	Bayesian networks	Contamination risk assessment of drinking water supply networks
Sun et al. (2019)	Recurrent Neural Networks (RNN)	River flow forecasting to optimise water management
Mohammadi et al. (2020)	LSTM neural networks	Predicting water levels in reservoirs with historical and climatological data
Ahmed et al. (2018)	Reinforcement algorithms	Optimisation of valve operation in water distribution systems
Hasan et al. (2020)	Regression algorithms	Predicting water quality in rural water supply systems
Chen et al. (2016)	Anomaly detection algorithms	Identifying leaks and failures in drinking water infrastructure
Smith et al. (2019)	Artificial Neural Networks (ANN)	Real-time water quality monitoring using distributed sensors
Park et al. (2021)	Self-organising neural networks	Optimising water use in smart buildings
Liu et al. (2020)	Hybrid AI algorithms	Combining deep learning and optimisation for sustainable water management

Predictive techniques are a set of statistical methods and algorithms that allow forecasting future outcomes based on historical data and current patterns. These techniques can include regression models, machine learning, neural networks, and other approaches that identify hidden trends and relationships within data. Data analytics involves collecting, processing, and analysing data sets to extract meaningful information to support decision-making (Hastie et al., 2009).

In water management, these predictive techniques are used to predict (Table 2):

- **Usage Patterns:** Water consumption over time, considering different factors such as seasonality, weather events, and user habits.
- **Climate:** Prediction of climatic events (rainfall, droughts) that can directly influence water availability and demand.
- **Future Water Demand:** Forecasts of how much water will be needed in the future, which helps to plan resource management and avoid shortages or wastage.

Predictive techniques allow water managers to adjust distribution systems according to anticipated demand, optimising supply and reducing losses.

Table 2. Predictive techniques and data analysis for water management

Research	Algorithms/Models	Contribution
Adamowski et al. (2012)	Artificial Neural Networks (ANN)	Predicting peak water use in urban environments using different learning algorithms
Jain et al. (2001)	Neural network models	Long-term drinking water demand forecasting using neural networks
Ghiassi et al. (2008)	Dynamic neural network models	Water consumption prediction by dynamic modelling based on neural networks
Herrera et al. (2010)	Clustering algorithms and Bayesian networks	Rural water demand forecasting using clustering and probabilistic analysis
House-Peters & Chang (2011)	Multiple regression models	Assessing climatic and socio-economic factors on future water demand
Zhou et al. (2012)	Support Vector Machine (SVM) algorithms	Residential water demand forecasting using SVM and time series analysis
Bougadis et al. (2005)	Regression and time series models	Analysing the effects of climate change on water use in Canada
Makropoulos et al. (2008)	Neural networks and fuzzy logic	Short-term forecasting model for water demand in urban systems
Brentan et al. (2017)	ARIMA and hybrid models	Water demand forecasting in distribution networks using a hybrid approach
Qi et al. (2016)	Machine learning algorithms	Real-time water demand forecasting using supervised machine learning
Lobbrecht & Solomatine (2002)	Neural network algorithms	Water demand forecasting using neural networks and hydrological modelling
Jothiprakash & Sasikumar (2009)	Optimisation algorithms and regression models	Predictive modelling of water use in agricultural areas using optimisation
Herrick et al. (2015)	Logistic regression algorithms	Assessing the relationship between climate and water use in arid environments
Kisi (2013)	Feedforward neural networks	Predicting water flow in watersheds using feedforward neural networks

continued on following page

Table 2. Continued

Research	Algorithms/Models	Contribution
Bae et al. (2020)	Convolutional Neural Network Models (CNN)	Predicting the demand for drinking water in urban areas using CNN
Canteli et al. (2019)	Deep learning algorithms	Modelling residential water consumption using deep learning techniques
Gato et al. (2007)	Regression models and decision trees	Predicting water demand as a function of climatic factors in Melbourne, Australia
Hutton et al. (2014)	Real-time forecasting models and data analysis	Developing predictive systems for real-time water management using sensor data
Adamowski & Karapataki (2010)	Learning models based on neural network theory	Short-term forecasting of water demand in arid regions
Guo et al. (2015)	Non-linear regression models	Predicting water demand as a function of meteorological variables in urban areas of China

4. INTERNET OF THINGS (IOT) IN SMART WATER SYSTEMS

The Internet of Things (IoT) has emerged as a fundamental tool in water resource management, enabling the implementation of Smart Water Systems that optimise water use and monitoring. In this context, essential IoT components include sensors, actuators and communication networks, each playing a crucial role in the collection and management of water-related data.

- Sensors are devices that allow the measurement of various physical and chemical variables in water, such as temperature, pH and turbidity. For example, in a study conducted in Cartagena, a monitoring system was developed that uses sensors to continuously measure water quality, allowing for the detection of pollutants and real-time adjustments (Álvarez, 2023). In addition, the implementation of smart sensor networks has proven to be effective in monitoring critical infrastructure, such as water pipes, facilitating leak detection and optimising maintenance (Valdez et al., 2018).
- Actuators, on the other hand, are devices that execute actions based on signals received from sensors. In automated irrigation systems, for example, actuators can control valves and water pumps, allowing for more efficient irrigation tailored to specific crop needs (Sanchez et al., 2021). A study in the Tacna region showed that the implementation of an IoT system optimised water use in agriculture, achieving significant savings in irrigation (Catacora & César, 2021). This highlights how actuators, when integrated into IoT systems, can contribute to sustainability and efficiency in water use.

- Communication networks are the third essential component of the IoT, enabling data transmission between sensors, actuators and management platforms. These networks can be of different types, including WiFi, LoRa, and cellular networks, depending on the application and environment. Telemetry, for example, has been used to monitor water consumption in commercial facilities, providing real-time data that helps manage water use more effectively (Lima et al., 2021). The connectivity provided by these networks is crucial for the creation of Smart Water Management systems that can operate autonomously and efficiently.

IoT in water management is based on the integration of sensors, actuators and communication networks, enabling more effective monitoring and control of water resources. This technology not only improves water use efficiency, but also contributes to environmental sustainability and the prevention of water crises in various regions.

The integration of the Internet of Things (IoT) with Artificial Intelligence (AI) is revolutionising the way real-time data is managed and analysed, especially in the context of environmental monitoring and water resource management. This synergy enables the use of smart sensors that not only collect data, but also analyse and process it to provide valuable and timely information.

Smart sensors are devices that, when connected to IoT networks, can measure environmental parameters such as water quality, temperature and other relevant indicators. These sensors, when integrated with AI, allow for real-time analysis of the collected data. For example, water quality sensors using AI algorithms have been shown to be essential for monitoring physicochemical contamination in aquifers, facilitating more efficient management of water resources in Latin America (Franco, 2023). This translates into the ability to detect anomalies or changes in water quality immediately, allowing a rapid response to possible contamination.

AI plays a crucial role in processing large volumes of data generated by these sensors. Through machine learning techniques, AI can identify patterns and trends in the data that would be difficult to detect manually. This not only improves the accuracy of the analysis, but also enables the prediction of future events, such as the occurrence of pollutants in water (Hernández et al., 2022). Furthermore, the combination of IoT and AI facilitates the creation of predictive models that can be used for the proactive management of water resources, thus optimising water use and minimising waste (Álvarez, 2023).

The connectivity provided by IoT allows data collected by sensors to be sent to cloud platforms, where it can be analysed using AI tools. This not only improves data accessibility, but also enables collaboration between different actors in water management, from local authorities to researchers. For example, in the field of public health, the analysis of wastewater using omics tools and AI algorithms has enabled the detection of patterns of pathogenic microorganisms, which is crucial for the prevention of epidemics (Hernández et al., 2022).

The Internet of Things (IoT) has significantly transformed water management, enabling the implementation of innovative solutions for water quality monitoring, leak detection and supply optimisation. These use cases are critical to improve efficiency in water resource management and to address challenges related to water scarcity (Table 3).

Table 3. Use case

Use Case	Description	IoT Technology Used	Research
Real-time water quality monitoring	Use of distributed sensors to measure water quality parameters (pH, turbidity, conductivity) in rivers and lakes.	Cloud-connected smart water quality sensors; real-time analysis through IoT platforms.	Li et al. (2018)
Leak detection in urban networks	IoT sensors installed in pipes to detect changes in flow that indicate water leaks.	Pressure and flow sensors, wireless communication and anomaly detection algorithms.	Huang et al. (2018)
Irrigation optimisation in agricultural systems	Soil and climate monitoring by sensors to automatically adjust water supply in irrigation systems.	Soil moisture sensors, IoT weather stations, automated actuators in irrigation systems.	Wang et al. (2015)
Intelligent control of valves in water networks	Remotely controlled valves to adjust water flow according to demand or network maintenance.	Smart valves connected to the cloud, remote control via IoT applications.	Ahmed et al. (2018)
Management of reservoirs and water reserves	IoT sensors monitor water levels in reservoirs to optimise water release and prevent flooding.	Water level sensors, wide area network (LoRa) communication, and IoT-based predictive analytics.	Riahi et al. (2020)
Detection of industrial pollutants	IoT systems monitor the presence of pollutants in wastewater from factories in real time.	Chemical water quality sensors, 5G connectivity, cloud-based analysis platforms.	Zhang et al. (2018)
Water demand forecasting	Predictive analysis of water consumption patterns based on historical data collected by IoT sensors.	Water flow sensors, machine learning for predictive analytics in IoT platforms.	Qi et al. (2016)
Water management in smart cities	Consumption monitoring in smart buildings, adjusting water supply according to usage patterns.	Grid-connected smart water meters, analytical applications for supply management.	Al-Juaidi et al. (2017)
Monitoring of sewerage systems	IoT sensors monitor the flow and quality of water in sewer systems to detect blockages or overloads.	Flow and quality sensors in sewers, wireless connectivity, and automated alerts.	Wu et al. (2019)
Optimisation of water supply in rural areas	Efficient management of water supply in remote areas using IoT-based sensors and remote control systems.	Water tank level sensors, wide area networks (LPWAN), and remote management through IoT platforms.	Li et al. (2019)

5. CYBER-PHYSICAL SYSTEMS APPLIED TO WATER

Cyber-physical systems applied to water represent an innovative convergence between physical infrastructure and advanced computational capabilities. These systems integrate Information and Communication Technologies (ICT) with physical processes to improve the management of water resources, optimising both water quality and its distribution and use. In this context, water planning and management benefit from the implementation of technological tools that enable real-time monitoring and rapid response to risk situations, such as pollution or water scarcity.

Cyber-physical systems applied to water are fundamental to the integration of physical infrastructure with advanced computational capabilities. These systems not only optimise water management and quality, but also promote better public health and sustainable development in communities. The combination of information technologies, risk assessment methodologies and integrated approaches to water planning is essential to meet today's challenges in water resources management.

Interaction models between physical and virtual systems in the context of Smart Water Management are essential for real-time monitoring and control, as well as for failure prediction. The integration of technologies such as the Internet of Things (IoT) and Digital Twins enables more efficient and sustainable management of water resources, facilitating informed decision-making and process optimisation.

Digitalisation in water management has proven to be a catalyst for improving water use efficiency and reducing pollution. For example, the use of digital technologies, such as IoT sensors and big data analytics, enables the creation of a Smart Water Management layer, which supports circular economy practices and accelerates sustainable development in this sector (Liu et al., 2021).

Cyber-physical systems (CPS) in both urban and rural water networks are prominent examples of how the integration of digital and physical technologies can improve water management. These systems enable more efficient monitoring and control, as well as optimisation of water resources in various applications, including wastewater infrastructure, smart irrigation and distribution networks.

- In the field of wastewater infrastructure, SCPs have been implemented to optimise wastewater treatment and management. For example, non-linear predictive controllers have been shown to be effective in maintaining the operation of treatment plants despite failures, highlighting the importance of resilience in water infrastructure Fernandez et al. (2022).
- In the context of smart irrigation, SCPs enable the implementation of automated irrigation systems that use real-time data to optimise water use. Methodologies combining artificial neural networks and genetic algorithms have been developed to predict water demand in irrigation communities, fa-

cilitating more efficient management of water resources (Perea et al., 2016). These systems not only improve irrigation efficiency, but also contribute to agricultural sustainability by reducing water consumption and minimising environmental impact.

- Water distribution networks also benefit from the implementation of SCP. The integration of real-time monitoring technologies allows early detection of leaks and infrastructure failures, which is crucial to minimise water waste. In the city of Guayaquil, the infrastructure for industrial wastewater treatment has been assessed, leading to the identification of areas for improvement and the implementation of more efficient technologies (Chere-Quiñónez et al., 2022). Furthermore, the use of geoportals for spatial data management in drinking water systems allows for a clearer visualisation of existing infrastructure and facilitates informed decision-making (Morejón et al., 2023).

6. CONCLUSION

The implementation of Smart Water Management systems (Smart Water) that integrate artificial intelligence (AI) faces multiple challenges. These challenges can be classified into three main categories: technical, economic and social, and ethical. This article discusses each of these challenges, providing a framework for understanding the complexities involved in adopting advanced technologies in water resource management.

The integration of AI into Smart Water Systems presents several significant technical challenges:

- System Interoperability. Interoperability between different systems and platforms is crucial to the success of Smart Water initiatives. The lack of common standards can hinder the integration of diverse technologies, limiting the effectiveness of implemented solutions Llanos-Zavalaga et al. (2022). The need to establish standardised communication protocols and data formats is essential to facilitate collaboration between different actors and technologies.
- Handling Large Data Volumes. The collection and analysis of large volumes of data generated by sensors and IoT devices is essential for informed decision making. However, efficient processing of this data poses challenges in terms of storage, analysis and visualisation (Alvarez-Cano et al., 2022). The capacity of current infrastructures to handle these volumes of data is a critical factor that may limit the effective implementation of AI-based solutions.
- Cybersecurity. Cybersecurity is a growing concern in the implementation of smart technologies. The interconnection of devices and systems increas-

es vulnerability to cyber attacks, which can compromise data integrity and availability (Gozzer et al., 2021). Implementing robust security measures is essential to protect critical infrastructure and ensure public trust in these systems.

Economic and social challenges also play a crucial role in the implementation of Smart Water:

- Cost of Implementation. The initial cost of implementing Smart Water technologies can be prohibitive for many communities and local governments. Investment in infrastructure, training and maintenance can be a disincentive to adopt these technologies, especially in resource-constrained regions (Schmidt, 2022). It is essential to develop financing models that facilitate investment in sustainable technologies.
- Public Acceptance. Public acceptance of Smart Water technologies is vital to their success. Lack of understanding of the benefits of these technologies can lead to resistance among end-users. Education and awareness raising are essential to foster acceptance and use of smart systems in water management (Zapata & Tejeda, 2018).
- Regulation and Policy. Government regulation and policies play a crucial role in the implementation of Smart Water technologies. The lack of a clear regulatory framework can hinder innovation and the adoption of new technologies (Freier et al., 2020). Policies that promote public-private collaboration and research and development in the field of water management need to be put in place.

The ethical challenges associated with the implementation of Smart Water are equally important:

- Data Privacy. The collection of real-time data on water use and water quality raises concerns about data privacy. Proper management of personal information and transparency in the use of this data are essential to maintain public trust (Holguín & Dutrénit, 2021). Privacy policies should be clear and accessible to users.
- Justice in Water Resource Allocation. The implementation of smart technologies can exacerbate existing inequalities in access to water resources. It is critical to ensure that Smart Water solutions benefit all communities, especially those that have been historically marginalised (Elenes et al., 2023). Equity in the distribution of water resources must be a priority in the design and implementation of smart systems.

Future trends in Smart Water Management are marked by advances in AI and sensor technologies, the development of resilient infrastructure and a focus on integrated global water management. Addressing these aspects is essential to ensure sustainable and efficient use of water resources in a world facing growing environmental and social challenges.

The integration of artificial intelligence into Smart Water Systems represents a significant opportunity to improve water resource management. Operational benefits, sustainability and cost reduction are just some of the positive aspects that these systems can offer. As we move towards a more interconnected and sustainable future, it is crucial to address existing challenges and foster collaboration between governments, businesses and communities to ensure efficient and equitable water management.

REFERENCES

Abrahart, R. J., & See, L. M. (2000). Neural network modeling of nonlinear hydrological relationships. *Hydrology and Earth System Sciences*, 4(3), 551–560. DOI: 10.5194/hess-4-551-2000

Adamowski, J., Chan, H. F., & Prasher, S. O. (2012). Comparison of multivariate regression and artificial neural networks for peak urban water demand forecasting: Evaluation of different ANN learning algorithms. *Journal of Hydrologic Engineering*, 17(5), 498–511. DOI: 10.1061/(ASCE)HE.1943-5584.0000492

Adamowski, J., Fung Chan, H., & Prasher, S. O. (2012). Comparison of multivariate regression and artificial neural networks for peak urban water demand forecasting: Evaluation of different ANN learning algorithms. *Journal of Hydrologic Engineering*, 17(5), 498–511. DOI: 10.1061/(ASCE)HE.1943-5584.0000492

Adamowski, J., & Karapataki, C. (2010). Water demand forecasting in Mediterranean regions using socio-economic and climatic variables. *Water Resources Management*, 24(11), 2417–2432. DOI: 10.1007/s11269-009-9566-4

Adewusi, A. O., Asuzu, O. F., Olorunsogo, T., Olorunsogo, T., Adaga, E., & Daraojimba, D. O. (2024). Ai in precision agriculture: A review of technologies for sustainable farming practices. *World Journal of Advanced Research and Reviews*, (1), 2276–2285. DOI: 10.30574/wjarr.2024.21.1.0314

Ahmed, A., Salih, A., & Zhang, Z. (2018). Reinforcement learning for optimizing valve operations in water distribution networks. *Journal of Hydroinformatics*, 20(6), 1235–1248. DOI: 10.2166/hydro.2018.098

Ahmed, A., Salih, A., & Zhang, Z. (2018). Reinforcement learning for optimizing valve operations in water distribution networks. *Journal of Hydroinformatics*, 20(6), 1235–1248. DOI: 10.2166/hydro.2018.098

Aivazidou, E., Banias, G., Lampridi, M., Vasileiadis, G., Anagnostis, A., Papageorgiou, E. I., & Bochtis, D. (2021). Smart technologies for sustainable water management: An urban analysis. *Sustainability (Basel)*, 13(24), 13940. DOI: 10.3390/su132413940

Al-Juaidi, A., Kharabsheh, S., & Al-Taani, A. (2017). Real-time water usage management in smart cities using clustering algorithms. *Journal of Water Resources Planning and Management*, 143(10), 04017054. DOI: 10.1061/(ASCE)WR.1943-5452.0000814

Al-Juaidi, A., Kharabsheh, S., & Al-Taani, A. (2017). Real-time water usage management in smart cities using clustering algorithms. *Journal of Water Resources Planning and Management*, 143(10), 04017054. DOI: 10.1061/(ASCE)WR.1943-5452.0000814

Aldegheishem, A., Alrajeh, N., García, L. R., & Lloret, J. (2022). Swap: Smart water protocol for the irrigation of urban gardens in smart cities. *IEEE Access : Practical Innovations, Open Solutions*, 10, 39239–39247. DOI: 10.1109/ACCESS.2022.3165579

Alexander Obaigbena, Michael Ayorinde DadaMichael Tega MajemiteOnyeka Henry DaraojimbaJohnson Sunday OlihaZamathula Queen Sikhakhane Nwokediegwu. (2024). Review of Smart Water Management: Iot and ai in water and wastewater treatment. *World Journal of Advanced Research and Reviews*, 21(1), 1373–1382. DOI: 10.30574/wjarr.2024.21.1.0171

Alghadeer, S. G. (2023). The use of artificial intelligence in water management projects in the kingdom of saudi arabia. *American Journal of Technology*, 2(1), 54–65. DOI: 10.58425/ajt.v2i1.208

Alshehri, F., & Rahman, A. (2023). Coupling machine and deep learning with explainable artificial intelligence for improving prediction of groundwater quality and decision-making in arid region, saudi arabia. *Water (Basel)*, 15(12), 2298. DOI: 10.3390/w15122298

Alvarez-Cano, J., Cavero, V., & Diez-Canseco, F. (2022). Idas y venidas del diseño de la política de alimentación saludable en el perú: Análisis comparativo de sus documentos regulatorios. *Revista Peruana de Medicina Experimental y Salud Pública*, •••, 480–488. DOI: 10.17843/rpmesp.2022.394.11896 PMID: 36888812

Amorós, A. M. R., Cantos, J. O., & Castiñeira, C. J. B. (2014). Competencias por el uso del agua en la provincia de alicante: experiencias de gestión en la armonización de usos urbano-turísticos y agrícolas - (competències per l'ús de l'aigua a la província d'alacant: experiències de gestió en l'harmonització...). *Documents d'Analisi Geografica*, 60(3), 523–548. Advance online publication. DOI: 10.5565/rev/dag.136

Bae, K., Kwon, S., & Hwang, M. (2020). CNN-based urban water demand forecasting. *Sustainable Cities and Society*, 60, 102275. DOI: 10.1016/j.scs.2020.102275

Bertel, M. E. C., Morales, J. D. C. J., Barrios, R. L. A., & Sierra, C. A. S. (2014). Precisión y exactitud para las determinaciones de la demanda bioquímica de oxigeno en aguas por luminiscencia. *Ingenium*, 8(19), 55. DOI: 10.21774/ing.v8i19.386

Bougadis, J., Adamowski, K., & Diduch, R. (2005). Short-term municipal water demand forecasting. *Hydrological Processes*, 19(1), 137–148. DOI: 10.1002/hyp.5763

Brentan, B., Meirelles, G., Manzi, D., & Izquierdo, J. (2017). Hybrid regression model for near real-time urban water demand forecasting. *Journal of Hydroinformatics*, 19(2), 293–306. DOI: 10.2166/hydro.2016.059

Cabarcas Álvarez, A., Puello Marrugo, P., & Lechuga Sierra, J. E. (2023). *Sistema de monitoreo para cuerpos de agua en la ciudad de cartagena basado en internet de las cosas*. Encuentro Internacional De Educación en Ingeniería., DOI: 10.26507/paper.3383

Cabarcas Álvarez, A., Puello Marrugo, P., & Lechuga Sierra, J. E. (2023). *Sistema de monitoreo para cuerpos de agua en la ciudad de cartagena basado en internet de las cosas*. Encuentro Internacional De Educación en Ingeniería., DOI: 10.26507/paper.3383

Canteli, S., García, J., Sanz, D., & Hernández, F. (2019). Deep learning for residential water consumption forecasting. *Procedia CIRP*, 80, 259–264. DOI: 10.1016/j.procir.2019.01.104

Carvajal, L. J. O. and Mazo, W. H. A. (2019). Oportunidades y desafíos: una mirada holística a las iniciativas y proyectos en curso para ciudades inteligentes en colombia y medellín. *Unaciencia Revista De Estudios E Investigaciones*, 12(22), 53-65. .DOI: 10.35997/runacv12n22a6

Catacora, A., & César, C. (2021). Implementación de un sistema de internet de las cosas para optimizar la gestión del agua en la agricultura de la región tacna. *Ingenieria e Investigacion*, 3(1), 519–533. DOI: 10.47796/ing.v3i1.478

Chang, F., Chang, L., & Chen, J. (2023). Artificial intelligence techniques in hydrology and water resources management. *Water (Basel)*, 15(10), 1846. DOI: 10.3390/w15101846

Chang, F., & Guo, S. (2020). Advances in hydrologic forecasts and water resources management. *Water (Basel)*, 12(6), 1819. DOI: 10.3390/w12061819

Chen, L., Yao, Y., & Zhang, W. (2016). Anomaly detection in urban water distribution systems using machine learning algorithms. *Procedia Engineering*, 154, 627–632. DOI: 10.1016/j.proeng.2016.07.530

Chere-Quiñónez, B. F., Souza, R. C. U., Reyna-Tenorio, L. J., Canchingre-Bone, M. E., & Mosquera-Quintero, G. A. (2022). Energía eléctrica a partir de aguas residuales industriales en guayaquil, ecuador. *Sapienza: International Journal of Interdisciplinary Studies*, 3(6), 202–210. DOI: 10.51798/sijis.v3i6.509

Codd, E. F. (1970). A relational model of data for large shared data banks. *Communications of the ACM*, 13(6), 377–387. DOI: 10.1145/362384.362685

Dasallas, L., Lee, J., Jang, S., & Jang, S. (2024). Development and application of technical key performance indicators (kpis) for Smart Water Cities (swcs) global standards and certification schemes. *Water (Basel)*, 16(5), 741. DOI: 10.3390/w16050741

Di Nardo, A., Boccelli, D. L., Herrera, M., Creaco, E., Cominola, A., Sitzenfrei, R., & Taormina, R. (2021). Smart urban water networks: Solutions, trends and challenges. *Water (Basel)*, 13(4), 501. DOI: 10.3390/w13040501

Drogkoula, M., Kokkinos, K., & Samaras, N. (2023). A comprehensive survey of machine learning methodologies with emphasis in water resources management. *Applied Sciences (Basel, Switzerland)*, 13(22), 12147. DOI: 10.3390/app132212147

Elenes, J. R. F., Escobar, R. R., & Urbano, P. M. (2023). La gestión de los recursos hídricos en el municipio de culiacán, sinaloa, méxico. Revistas Cuadernos De Trabajo De Estudios Regionales en Economía, Población Y Desarrollo, 13(73), 3-42. .DOI: 10.20983/epd.2023.73.1

Fernández, A. S., Sutil, M. F., Aparicio, M. J. d. l. F., & Cruz, P. V. (2022). Control predictivo no lineal tolerante a fallos en una planta de tratamiento de aguas residuales. Actas De Las XXXVII Jornadas De Automática 7, 8 Y 9 De Septiembre De 2016, Madrid. .DOI: 10.17979/spudc.9788497498081.0623

Franco, E. F., Ramos, R., Ovando-Javier, A., Montero-Espaillat, E., Bonilla, S., & Veda, A. (2023). Sensores de calidad de agua para el control de la contaminación fisicoquímica en los acuíferos de latinoamérica: una revisión. *Ciencia, Ambiente Y Clima*, 6(1), 45-70. .DOI: 10.22206/cac.2023.v6i1.pp45-70

Freier, A., Mazzalay, V. H., & Rolando, A. (2020). Calentamiento global y política energética. incidencias de la gobernanza multinivel en el caso argentino. *Td. Temas y Debates (Rosario)*, (40), 33–58. DOI: 10.35305/tyd.v0i40.471

Fu, G., Sun, S., Hoang, L., Yuan, Z., & Butler, D. (2023). Artificial intelligence underpins urban water infrastructure of the future: A holistic perspective. Cambridge Prisms. *Water (Basel)*, 1, e14. Advance online publication. DOI: 10.1017/wat.2023.15

Gao, Y., Liu, X., Zhang, Y., & Wang, L. (2020). Deep learning-based real-time water demand prediction model. *Journal of Hydrology (Amsterdam)*, 588, 124892. DOI: 10.1016/j.jhydrol.2020.124892

Garzón, A., Kapelan, Z., Langeveld, J., & Taormina, R. (2022). Machine learning-based surrogate modeling for urban water networks: Review and future research directions. *Water Resources Research*, 58(5), e2021WR031808. Advance online publication. DOI: 10.1029/2021WR031808

Gato, S., Jayasuriya, N., & Roberts, P. (2007). Forecasting residential water demand: Case study of Melbourne, Australia. *Journal of Water Resources Planning and Management*, 133(4), 309–319. DOI: 10.1061/(ASCE)0733-9496(2007)133:4(309)

Geetha, S., & Gouthami, S. (2016). Internet of things enabled real time water quality monitoring system. *Smart Water*, 2(1), 1. Advance online publication. DOI: 10.1186/s40713-017-0005-y

Ghiassi, M., & Zimbra, D. K. (2008). Modeling and forecasting short-term water demand using neural networks and dynamic regression. *Journal of Water Resources Planning and Management*, 134(2), 119–128. DOI: 10.1061/(ASCE)0733-9496(2008)134:2(138)

Gozzer, E., Canchihuamán, F., Chalco, V., Huari, G., Cisneros-Dionisio, A., Quispe, W., & Espinoza, R. (2021). Duración de la gestión ministerial y logros en salud en el perú. *Revista Peruana de Medicina Experimental y Salud Pública*, 38(2), 326–336. DOI: 10.17843/rpmesp.2021.382.6721 PMID: 34468584

Guo, Y., Wang, J., & Cheng, J. (2015). Prediction of water demand in China's urban areas based on nonlinear regression and meteorological data. *Journal of Water Resources Planning and Management*, 141(7), 04014083. DOI: 10.1061/(ASCE)WR.1943-5452.0000493

Hasan, M., Khatun, R., & Ahmed, F. (2020). Regression analysis for predicting rural water quality in developing countries. *Water Research*, 170, 115373. DOI: 10.1016/j.watres.2019.115373

Hastie, T., Tibshirani, R., & Friedman, J. (2009). *The elements of statistical learning: Data mining, inference, and prediction* (2nd ed.). Springer. DOI: 10.1007/978-0-387-84858-7

Hernández, E. V., Lalaleo, L., & Cáceres, W. C. (2022). ¿se pueden prevenir las pandemias? epidemiología basada en aguas residuales. *Investigación Y Desarrollo*, 12(1), 88-90. .DOI: 10.31243/id.v12.2020.991

Herrera, M., Torgo, L., Izquierdo, J., & Pérez-García, R. (2010). Predictive models for forecasting hourly urban water demand. *Journal of Hydrology (Amsterdam)*, 387(1-2), 141–150. DOI: 10.1016/j.jhydrol.2010.04.005

Herrick, C., Dymond, R. L., & Anderson, M. R. (2015). Climate variability and water demand: Evidence from a southwestern United States urban setting. *Urban Climate*, 12, 1–15. DOI: 10.1016/j.uclim.2015.01.001

Holguín, H. M., & Dutrénit, G. (2021). La regulación como instrumento de fomento a la innovación para la atención de la diabetes en méxico. *Brazilian Journal of Business*, 3(3). Advance online publication. DOI: 10.34140/bjbv3n3-032

House-Peters, L. A., & Chang, H. (2011). Urban water demand modeling: Review of concepts, methods, and organizing principles. *Water Resources Research*, 47(5), W05401. Advance online publication. DOI: 10.1029/2010WR009624

Huang, Y., Yang, J., & Li, M. (2018). Leak detection in urban water pipelines using support vector machines. *Journal of Water Resources Planning and Management*, 144(9), 04018062. DOI: 10.1061/(ASCE)WR.1943-5452.0000984

Huang, Y., Yang, J., & Li, M. (2018). Leak detection in urban water pipelines using support vector machines. *Journal of Water Resources Planning and Management*, 144(9), 04018062. DOI: 10.1061/(ASCE)WR.1943-5452.0000984

Hutton, C. J., Veldhuis, M.-C., & Allsop, W. (2014). Real-time demand forecasting for urban water systems. *Water Resources Research*, 50(5), 3723–3743. DOI: 10.1002/2014WR015563

Inalegwu, O. C., Roman, M., Mumtaz, F., Zhang, B., Nambisan, A., & Huang, J. (2024). Optical fiber specklegram sensor for water leak detection and localization. *Optical Waveguide and Laser Sensors*, III, 15. Advance online publication. DOI: 10.1117/12.3014056

Jain, A., Varshney, A. K., & Joshi, U. C. (2001). Short-term water demand forecast modeling at IIT Kanpur using artificial neural networks. *Water Resources Management*, 15(5), 299–321. DOI: 10.1023/A:1014415503476

Jothiprakash, V., & Sasikumar, K. (2009). Water demand forecasting for a water supply project using optimization techniques. *Journal of Water Resources Planning and Management*, 135(5), 371–381. DOI: 10.1061/(ASCE)0733-9496(2009)135:5(371)

Kelley, C. T. (1960). *Iterative methods for optimization*. SIAM.

Kisi, O. (2013). Streamflow forecasting using different artificial neural network algorithms. *Journal of Hydrologic Engineering*, 18(1), 88–100. DOI: 10.1061/(ASCE)HE.1943-5584.0000625

Kuğuoğlu, B., Voort, H., & Janssen, M. (2021). The giant leap for smart cities: Scaling up smart city artificial intelligence of things (aiot) initiatives. *Sustainability (Basel)*, 13(21), 12295. DOI: 10.3390/su132112295

Labadie, J. W. (2004). Optimal operation of multireservoir systems: State-of-the-art review. *Journal of Water Resources Planning and Management*, 130(2), 93–111. DOI: 10.1061/(ASCE)0733-9496(2004)130:2(93)

Lassiter, A., & Leonard, N. D. (2022). A systematic review of municipal smart water for climate adaptation and mitigation. *Environment and Planning. B, Urban Analytics and City Science*, 49(5), 1406–1430. DOI: 10.1177/23998083211072864

Li, X., Zhang, Y., & Wang, L. (2018). Real-time water quality monitoring using IoT-enabled sensors in river ecosystems. *Water Science and Technology*, 78(1), 185–194. DOI: 10.2166/wst.2018.091

Li, Y., Feng, L., & Wang, R. (2019). Deep learning models for drought prediction and water availability assessment in arid regions. *Journal of Hydrology (Amsterdam)*, 575, 370–383. DOI: 10.1016/j.jhydrol.2019.05.027

Li, Y., Feng, L., & Wang, R. (2019). Deep learning models for drought prediction and water availability assessment in arid regions. *Journal of Hydrology (Amsterdam)*, 575, 370–383. DOI: 10.1016/j.jhydrol.2019.05.027

Li, Z., Wang, J., & Chen, X. (2017). Water quality classification using k-means clustering algorithm: A case study of the Pearl River, China. *Water Science and Technology*, 76(8), 2193–2203. DOI: 10.2166/wst.2017.379

Liang, B., Verma, S., Xu, J., Liang, S., Liu, Z., Wang, Y., . . . Chen, F. (2020). A data driven approach for leak detection with smart sensors. 2020 16th International Conference on Control, Automation, Robotics and Vision (ICARCV). DOI: 10.1109/ICARCV50220.2020.9305403

Lima, M. F. d., Freitas, L. L. G., Henning, E., & Kalbusch, A. (2021). *Avaliação da utilização de hidrômetros com aquisição de dados por telemetria para monitoramento do consumo de água de uma agência bancária em joinville/sc*. Simpósio Nacional De Sistemas Prediais., DOI: 10.46421/sispred.v1i.1578

Liu, F., Zhang, Y., & Zhao, X. (2020). Hybrid AI algorithms for sustainable water management in urban environments. *Sustainability*, 12(4), 1452. DOI: 10.3390/su12041452

Liu, Q., Yang, L., & Yang, M. (2021). Digitalisation for water sustainability: Barriers to implementing circular economy in Smart Water Management. *Sustainability (Basel)*, 13(21), 11868. DOI: 10.3390/su132111868

Liu, Y., Ma, X., Li, Y., Tie, Y., Zhang, Y., & Gao, J. (2019). Water pipeline leakage detection based on machine learning and wireless sensor networks. *Sensors (Basel)*, 19(23), 5086. DOI: 10.3390/s19235086 PMID: 31766356

Llanos-Zavalaga, F., Quiroz, J. A. C., & Silva, C. C. (2022). Análisis de políticas públicas en salud: El caso de recursos humanos en el perú. *Revista Médica Herediana*, 33(3), 178–186. DOI: 10.20453/rmh.v33i3.4338

Lobbrecht, A. H., & Solomatine, D. P. (2002). Machine learning in real-time control of water systems. *Urban Water Journal*, 4(3), 257–263. DOI: 10.1080/15730620500246236

Maestre-Góngora, G. (2015). Revisión de literatura sobre ciudades inteligentes: Una perspectiva centrada en las tic. *Ingeniare*, (19), 137–149. DOI: 10.18041/1909-2458/ingeniare.19.531

Makropoulos, C. K., & Butler, D. (2008). A neurofuzzy modeling framework for predicting urban water demand. *Journal of Water Resources Planning and Management*, 134(2), 112–118. DOI: 10.1061/(ASCE)0733-9496(2008)134:2(112)

McCulloch, W. S., & Pitts, W. (1943). A logical calculus of the ideas immanent in nervous activity. *The Bulletin of Mathematical Biophysics*, 5(4), 115–133. DOI: 10.1007/BF02478259

Mitchell, T. M. (1997). *Machine Learning*. McGraw-Hill.

Mohammadi, A., Azmi, M., & Araghinejad, S. (2020). Long short-term memory neural networks for predicting reservoir water levels. *Journal of Hydrology (Amsterdam)*, 581, 124412. DOI: 10.1016/j.jhydrol.2019.124412

Mohammed, M., K. Alqahtani, N., Munir, M., & A. Eltawil, M. (2024). Applications of ai and iot for advancing date palm cultivation in saudi arabia. *Internet of Things - New Insights*. .DOI: 10.5772/intechopen.113175

Morejón, M. A. E., Manobanda, M. E. B., Velasco, S. R. M., & Ballesteros, H. F. V. (2023). Implementación del geoportal para la junta administradora de agua potable y saneamiento – jaapys - santa fé centro, año 2023. *ConcienciaDigital*, 6(2.1), 6-23. .DOI: 10.33262/concienciadigital.v6i2.1.2553

Mounce, S. R., Boxall, J. B., & Khan, A. (2016). Novel application of genetic algorithms for optimization of water distribution network design. *Water Resources Research*, 52(6), 4379–4394. DOI: 10.1002/2016WR018896

Nikoloska, R., Bykerk, L., Vitanage, D., Miró, J. V., Chen, F., Wang, Y., & Liang, B. (2020). Enhancing sydney water's leak prevention through acoustic monitoring. *Water E-Journal*, 5(2), 1–15. DOI: 10.21139/wej.2020.014

Nishant, R., Kennedy, M., & Corbett, J. (2020). Artificial intelligence for sustainability: Challenges, opportunities, and a research agenda. *International Journal of Information Management*, 53, 102104. DOI: 10.1016/j.ijinfomgt.2020.102104

Okoli, N. J. (2023). Smart water metering system (swms) adoption: a systematic literature review. *International Conference on Artificial Intelligence and Its Applications*, 2023, 172-180. DOI: 10.59200/ICARTI.2023.025

Okoli, N. J., & Kabaso, B. (2024). Building a smart water city: Iot smart water technologies, applications, and future directions. *Water (Basel)*, 16(4), 557. DOI: 10.3390/w16040557

Park, S., Lee, J., & Kang, D. (2021). Self-organizing neural networks for water usage optimization in smart buildings. *Automation in Construction*, 122, 103458. DOI: 10.1016/j.autcon.2020.103458

Perea, R. G., Poyato, E. C., Montesinos, P., & Díaz, J. A. A. (2016). Optimización de la predicción de demanda de agua mediante algoritmos neuro-genéticos para un conjunto de datos reducido. *XXXIV Congreso Nacional De Riegos*. DOI: 10.21151/CNRiegos.2016.C18

Popartan, L. A., Cortés, À., Garrido-Baserba, M., Verdaguer, M., Poch, M., & Gibert, K. (2022). The digital revolution in the urban water cycle and its ethical–political implications: A critical perspective. *Applied Sciences (Basel, Switzerland)*, 12(5), 2511. DOI: 10.3390/app12052511

Qi, J., Chang, N.-B., & Yang, Y. J. (2016). Development of a web-based decision support system for improved urban water demand forecasting and management. *Environmental Modelling & Software*, 76, 3–14. DOI: 10.1016/j.envsoft.2015.11.005

Qi, J., Chang, N.-B., & Yang, Y. J. (2016). Development of a web-based decision support system for improved urban water demand forecasting and management. *Environmental Modelling & Software*, 76, 3–14. DOI: 10.1016/j.envsoft.2015.11.005

Riahi, S., Nasrabadi, T., & Shahsavari, P. (2020). Multi-objective optimization of wastewater treatment plants using evolutionary algorithms. *Journal of Cleaner Production*, 246, 118964. DOI: 10.1016/j.jclepro.2019.118964

Riahi, S., Nasrabadi, T., & Shahsavari, P. (2020). Multi-objective optimization of wastewater treatment plants using evolutionary algorithms. *Journal of Cleaner Production*, 246, 118964. DOI: 10.1016/j.jclepro.2019.118964

Şahin, E., & Yüce, H. (2023). Prediction of water leakage in pipeline networks using graph convolutional network method. *Applied Sciences (Basel, Switzerland)*, 13(13), 7427. DOI: 10.3390/app13137427

Sanchez, E., Sanchez, E., Acosta, A., & González, E. (2021). Sistema de riego automatizado iot aplicado en el cultivo del pepino (cucumis sativus). Congreso Nacional De Ciencia Y Tecnología – APANAC, 523-529. .DOI: 10.33412/apanac.2021.3234

Schmidt, M. (2022). El cañito de la red llega, lo que no llega es el agua. Hacia una ecología política del agua en el chaco salteño, argentina. *Territorios*, (46). Advance online publication. . DOI: 10.12804/revistas.urosario.edu.co/territorios/a.9250

Shaikh, M., & Birajdar, F. (2024). Artificial intelligence in groundwater management: Innovations, challenges, and future prospects. *International Journal of Science and Research Archive*, 11(1), 502–512. DOI: 10.30574/ijsra.2024.11.1.0105

Shaikh, M., & Birajdar, F. (2024). Harmony in hydroinformatics: Integrating ai and iec for sustainable groundwater conservation in solapur. *International Journal of Science and Research Archive*, 11(1), 2163–2175. DOI: 10.30574/ijsra.2024.11.1.0294

Smith, A., Johnson, M., & Turner, C. (2019). Real-time water quality monitoring using distributed sensor networks and artificial neural networks. *Journal of Water Resources Planning and Management*, 145(6), 04019030. DOI: 10.1061/(ASCE) WR.1943-5452.0001060

Somayaji, S. R. K., Nallakaruppan, M. K., Chengoden, R., Koppu, S., Iyapparaja, M., Sadhasivam, J., & Sethuraman, S. (2022). Smart water resource management using artificial intelligence—A review. *Sustainability (Basel)*, 14(20), 13384. DOI: 10.3390/su142013384

Sun, Z., Zhang, Q., & Wang, X. (2019). River flow prediction using recurrent neural networks for enhanced water management. *Water Resources Research*, 55(3), 2547–2563. DOI: 10.1029/2018WR024567

Tsai, Y., Chang, H., Lin, S., Chiou, A., & Lee, T. (2022). Using convolutional neural networks in the development of a water pipe leakage and location identification system. *Applied Sciences (Basel, Switzerland)*, 12(16), 8034. DOI: 10.3390/app12168034

Tyagi, V., Pandey, P., Jain, S., & Ramachandran, P. (2023). A two-stage model for data-driven leakage detection and localization in water distribution networks. *Water (Basel)*, 15(15), 2710. DOI: 10.3390/w15152710

Valdez, J. E., Pandolfi, D., & Villagra, A. (2018). Red de sensores inteligentes para adquisición de datos de una planta de desalinización de agua. Informes Científicos Técnicos - UNPA, 10(2), 83-95. .DOI: 10.22305/ict-unpa.v10i2.272

Wang, F., Zhu, X., & Jiang, Q. (2015). IoT-based agricultural water management using evolutionary optimization algorithms. *Computers and Electronics in Agriculture*, 118, 56–63. DOI: 10.1016/j.compag.2015.08.007

Wang, F., Zhu, X., & Jiang, Q. (2015). IoT-based agricultural water management using evolutionary optimization algorithms. *Computers and Electronics in Agriculture*, 118, 56–63. DOI: 10.1016/j.compag.2015.08.007

Wu, H., He, S., & Zhang, W. (2019). Prediction of industrial wastewater contamination using decision trees. *Environmental Monitoring and Assessment*, 191(11), 679. DOI: 10.1007/s10661-019-7884-4 PMID: 31655913

Wu, H., He, S., & Zhang, W. (2019). Prediction of industrial wastewater contamination using decision trees. *Environmental Monitoring and Assessment*, 191(11), 679. DOI: 10.1007/s10661-019-7884-4 PMID: 31655913

Zapata, G., & Tejeda, I. (2018). Publicidad engañosa en educación superior: Definiciones, regulación y una revisión de la publicidad en prensa escrita. *Calidad en la Educación*, (44), 197–242. DOI: 10.31619/caledu.n44.37

Zhang, J., Li, H., & Huang, Z. (2021). Modeling water flow in intelligent irrigation systems using artificial neural networks. *Journal of Water Resources Management*, 35(2), 451–466. DOI: 10.1007/s11269-020-02702-w

Zhang, Q., Zhang, W., & Fang, L. (1998). Application of artificial neural networks in river flow forecasting. *Water Resources Management*, 12(3), 185–198. DOI: 10.1023/A:1007900110595

Zhang, X., Wang, H., & Li, J. (2018). Remote sensing-based detection of water contaminants using convolutional neural networks. *Remote Sensing of Environment*, 213, 382–392. DOI: 10.3390/rs10030382

Zhang, X., Wang, H., & Li, J. (2018). Remote sensing-based detection of water contaminants using convolutional neural networks. *Remote Sensing of Environment*, 213, 382–392. DOI: 10.3390/rs10030382

Zhou, S., McMahon, T. A., Walton, A., & Lewis, J. (2012). Forecasting operational demand for an urban water supply zone. *Water Resources Management*, 26(14), 4147–4163. DOI: 10.1007/s11269-012-0130-8

Zhou, Y., Guo, W., & Li, Z. (2020). Bayesian networks for risk assessment in potable water supply systems. *Journal of Risk Research*, 23(5), 655–672. DOI: 10.1080/13669877.2019.1626002

Chapter 2
Challenges and Opportunities in Implementing AI for Smart Water

Alejandro Fuentes-Penna
https://orcid.org/0000-0002-4303-3852
El Colegio de Morelos, Mexico

Raúl Gómez Cárdenas
El Colegio de Morelos, Mexico

Anayeli Silva Aguilar
El Colegio de Morelos, Mexico

ABSTRACT

Adopting emerging technologies based on Artificial Intelligence in water management is essential to address the challenges of drinking water scarcity. Artificial Intelligence proposes new advantages over current technologies, with the main objective of optimizing the drinking water supply and ensuring that future generations have access to safe and clean water. The current issues detected from the Sustainable Development Goals (SDGs), in particular SDG 6 - clean water and sanitation, seek to reduce water scarcity, optimize efficiency and sustainability in water use, improve the quality of water for human consumption, maintain real-time monitoring and management, among other aspects relevant to everyday life. This chapter aims to identify opportunities and challenges in water management through the ethical and appropriate use of Artificial Intelligence to achieve SDG 6 and propose viable alternative solutions.

DOI: 10.4018/979-8-3693-8074-1.ch002

1. INTRODUCTION

1.1 Importance of Water for Human Consumption

Water is fundamental to life; in fact, it is the origin of life. Playing a vital role in all physical and chemical reactions, including those that occur internally in living organisms, water is the most crucial element on our planet and has enabled the emergence and development of species, according to the National Institute of Public Health (INSP, 2022).

Suffice it to say that 80% of the chemical composition of living beings is water (70% in the case of humans, which is stored especially in cells), which allows metabolic processes to occur; for example, digestive processes require water to dissolve nutrients from food. Water is part of almost all cellular chemical reactions (UNAM, 2024).

For people, water enables the development of distinct functions, among them:

- Transports nutrients to the cells.
- Collaborates in blood circulation, respiration, and digestion.
- Enables the proper functioning of the nervous system and brain.
- Forms part of muscle contraction and allows tissues to have flexibility and elasticity.
- Helps regulate body temperature.
- Provides basic minerals such as magnesium, calcium, and fluoride.
- Aging dose.

In fulfilling such principal functions, a lack of water can prevent the body from adequately developing its activities, such that chronic dehydration can be fatal (Secretaría de Salud, 2016). Therefore, daily water consumption, between six and eight glasses a day, is essential to stay hydrated and for the body to function efficiently.

Water is an essential resource for all forms of life. It is a common good (Zamagni, 2014), which has no substitute in the market, a renewable but finite and exhaustible natural resource, which has social value and is considered a third-generation human right (Shiva, 2002).

Recent studies by the United Nations (UN) show the existence of a serious water crisis caused both by climatic factors and by human or anthropogenic actions. It is estimated that 40% of the world's population suffers from water scarcity and that 50% of humanity could live in water-scarce areas by 2025 because global demand for freshwater will exceed supply by approximately 40%.

It should be noted that, in addition to water consumption for the sustainability of human life and biodiversity, there are other types of consumption. The National Water Commission (2020) points out that, in Mexico, thirteen uses of this vital liquid have been identified. These include recreational and productive activities such as tourism, agriculture, livestock, mining, industry, and hydroelectric power, which affect the balance of nature by causing distinct levels of global warming, desertification, and/or contamination of oceans and rivers.

These effects are caused by activities that provoke carbon emissions, as well as by human actions in construction and urbanization, indiscriminate felling of forests, and irresponsible waste treatment.

Scientific research is being carried out to find out if life exists beyond Earth and is focused on the search for carbon and pure water on other planets. The Nuclear Research Institute of the UNAM has explained that carbon is a chemical element capable of forming very large and complex structures, thanks to its ability to bind to other atoms of the same element, but it requires a liquid medium that facilitates the molecules to meet, react and accumulate (Suarez, 2024). On Earth, that liquid medium is water.

1.2 The Current Situation of Drinking Water in the World

According to the United Nations (UN, 2024), water is fundamental for sustainable development as it enables, among other processes, socioeconomic growth, food production, energy generation, and the proper functioning of ecosystems, in addition to its vital contribution to the human species. Water links society to the environment, so we need it to adapt to climate change.

As the world's population has grown, it has become increasingly necessary to reconcile the competing demands of industry, commerce, and the economy in general, with those of human consumption in cities and communities. Water is needed not only for drinking and hydration but also for the existence and operation of clean sanitation facilities for human use, for example, for girls and women to take care of their menstruation and to exercise their maternity in a dignified and safe manner, with privacy.

In this regard, it should be noted that the need for water for human consumption and that required for drainage and sanitation systems should be addressed separately. Both uses are particularly important for health care, as they make it possible to reduce the number and type of diseases. Healthy people are a vital part of the sustainable development to which we all aspire.

In 2010, the United Nations General Assembly recognized the human right to water supply and sanitation. This means that everyone has the right to sufficient and continuous access to safe drinking water of acceptable quality for personal and domestic use.

According to the World Health Organization (WHO, 2022), water contaminated with microbes harmful to humans can transmit diarrheal diseases (cholera, dysentery, typhoid fever) as well as polio. Every year, 505,000 people worldwide die from diarrheal diseases alone. Global statistics indicate that, in 2022, 73% of the world's population (six billion people) had continuous, safe, and quality access to safe drinking water supply. This implies that more than a quarter of the total number of people in the world do not have access, in these conditions, to the vital liquid, a lack that threatens their health and their lives.

1.3 Ethical and Privacy Considerations in the Use of AI in Water Management

Given the needs and shortcomings already described, not only present but also future, water supply and sanitation management are relying on artificial intelligence (AI) to make these vital processes more efficient, sustainable, and cost-effective, as well as to cope with extreme weather events that cause floods or droughts (Sener, 2024).

However, the ethical and privacy implications of using Artificial Intelligence (AI) in any activity or process must be considered, which is even more delicate in the case of a vital issue for humanity such as water. Much of the colossal utility of AI applied to water management relates to its ability to manage and predict numerous problems such as flooding, networks, water availability and demand, and water treatment, among others.

Having access to data as important as those mentioned above necessarily implies their ethical management. That is, maintaining the approach that water is a human right and a determining factor for sustainable development, and that the AI used in water management must safeguard this fundamental motive, avoiding bias or misuse to favor companies or regions, or to prioritize uses other than human consumption in water management.

Romero (2024) points out that, in this sense, it is essential to use AI with a focus on human and social development, seeking to ensure equitable, egalitarian, and non-discriminatory access to water and its technologies; to give privacy to the management of available data; with transparency and full understanding and traceability of the algorithms used, as well as ensuring the cybersecurity of strategic infrastructures associated with water resource management.

Regarding this last point, the pillars of information security (confidentiality, availability, and integrity) are essential to ensure the management of water resources by avoiding cyber-attacks, vulnerability, disinformation, or misuse of sensitive data.

2. IA IN WATER MANAGEMENT

Based on the United Nations 2030 Agenda (UN, 2024), access to safe drinking water, sanitation, and hygiene is a priority human need for health care and well-being. Water is essential not only for health, but also for reducing poverty and ensuring food security, peace, decent work, human rights, ecosystems, and education.

In this regard, countries face increasing challenges related to water scarcity, water pollution, degradation of water-related ecosystems, and transboundary water cooperation.

Among the measures needed to ensure universal access to safe and affordable drinking water by 2030 are investments in sanitation technology, infrastructure and facilities, protection, and restoration of water-related ecosystems, as well as hygiene education. Among these measures, the role of AI in water management can be of paramount importance.

For example, Big Data or Machine Learning is especially useful for storing, managing, and taking advantage of a large amount of historical data and new data to optimize the management of water resources in a region, as well as to define the needs and characteristics of water infrastructure or improve the infrastructure associated with them.

Likewise, algorithms based on artificial neural networks (ANNs) can solve complex problems to optimize the planning and management of water resources, due to their ability to adapt to discontinuous problems, handle multiple objectives, and cope with uncertainty and variability in supply and distribution systems.

2.1 Energy, Technology, and Drinking Water Management

Environmental management presents several challenges, mainly in the design of methodological procedures linked to environmental information. The aspects that have been addressed by different researchers provide a basis for the problematic situation, considering that the theoretical elements that define environmental management show that, despite the contributions made, there are still shortcomings.

In the global environmental strategy, the scarcity of drinking water is established as one of the main problems, being this a central element for sustainable development it should be taken into account that, according to Pichs (2012), although 70% of the earth's surface is covered with water, only 2.5% is fresh, only 0.26% is accessible to man in rivers, lakes and other reservoirs, while 2.24% is frozen in the polar areas.

Environmental management systems allow for the prevention, reduction, and elimination of environmental impacts and are increasingly focused on preventive measures, where knowledge and cost control are the most important tools to obtain benefits if they respect the environment. This is why it is necessary to calculate the costs of environmental activities, to enrich the environmental management of the entity and favor decision-making based on the achievement of sustainability of natural resources.

All this allows formulating the following research problem: how to quantify the environmental costs to contribute to the sustainability of the drinking water resource and which technologies are necessary to optimize its use?

The evaluation of the Millennium Development Goals concluded in 2015 and a new development agenda for 2030 was proposed. This agenda is composed of 17 Sustainable Development Goals (SDGs) that represent an expansion of measures to help end poverty by 2030.

Goal six is a target dedicated to ensuring the availability and sustainable management of water and sanitation for all. The 2030 agenda includes capacity development, financing, institutions, policies, partnerships, and technology as catalysts for this change.

Maestu (2015) presents some key challenges for the implementation of the different Sustainable Development Goals related to water which are:

- Water, Sanitation and Hygiene
  - Increase and adequately locate funding sources, institutional capacity, political support, and management of inconsistencies in the implementation of legislation in this area.
  - Expand appropriate technologies and improve capacity to address sanitation and hygiene issues in schools and health facilities.
- Water Resources Management
  - Funding for water resources management is insufficient.
  - There is an infrastructure deficit.
  - There is no increase in investment in water infrastructure.
  - Lack of payment for water resource services and services for ecosystem protection.
  - Challenges in implementing appropriate technologies.
  - Water efficiency has not been integrated into water resources management.
  - Little progress in the implementation of techniques to improve irrigation and rainwater harvesting.

- Lack of implementation of new technologies as a benefit to ensure that they become an effective means to achieve socially and ecologically sustainable development.
- Lack of human capacity in knowledge, planning, and management of drinking water.

- Water quality
    - Challenges in implementing water quality systems and protecting ecosystems.
    - Improving financing for soft and intangible infrastructure.
    - Institutional capacity building for the development of standards and regulations, as well as for their monitoring and enforcement.
- Risks
    - Inadequacy of current technologies.
    - Lack of capacity and limited use of traditional knowledge or Lack of capacity and limited use of traditional knowledge.
    - Improve water governance with greater emphasis on the resource itself.
    - Ensure participation of all relevant sectors in climate action.
    - Lack of discussion of institutional barriers.
    - Difficulty in designing and implementing strategies to access local government resources.
    - Untapped knowledge, unreliability of data presented, and inaccessibility or lack of hydrological information.

2.1.1 The Importance of Technology in Driving Water Action

Technology plays a vital role in the implementation of the 2030 Sustainable Development Goals. The transition to a developed society requires secure and adequate access to water and energy for its citizens, as virtually all goods and services involve water and energy as factors of production, and for the environment itself, as many energy sources depend on a continuous supply of freshwater.

It is difficult to overestimate the importance of technology for economic development. Despite recent advances, for developing countries, foreign sources of technology can account for up to 90% of national productivity growth. In general, technology opens the following development opportunities:

- **Economic Development:** Technology diffusion is a strong catalyst for economic growth, job creation, human capital accumulation, and efficiency.
- **Environmental Improvement:** Resource-efficient technologies enable societies to reduce their environmental impact, reduce risks, degrade, or collapse, and enable adaptability.

- **Poverty Reduction:** Some innovations bring crucial social benefits, such as improved access to water, better sanitation, reduced energy poverty, and reduced health risks because of replacing polluting products and processes with cleaner technologies.
- **Competitiveness:** Technologies enable meeting stringent social and environmental needs through technological innovation and economic empowerment.
- **Resource Productivity:** Environmental technologies enable societies to improve the productivity of their resources, resulting in actual cost savings for the economy and the environment.
- **Social Capital:** Technology development and diffusion often take a multi-stakeholder approach that should build trust and legitimacy.

To meet the SDGs, the world needs recent technologies and ways of organizing human activity to combine better human development and environmental goals. However, the technological challenges are not a lack of know-how, or even a lack of innovation, but mainly about putting technology at the service of achieving human development goals. This application challenge is about bridging the gap between knowledge and action.

In moving from knowledge to practice, the set of available options is reduced to those that are socially acceptable and if the involvement of businesses and individuals is required, the set is reduced to those that are economically profitable or can be made financially sustainable in the long term.

In many areas of development, but particularly in those related to water, this implies an unprecedented mobilization of knowledge applied to multiple sectors in different regions. Governments, international institutions, private companies, academia, and civil society will need to work together to identify pathways to success in ways that combine technical expertise and democratic representation.

Green technologies, which increase the amount of water available, and the efficiency of the resource and contribute to the achievement of development goals, can become options for creating new business opportunities, markets, and jobs. The development of technology, science, and innovation, when combined with public awareness, can greatly contribute to efficiency and sustainable growth in most water-consuming sectors.

2.1.2 Monitoring

Mobile and web-enabled platforms have become a key asset in aiding the long-term sustainability of water services. Mobile technologies collect data on the type, location, and functionality of water points, display real-time distribution maps, and monitor the status of water infrastructure nationwide.

The information collected can provide valuable findings, which can be used as a basis for informed decision-making making, program planning, private companies, academic institutions, and civil society should work together to identify pathways to success in a way that combines technical expertise and democratic representation. However, technology transfer requires both local knowledge and local capabilities to make these options adaptable to local conditions.

Effective adaptation and use of these technologies depend critically on local knowledge, human resources, and social capacities, as well as existing institutions and policies.

Today, there are several innovative and low-cost technologies, behavioral change approaches to sanitation, water supply and management, as well as technical alternatives to increase efficiency in water supply and use. There are also many alternatives for adapting to climate change and reducing the risks of extreme water events such as floods and droughts. Global problem-solving networks for sustainable water development will become crucial options in the coming years.

2.1.3 Information Exchange: The Role of Social Networks

AI and social networks represent a real opportunity to reduce the transaction cost of choosing between existing alternatives and their final selection. Scientists, technologists, civil society activists, and others are increasingly turning to virtual networks for collaboration, crowdsourcing, group problem-solving, and open-source solutions such as software and its applications.

Pathways to sustainable development will not be determined through a top-down approach but through a highly energized era of networked problem-solving involving universities, businesses, non-governmental organizations, governments, and especially the world's youth, who must become the experts and leaders of a profoundly challenging new era. Knowledge sharing, through global technology platforms, is a means to improve water decisions, including not only the dissemination of techniques but also the conditions to facilitate their transfer and adaptation and the skills needed to make them work.

2.2 Technical and Interoperability Issues in the Integration of IA Systems

Boutin Mesa (2018) describes drinking water supply management as part of the urban water cycle, which cannot be isolated from the context of integrated water resource management, being evident in its importance and connection with aspects related to water security in the medium and long term. For this reason, poor management of drinking water supply, together with climate change, results in a scenario in which it is essential to advance in the exploration of new tools and techniques that help decision-making at diverse levels, from the management of water sources to the management of drinking water supply.

The adoption in the drinking water sector of new technological solutions aimed at ensuring the integration and linking of all system information, adjusting demand forecasting, detecting and identifying losses in the network, planning and anticipating the management of operational scenarios, or promoting and supporting transparency policies to encourage savings among consumers, will enable more efficient use of resources and, in turn, will turn supply systems into infrastructures with greater resilience to adverse scenarios and contingencies.

Fragoso Sandoval, Roberto Ruiz, Flores, and Juárez León (2013) presented a project that developed a geographic information system (GIS) to simulate the operation of a drinking water distribution network, which will allow the management, evaluation, and distribution of good quality water, in addition to automating the system to implement and develop mechanisms in the future that will allow it to be at the technological forefront. This GIS was applied in two housing units in Tehuacán, Puebla, Mexico. The hydraulic calculations were carried out with Epanet software. The management of water distribution systems is a subject of study, updating, and innovation in various governmental and private organizations and the sectors involved in society. Actions aimed at building management and regulatory capacities in water management are priority issues for the development of rural communities and urban society in general.

Boutin Mesa (2018) presents the technological innovation called WatEner as a set of algorithms and artificial intelligence techniques (machine learning, pattern recognition) that allows incorporating and storing the existing expert knowledge of the technicians and operators of the supplying company (the people) to subsequently manage it and generate, through a learning system, operation recommendations when planning the tasks to be performed. The benefits of this methodology are enhanced with the integration of data and information from the different areas of the supply service (hydraulic, energy, geographic, economic, etc.), increasing its capacity to generate recommendations and evaluate strategies and making available all the necessary information for the total management of the service.

In conclusion, Boutin Mesa (2018) states that drought events must be faced with new technological solutions. For this, it is necessary to enhance the incorporation of advanced management models and the implementation of actions in the field of innovation-oriented to the implementation of new technologies in supplier companies. Tools such as WatEner, which guarantees the global vision of the system by incorporating innovative solutions such as artificial intelligence, offer an excellent response to the current challenges in the operation of supply services that result in benefits in the field of integrated water resource management, also favoring proactivity in organizations, and promoting the development of environmental education policies and citizen participation and awareness in the efficient use of resources.

Godo Pla et al. (2018) developed an Artificial Intelligence Application for drinking water treatment called DrinkIA Project. DrinkIA constitutes a significant advance in coping with the high complexity of the operation of Drinking Water Treatment Stations (DWTS). This proposal integrates a decision support system (DSS) that automates this process by increasing the efficiency and robustness of decisions while reducing decision-making time and associated costs.

Drinking water treatment by DWTP is complex due to the large multiplicity of technological, environmental, and economic parameters that generate uncertainty and subjectivity in decision-making (Figure 2). This complexity will increase due to climate change, increased waste and garbage, demographic changes, and the misuse of drinking water, among others, which are important evidence of the vulnerability of the infrastructure in the face of an increasingly dynamic regime of the input resource. In the last 30 years, approximately one hundred patents have been registered on ETAP monitoring and control methodology and tools (WIPO, 2024). These patents indicate how to manage the unitary operations of DWTPs, such as disinfection, activated carbon filtration, pre-oxidation, flocculation, coagulation, decantation, membrane filters, etc. However, none of them presents an integrated management.

Contreras Clavijo and Rojano Mueses (2023) presented the project entitled "Intelligent Monitoring and Control System for the El Carrizal Salcedo Drinking Water Treatment Plant based on IOT and Artificial Intelligence". The authors designed a prototype to monitor important variables for regulatory compliance such as pH, temperature, conductance, and flow, among others, and took control of the pumps and solenoid valves that control the water supply. The prototype was based on a three-layer Internet of Things (IoT) architecture, Arduino and ESP8266 microprocessors were used to collect and store information from the sensors. The control of solenoid valves and pumps is managed under the MQTT protocol, a web application design was implemented with the combination of the Grafana service and programming languages such as PHP, JavaScript, and HTML, and a Python-based control system was developed under a machine learning and polynomial regression model to process temperature data.

Based on WIPO (2024), Table 1 presents the patents that have been developed using Artificial Intelligence, Big Data, cloud computing, or the Internet of Things, applied to water use optimization.

Table 1. AI applied to smart water

Authors	Title	Description
Vijila & Sahaya Vasanthi (2022).	Artificial intelligence-based drinking water purification system with Graphene Nano-material using Machine Learning and Wireless Sensor Networks	The authors present a Drinking Water Purification System based on AI, ML, and wireless sensor networks with graphene nanomaterial. AI is used to monitor the performance of graphene membranes and contributes to the increased versatility of these membranes.
Pushpalatha, et al. (2022).	Artificial Intelligence-Based Smart Water Bottle with Drinking Water Consumption Remainder and Water Quality Monitor	The authors present as a patent an AI-based smart water bottle with a drinking water consumption remnant and water quality monitor that guarantees adequate water intake and ensures water quantity and quality. In turn, the proposed water container is reusable to reduce risks of dehydration and waterborne diseases.
Xiao Chuyu, et al. (2018)	Drinking water distribution system and method based on artificial intelligence	A system based on a drinking water distribution method with AI is presented. The system is composed of a user's water intake that sends water orders with a terminal to record the amount of water taken; and a water distribution management unit to plan the amount of drinking water to be produced.
Liu Renwu (2019)	Artificial intelligence system for water dispenser	The utility model is based on an IA system for the management of a water dispenser integrated by data comparison, storage, control, water flow measurement, and water temperature control modules.
Vishnupriyan Marimuthu et al. (2022)	Artificial Intelligence-Based Techniques to Monitor and Maintain the Quality of Drinking Water Supplied to Households	The patent is based on AI and IOT to monitor, control, and corroborate the quality of drinking water supplied in homes for domestic purposes.
Lei Feiy (2021)	Intelligent water drinking equipment based on artificial intelligence	The patent is a smart equipment for drinking water management integrated by an AI system and composed of several modules aimed at making potable water stored in a container to meet the collaborative management requirement of a smart home for domestic equipment.
Ashwini et al. (2022)	Artificial Intelligence-Based Techniques to Monitor and Maintain the Quality of Drinking Water Supplied to Households	The patent presents the use of AI, Big data, machine learning, and NR (Neural Networks) to predict drinking water quality.

continued on following page

Table 1. Continued

Authors	Title	Description
Wan Xinlei et al. (2023)	Rural safe drinking water monitoring system	A rural safe drinking water monitoring system based on water quality detection equipment operating with the communication base station, cloud servers, intelligent query system, information display system, and finally anomaly feedback system is presented. Water quality detection is carried out in real-time from a drinking water source or directly from the supply pipe for further analysis on the cloud server.
Zhong Caohui (2017)	Domestic smart water dispenser based on artificial intelligence	An intelligent domestic water dispenser composed of the following modules is presented: a control processor, indoor humidity sensor, indoor water temperature calculation module, and heater control.
Sajjan Singh, Amarjeet kaur & Ashmeet kaur (2022).	Automatic Tank Cleaning Device	An automatic tank cleaning device consisting of a hollow body for storage, an odor sensor, a water pH sensor, an AI-based camera that communicates with the microcontroller to determine water quality, an inflatable unit with a primary pump and tubing for pumping water, the water level sensor, a unit with a sensor for removing algae, and a drainage unit is disclosed.

2.3 Future of Water Management: Risks, Opportunities and Challenges

Water uses can be natural or anthropogenic. The former includes the maintenance of rivers, plants, ecosystems, sediment transport, and natural reserves. The latter include domestic supply, recreation, agriculture, livestock, mining, industry, hydroelectric power, indiscriminate logging, and all other human intervention.

These activities affect the balance of nature, among other reasons because they accelerate the loss of biodiversity, desertification, and/or the pollution of oceans and rivers. In general terms, the types of consumption are:

1. Water consumption for the sustainability of human life and biodiversity. The water demand has increased six-fold in the last century, while the world's population has tripled during the same period. Population growth generates increased consumption of food, hydropower, and other material goods. This increased demand for water leads to a faster depletion of groundwater and surface water reserves than the time needed to replenish them.
2. Water is used in production processes, services, and for domestic use. Where legal requirements exist, the focus of measurement is on emission levels, and quality indicators of production processes are also required. In short, different activities require different measurement parameters.

Specific water resource indicators must consider the problem in a comprehensive and comprehensible manner and include comparative criteria to analyze in real-time the hydrological, social, political, and economic variables that affect the use of the resource.

Current emerging technologies such as the IOT methodology allow the intelligent management of water distribution, through an open platform validated as a feasible solution for the needs of the sector, called the Intelligent Water Supply and Consumption Management System.

This solution allows online metering, processing the data collected with ICTs and establishing automatically measurable parameters, defining the appropriate communication network for data transmission through information technology. The minimum requirements for this type of project include:

- Wireless Communications Network, analysis of maps and coverage area to estimate the number of base stations (Gateway) to install, and monitoring of wells (flow, pressure, power consumption, etc.).
- Analysis of tools for level measurement in wells and storage tanks: Detection of leaks in distribution networks, study of plans with lengths, measurements, and manholes.
- Evaluation of different scenarios for leak detection, consumption, and pressure measurement in dwellings.
- Determination of the number of control points to be considered.
- Survey of occupational health and safety suppliers, type, and quantity of equipment to be purchased.
- Integration with industrial protocols, e.g., software platform, e.g., monitoring, alerts, and possible trends.
- Detailed technical specifications and economic-financial budget to acquire the IT infrastructure and perform a proof of concept before project implementation.

In the face of climate change and without a significant improvement in the management of water resources worldwide, it is not possible to guarantee sustainable development by 2030, and this horizon extends to 2050. The growing scarcity of the resource makes it essential to know which places have water, which do not, and how it is consumed and used. Reducing poverty and promoting access to quality education and affordable healthcare also requires the equitable and sufficient availability of two fundamental resources: water and energy.

In a context where the COVID-19 pandemic highlighted the importance of water in disease prevention and containment, it is essential to understand that the real crisis is essentially due to human beings' inappropriate methods of managing water resources. The effects of these practices have grave consequences on the daily lives of the poorest communities, who suffer from diseases related to lack of water or untreated water.

Understanding the issue requires a broader perspective of analysis, in the context of which it is necessary to have reliable information in real-time. What is not measured cannot be improved. It is necessary to understand the links between the different dimensions that compromise the preservation of and access to water resources so that communities can face these challenges in the coming decades.

In terms of governance, it is essential to develop public policy instruments that promote the proper management of water resources and the definition of projects with a triple economic, social, and environmental impact.

It is also necessary to safeguard the status of watersheds and groundwater, as well as to treat and dispose of wastewater or sewage. Of course, the Nation States must assume a tutelary role in the care of occupational safety and health, which implies the availability and allocation of the necessary budgetary resources for the sustainability of these actions, an aspect not addressed at present.

In terms of sustainable practices, it is essential to raise awareness among decision-makers in companies and organizations of the importance of applying good practices in the use of water resources both internally and about the actors in their value chains. It is therefore important for countries to raise awareness and provide training and information to their communities, promoting practices for the care of water resources.

From the fiscal point of view, environmental goods must be priced, and, from the financial point of view, new instruments must be put in place to facilitate risk sharing between governments and investors, using technologies with low socio-environmental impact.

Although various actions and measures have been taken over the last thirty years to overcome the water problem, the progress achieved to date is not encouraging. Nor are expectations for the hundreds of millions of people in poor countries or for the environment in general auspicious.

Making the exercise of this third-generation human right to a dignified life a reality entails processes of cultural change towards responsible consumption habits within the framework of quality education (SDG4) and sustainable production by companies and organizations (SDG12), taking advantage of the benefits offered by emerging technologies. Otherwise, it makes no sense to speak of security of access to water when it is overexploited, polluted, and/or its catchment sources are neglected.

Ultimately, only a multidimensional approach that considers all aspects of the water crisis will make it possible to move towards the sustainability of life for generations to come.

3. CONCLUSION

Water is the origin of life. Playing a significant role in all physical and chemical reactions, including those that develop internally in living organisms, water is the most crucial element of our planet and has allowed the emergence and development of species. It is a resource that has no substitute in the market, is finite and exhaustible, and is considered a third-generation human right.

Water is used for human and living beings' consumption, as well as for different productive and recreational purposes. According to the United Nations (2024), water is fundamental for sustainable development as it favors socioeconomic growth, food production, energy generation, and the proper functioning of ecosystems, in addition to its vital contribution to the human species. Water also links society to the environment, which is why we need it to adapt to climate change.

In 2010, the United Nations General Assembly recognized the human right to water supply and sanitation. This means that everyone has the right to sufficient and continuous access to safe drinking water of acceptable quality for personal and domestic use.

Today we are facing a serious water crisis caused by both climatic factors and human or anthropogenic actions. It is estimated that 40% of the world's population suffers from water scarcity and that 50% of humanity could live in water-scarce areas by 2025 because global demand for freshwater will exceed supply by approximately 40%.

This is both a quantitative and qualitative problem because the lack of water quality can transmit diarrheal diseases, the cause of 505,000 deaths worldwide each year. In 2022, only 73% of the world's population (six billion people) had continuous, safe, and quality access to safe drinking water supply.

Some key water-related challenges for implementing the various Sustainable Development Goals relate to sanitation and hygiene, water resources management, water quality, as well as the risks inherent in water management. All of them highlight the need for a joint effort and coordination between society and governments, as well as the involvement of local authorities in the processes inherent to the intelligent extraction, distribution, and consumption of water.

In this framework, artificial intelligence (AI) is a crucial tool to address this crisis. There are several innovative and low-cost technologies, behavioral change approaches for sanitation, water supply, and management, as well as technical alternatives to increase efficiency in water supply and use.

There are also many alternatives to adapt to climate change and reduce the risks of extreme water events such as floods and droughts. Global networks, and the global exchange of information through them, will become crucial elements in the coming years to solve problems in sustainable water management.

In this paper, we highlight some of the AI applications that are already being applied in this important topic and seek to contribute to the solution of its different problems and challenges. For example, mobile and web-enabled platforms have become a key asset to help the long-term sustainability of water services by collecting data and displaying important operational aspects in real-time.

Geographic information systems (GIS) allow the management, evaluation, and distribution of excellent quality water, operating through Epanet software. Other tools, such as WatEner, also offer an excellent response to current challenges in the operation of supply services. There are already ten patents for AI applications (Big Data, cloud computing, or the Internet of Things), which we present here, which optimize the use and management of water.

We also refer in this paper to the DrinkIA project, which constitutes a significant advance in addressing the high complexity in operational decision-making in Drinking Water Treatment Plants (DWTP), by addressing the large multiplicity of technological, environmental, and economic parameters involved.

Another AI application to efficiently monitor and control treatment plants has been the Internet of Things (IoT), based on Python under a machine learning and polynomial regression model. The Intelligent Water Supply and Consumption Management System uses IOT technology, through a validated open platform, to facilitate the intelligent management of water distribution.

Significant improvement in the management of water resources worldwide is a key element in addressing climate change and ensuring future sustainable development. In this context, it is particularly important to understand that it is humanity itself that contributes to this crisis by inadequately managing scarce and poorly distributed water resources.

In short, only a multidimensional approach that takes into account all aspects of the water crisis will allow progress towards the sustainability to which the global community aspires with the support of AI, but it is necessary to focus this powerful tool towards the promotion of human and social development, ensuring equitable, egalitarian and non-discriminatory access to water and its technologies; giving privacy to the available data; with transparency and full understanding and traceability of the algorithms used, as well as ensuring the cybersecurity of strategic infrastructures associated with water resource management.

REFERENCES

Ashwini, R. M., Pallavi, C. V., Jayashree, S., Geetha, L. S., Jalaja, G., & Raghavendra, C. K & Santosh Reddy (2022). Artificial Intelligence-Based Techniques to Monitor and Maintain the Quality of Drinking Water Supplied to Households [Patent]. https://patentscope.wipo.int/search/es/detail.jsf?docId=IN379761934&_cid=P20-LZCWA8-96371-1

Boutin Mesa, J. M. (2018). Combatiendo la sequía con inteligencia artificial. *Tecnoaqua*, (30), 98–101.

Collado, I. (2024), IA y Machine Learning en las disciplinas del agua y el medioambiente: del hidrólogo al "hidroinformático". Sitio Sener, https://www.group.sener/insights/ia-machine-learning-disciplinas-del-agua-medioambiente-del-hidrologo-al-hidroinformatico/

Comisión Nacional del Agua (Conagua). (2020) Usos del agua en México 2020. https://sinav30.conagua.gob.mx:8080/UsosAgua/

Contreras Clavijo, S. A., & Rojano Mueses, X. G. (2023). Sistema Inteligente de Monitoreo y Control para la Planta de Tratamiento de Agua Potable El Carrizal Salcedo basado en IOT e Inteligencia Artificial [Tesis]. Universidad Técnica de Ambato. https://repositorio.uta.edu.ec/handle/123456789/38419

Facultad de Medicina Veterinaria y Zootecnia UNAM. (2024) Estructura e importancia del agua en el funcionamiento molecular. https://fmvz.unam.mx/fmvz/p_estudios/apuntes_bioquimica/Unidad_2.pdf

Feiy, L. (2021). Intelligent water drinking equipment based on artificial intelligence [Patent]. https://patentscope.wipo.int/search/es/detail.jsf?docId=CN345721710&_cid=P20-LZCWA8-96371-1

Fragoso Sandoval, L., Roberto Ruiz, J., Flores, Z., & Juárez León, A. B. (2013). Sistema para control y gestión de redes de agua potable de dos localidades de México. Ingeniería Hidráulica y Ambiental, 34(1), 112-126. Recuperado en 31 de julio de 2024, de http://scielo.sld.cu/scielo.php?script=sci_arttext&pid=S1680-03382013000100009&lng=es&tlng=es

Godo Pla, L., Emiliano, P., Valero, F., Poch, M., & Monclús Sales, H. (2018). Aplicación de la Inteligencia Artificial en la potabilización: Proyecto DrinkIA. © RETEMA: revista técnica de medio ambiente, 2020, núm. 210, p. 96-102.

Instituto Nacional de Salud Pública (INSP). (2022) Importancia del agua para el buen funcionamiento de nuestro organismo. https://www.insp.mx/resources/images/stories/INSP/Docs/cts/210131_cts2.pdf

Liu, R. (2019). Artificial intelligence system for water dispenser [Patent]. https://patentscope.wipo.int/search/es/detail.jsf?docId=CN276502343&_cid=P20-LZCWA8-96371-1

Marimuthu, V., Khobragade, P. R., & Patil, M. B. (2022). Artificial Intelligence-Based Techniques to Monitor and Maintain the Quality of Drinking Water Supplied to Households [Patent]. https://patentscope.wipo.int/search/es/detail.jsf?docId=IN374031194&_cid=P20-LZCWA8-96371-1

Pichs, R. (2012). *Recursos naturales, economía mundial y crisis ambiental.* Editorial Científico-Técnica.

Pushpalatha, N., Gunapriya, D., Sri Sangeetha, R., Anbarasu, P., Rajan Babu, W., Premkumar, R., & Hemananth, B. (2022). Artificial Intelligence-Based Smart Water Bottle with Drinking Water Consumption Remainder and Water Quality Monitor [Patent]. https://patentscope.wipo.int/search/es/detail.jsf?docId=IN379762001&_cid=P20-LZCWA8-96371-1

Romero, G. (2024) Binomio IA - Agua: la revolución digital de la gestión hídrica, en Retema, Revista Técnica de Medio Ambiente https://drive.google.com/file/d/10dcNgc5YSGdDodxlslMjRLJ1Su8sqM4G/view

Secretaría de Salud. (2016) https://www.gob.mx/salud/articulos/sabes-por-que-es-importante-consumir-agua

Singh, S. Amarjeet kaur & Ashmeet kaur (2022). Automatic Tank Cleaning Device [Patent]. https://patentscope.wipo.int/search/es/detail.jsf?docId=IN389950475&_cid=P20-LZCW81-94862-1

Suárez. Patricia (2024), "¿Existe vida en otros planetas?" Gaceta UNAM (5489) https://www.gaceta.unam.mx/g20240725

UN. (2024). Goal 6: Ensure access to water and sanitation for all. https://www.un.org/sustainabledevelopment/water-and-sanitation/

Vijila, J., & Sahaya Vasanthi, S. (2022). Artificial intelligence-based drinking water purification system with Graphene Nano-material using Machine Learning and Wireless Sensor Networks [Patent]. https://patentscope.wipo.int/search/es/detail.jsf?docId=IN355391255&_cid=P20-LZCWA8-96371-1

Wan, X., Yang, X., Yu, L., Yang, Y., Hao, F., & Ting, T. (2023). Rural safe drinking water monitoring system [Patent]. https://patentscope.wipo.int/search/es/detail.jsf?docId=CN422248651&_cid=P20-LZCWA8-96371-1

WIPO. (2024). WIPO – Patentscope. https://patentscope.wipo.int/search/es/search.jsf

Xiao, C., Lu, B., Jing, X., Wu, W., & Zheng, H. (2018). Drinking water distribution system and method based on artificial intelligence [Patent]. https://patentscope.wipo.int/search/es/detail.jsf?docId=CN215328653&_cid=P20-LZCWA8-96371-1

Zamagni, S. (2014). Bienes comunes y economía civil. V32, No 87 (2014): Bien común, justicia y educación. https://erevistas.uca.edu.ar/index.php/CECON/issue/view/140

Zhong, C. (2017). Domestic smart water dispenser based on artificial intelligence [Patent]. https://patentscope.wipo.int/search/es/detail.jsf?docId=CN207671726&_cid=P20-LZCWA8-96371-1

Chapter 3
Water Quality Monitoring and Control With AI

Ashay Devidas Shende
https://orcid.org/0000-0002-2082-1311
K.D.K. College of Engineering, India

Shrikant A. Tekade
Government College of Engineering, Nagpur, India

Arpan Arunrao Deshmukh
G.H. Raisoni College of Engineering and Management, India

Sandeep Prabhudas Tembhurkar
https://orcid.org/0000-0003-2937-1019
SSVPS Bapusaheb Shivajirao Deore College of Engineering, Dhule, India

P. Selvakumar
https://orcid.org/0000-0002-3650-4548
Department of Science and Humanities, Nehru Institute of Technology, India

ABSTRACT

Water quality monitoring is a crucial process that involves the assessment and analysis of water characteristics to determine its suitability for various uses, from drinking and agriculture to recreation and aquatic life support. This monitoring plays a pivotal role in environmental stewardship and public health protection by identifying contaminants, tracking pollution sources, and guiding regulatory decisions. However, the practice is fraught with challenges stemming from technological limitations, resource constraints, and the dynamic nature of water systems. This article explores the significance of water quality monitoring, the challenges faced in implementing effective monitoring programs, and the advancements and

DOI: 10.4018/979-8-3693-8074-1.ch003

strategies employed to overcome these hurdles. In agriculture, water quality directly affects crop productivity and soil health, especially in irrigation-dependent regions. Poor water quality, characterized by high salinity or heavy metal contamination, can impair soil fertility and reduce crop yields, thereby jeopardizing food security and economic stability.

INTRODUCTION TO WATER QUALITY MONITORING: IMPORTANCE AND CHALLENGES

Water quality monitoring is a crucial process that involves the assessment and analysis of water characteristics to determine its suitability for various uses, from drinking and agriculture to recreation and aquatic life support. This monitoring plays a pivotal role in environmental stewardship and public health protection by identifying contaminants, tracking pollution sources, and guiding regulatory decisions. However, the practice is fraught with challenges stemming from technological limitations, resource constraints, and the dynamic nature of water systems. This article explores the significance of water quality monitoring, the challenges faced in implementing effective monitoring programs, and the advancements and strategies employed to overcome these hurdles. In agriculture, water quality directly affects crop productivity and soil health, especially in irrigation-dependent regions. Poor water quality, characterized by high salinity or heavy metal contamination, can impair soil fertility and reduce crop yields, thereby jeopardizing food security and economic stability. Furthermore, industries reliant on water for manufacturing processes must adhere to stringent quality standards to prevent contamination and ensure product safety. Beyond human and economic considerations, water quality monitoring is essential for preserving recreational opportunities and cultural practices associated with water bodies. Lakes, rivers, and coastal areas serve as recreational hubs for activities like swimming, boating, and fishing, contributing to tourism revenue and community well-being. Monitoring ensures these activities remain safe and enjoyable by detecting harmful algal blooms, bacterial contamination, or chemical pollutants that could pose risks to public health and the environment. In essence, water quality monitoring serves as a critical safeguard against the adverse impacts of pollution on both natural ecosystems and human societies. By providing timely data on water conditions, monitoring programs empower policymakers, regulators, and stakeholders to make informed decisions that promote sustainable water management and protect vital resources for future generations. Despite its importance, effective water quality monitoring faces numerous challenges that complicate the accurate assessment and management of water resources. One of the primary obstacles is the sheer diversity and variability of water systems, which range from small streams and urban waterways to large rivers and coastal estuaries. Each of these environments presents unique physical, chemical, and biological characteristics that influence water quality dynamics, posing challenges for standardized monitoring approaches. Furthermore, ensuring data accuracy and reliability is challenging due to natural variability in water quality parameters and the influence of transient events such as storms or seasonal fluctuations. Continuous monitoring efforts are essential to capture these temporal changes but require substantial financial investment in equipment and personnel training, particularly in resource-limited regions.

Another critical challenge is the identification and mitigation of emerging contaminants that pose risks to water quality and public health. These include pharmaceuticals, personal care products, and microplastics, which can accumulate in aquatic ecosystems and exert detrimental effects on aquatic organisms and human consumers alike. Monitoring programs must adapt to detect these emerging threats and establish regulatory frameworks to limit their environmental impact. Moreover, ensuring stakeholder engagement and data transparency is crucial for the success of water quality monitoring initiatives. Effective communication among scientists, policymakers, local communities, and industry stakeholders facilitates data sharing, promotes awareness of water issues, and fosters collaborative efforts to address pollution sources and improve water management practices (Abhilasha A., et al., 2022). In response to the growing threat of emerging contaminants, interdisciplinary research initiatives and collaborative partnerships are promoting innovative monitoring approaches and regulatory frameworks. This includes the development of analytical techniques capable of detecting trace levels of contaminants in water samples and assessing their potential risks to human and ecological health (Brown, E. C., et al.,2018). In parallel, efforts to enhance public awareness and participation in water quality monitoring are fostering community-driven initiatives and citizen science projects. By empowering local stakeholders to collect and interpret water quality data, these initiatives promote environmental stewardship, encourage sustainable water use practices, and strengthen resilience to climate change impacts. In conclusion, water quality monitoring is indispensable for safeguarding human health, supporting ecosystem vitality, and ensuring sustainable water use in the face of mounting environmental pressures. Despite the challenges posed by diverse water systems, technological limitations, and emerging contaminants, ongoing advancements and collaborative strategies are expanding the scope and efficacy of monitoring programs worldwide. By prioritizing innovation, data transparency, and stakeholder engagement, policymakers and practitioners can strengthen resilience to water quality threats and promote inclusive, science-based approaches to water resource management. As global efforts continue to evolve, the integration of cutting-edge technologies and interdisciplinary insights will be essential for achieving the shared goal of clean and accessible water for all.

AI AND MACHINE LEARNING IN WATER QUALITY MONITORING: OPPORTUNITIES AND APPLICATIONS

By providing early warning systems, decision-makers can implement timely interventions to mitigate risks, such as issuing advisories or adjusting water treatment protocols, thereby minimizing adverse consequences for downstream users and ecosystems. Another significant opportunity lies in optimizing monitoring networks and resource allocation through AI-driven analytics (Chen, Y., et al.,2021). Machine learning algorithms can prioritize sampling locations, sensor deployments, and monitoring frequencies based on spatial variability and pollution risk assessments. This targeted approach maximizes the efficiency of monitoring efforts, reduces operational costs, and enhances the overall effectiveness of water quality management strategies.

Applications of AI and Machine Learning in Water Quality Monitoring

AI and ML applications in water quality monitoring span a wide range of functionalities, from data processing and analysis to decision support and policy formulation. One notable application is in remote sensing and image analysis, where AI algorithms interpret satellite imagery to monitor water bodies for indicators of algal blooms, sedimentation, and pollution plumes. . This continuous monitoring approach ensures high data quality while reducing the need for manual intervention, thereby optimizing resource allocation and operational efficiency. Machine learning techniques also play a crucial role in data fusion and integration across disparate datasets, enabling comprehensive assessments of water quality parameters and ecosystem health. By combining data from biological surveys, chemical analyses, and hydrological models, AI algorithms identify complex interactions and correlations that influence water quality dynamics, providing holistic insights into ecosystem functioning and resilience. Furthermore, AI-driven decision support systems assist policymakers and water managers in evaluating management scenarios, predicting the outcomes of regulatory measures, and optimizing water resource allocation strategies (Cui, T., et al.,2020). These systems incorporate risk assessments, cost-benefit analyses, and stakeholder preferences to facilitate informed decision-making processes that balance environmental protection with socio-economic considerations.

Technical limitations, such as computational resource requirements and scalability of AI applications, also pose challenges for implementation in resource-constrained environments or remote regions with limited connectivity. Addressing these barriers through cloud-based solutions, edge computing technologies, and capacity building initiatives can extend the reach and effectiveness of AI-driven monitoring systems, ensuring equitable access to advanced water quality assessment tools (Garcia-Cuerva, L., et al.,2023). Furthermore, ethical considerations related to data privacy, security, and algorithmic bias require careful attention to safeguard sensitive information and uphold equity in decision-making processes. Implementing robust data governance frameworks, adhering to ethical guidelines, and promoting inclusivity in AI development and deployment are essential for fostering responsible innovation and maximizing societal benefits from AI-driven water quality monitoring solutions.

Water Quality Parameters: Physical, Chemical, and Biological Properties

These parameters serve as essential indicators of water quality status, reflecting natural processes, human activities, and environmental conditions influencing aquatic ecosystems. Understanding the significance, measurement methods, and implications of these parameters is fundamental to effective water resource management and environmental stewardship (Gupta, S. K., et al.,2019). This article explores the key physical, chemical, and biological properties used to assess water quality, highlighting their importance, measurement techniques, and implications for human health and ecosystem integrity.

Physical Parameters

Physical parameters describe the physical characteristics of water bodies, including temperature, turbidity, color, and conductivity. Temperature influences biochemical processes, dissolved oxygen levels, and aquatic species' metabolic rates. Rapid temperature fluctuations can stress aquatic organisms, affecting their growth and reproduction.

Chemical Parameters

Elevated nutrient levels stimulate algal blooms, depleting oxygen levels upon decomposition and harming aquatic organisms. Metals like mercury, lead, and arsenic can accumulate in water bodies through natural processes or industrial activities, posing health risks to humans and wildlife. Monitoring heavy metal concentrations is critical to assessing pollution impacts and implementing remediation measures. pH levels indicate water acidity or alkalinity, influencing chemical reactions and biological processes. Fluctuations in pH can affect aquatic species' survival and alter nutrient availability in aquatic ecosystems.

Biological Parameters

Biological parameters assess the health and biodiversity of aquatic ecosystems through the presence, abundance, and diversity of aquatic organisms. Biological indicators include macroinvertebrates, algae, fish, and microbial communities, reflecting ecosystem health and water quality conditions. Macroinvertebrates such as mayflies and stoneflies are sensitive to pollution levels and habitat alterations, making them valuable indicators of water quality. Algal communities vary in composition and abundance based on nutrient availability and light conditions, with harmful algal blooms posing risks to human health and aquatic ecosystems. Fish populations reflect habitat suitability, water quality, and ecosystem integrity, with declines indicating pollution impacts or habitat degradation (Hu, H., et al.,2020). Microbial indicators such as coliform bacteria and pathogens signal potential health risks from fecal contamination or inadequate water treatment, necessitating regular monitoring for safe drinking water supplies and recreational activities. Measurement techniques for water quality parameters vary depending on parameter type, monitoring objectives, and environmental conditions. Physical parameters like temperature and turbidity are typically measured using field instruments such as thermometers, turbidimeters, and secchi disks. Conductivity meters assess electrical conductivity to gauge dissolved ion concentrations in water (Huang, G., et al.,2022). Chemical parameters require laboratory analysis techniques such as spectrophotometry, chromatography, and atomic absorption spectroscopy for precise measurement of nutrient concentrations, heavy metals, and other contaminants. pH levels are measured using pH meters or colorimetric test kits, providing insights into water acidity or alkalinity variations. Biological parameters involve sampling and identification of aquatic organisms through field surveys, microscopy, and genetic analysis to assess biodiversity, species composition, and ecosystem health indicators.

Implications for Human Health and Ecosystem Integrity

Water quality parameters directly impact human health, ecosystem resilience, and socio-economic activities dependent on clean water resources (Johnson, L. E., et al.,2017). Poor water quality, characterized by high nutrient levels, toxic contaminants, or microbial pathogens, jeopardizes public health through waterborne diseases, gastrointestinal illnesses, and long-term exposure risks. Contaminated water sources compromise agricultural productivity, fisheries, and recreational opportunities, undermining local economies and community well-being. Ecosystem integrity suffers from habitat degradation, biodiversity loss, and impaired ecosystem services such as water purification and flood regulation. Monitoring and managing water quality parameters are essential to mitigating pollution impacts, preserving aquatic biodiversity, and ensuring sustainable water resource use for present and future generations. In conclusion, understanding water quality parameters encompasses a multidimensional approach to assessing environmental health, human well-being, and ecosystem sustainability. Integrating physical, chemical, and biological indicators enables comprehensive monitoring of water quality status, identifying pollution sources, and informing evidence-based management decisions (Kumar, A., et al.,2021). By prioritizing effective monitoring strategies, interdisciplinary collaborations, and adaptive management practices, stakeholders can safeguard water resources, mitigate pollution impacts, and promote resilient ecosystems capable of supporting diverse ecological functions and societal needs. Continued research, technological advancements, and policy interventions are essential for addressing emerging water quality challenges and achieving global water security goals in an era of rapid environmental change

Sensor Technologies for Water Quality Monitoring: Advances and Limitations

Sensor technologies play a pivotal role in modern water quality monitoring by providing real-time data acquisition, enhancing spatial coverage, and enabling continuous assessment of water parameters critical to environmental health and human well-being. These technologies range from traditional handheld devices to advanced automated systems integrated with remote sensing and wireless communication capabilities. While sensor advancements have revolutionized monitoring practices, they also face challenges related to accuracy, reliability, and operational limitations (Lee, J., et al.,2019). This article explores the evolution, applications, advancements, and limitations of sensor technologies in water quality monitoring, highlighting their role in environmental management and sustainable water resource utilization. The evolution of sensor technologies has significantly transformed water quality monitoring from labor-intensive sampling methods to automated, real-time data collection systems capable of monitoring diverse parameters simultaneously. Traditional sensors measure fundamental physical and chemical properties such as temperature, pH, dissolved oxygen (DO), turbidity, and conductivity using robust field instruments and handheld devices. These sensors are essential for routine monitoring activities, compliance assessments, and early detection of water quality fluctuations.

Recent advancements in sensor technologies have focused on improving accuracy, reliability, and versatility to address complex monitoring challenges and diverse environmental conditions. Miniaturization and sensor integration have reduced equipment size and power requirements, facilitating deployment in remote or inaccessible locations and enhancing operational efficiency. Multiparameter sensors combine multiple measurement capabilities into compact devices, reducing the need for multiple instruments and streamlining field operations (Li, J., et al.,2022). Moreover, sensor calibration techniques and quality assurance protocols ensure data accuracy and consistency over extended monitoring periods. Automated calibration routines, sensor validation procedures, and data quality checks mitigate measurement errors and drift, enhancing confidence in monitoring outcomes and decision-making processes. Continuous advancements in sensor materials, coatings, and sensing elements optimize sensor performance in harsh environments, such as saline waters or fluctuating temperature regimes, improving sensor longevity and reliability. Despite their transformative impact, sensor technologies in water quality monitoring encounter several limitations that affect their widespread adoption and operational effectiveness. Sensor accuracy and reliability can be compromised by environmental variability, biofouling, sensor drift, and interferences from natural and anthropogenic factors. Calibration drift over time necessitates regular maintenance and calibration checks to ensure data integrity and measurement consistency.

Sensor deployment and maintenance costs, including equipment procurement, installation, and personnel training, pose financial challenges for resource-constrained regions or monitoring programs with limited funding. Operational complexities associated with sensor networks, data management, and integration with existing monitoring infrastructure require technical expertise and logistical support to overcome implementation barriers effectively. Furthermore, sensor limitations may constrain their applicability for certain water quality parameters or environmental conditions. For instance, turbidity sensors may struggle to differentiate between organic and inorganic particles in highly turbid waters, affecting accuracy in sediment-laden environments. Sensors designed for specific chemical contaminants may exhibit cross-sensitivity or require customized calibration protocols to ensure reliable detection and quantification of target analytes. Data management and interpretation present additional challenges, particularly with large-scale sensor networks generating high-volume datasets. Ensuring data compatibility, standardization, and interoperability across different sensor platforms and monitoring systems requires robust data management frameworks and collaborative efforts among stakeholders, researchers, and policymakers (Li, W., et al.,2022).Looking ahead, the future of sensor technologies in water quality monitoring hinges on advancing sensor performance, data analytics capabilities, and integration with emerging technologies. Continued research and development efforts are focused on enhancing sensor sensitivity, selectivity, and operational robustness to expand monitoring capabilities for emerging contaminants, microbial pathogens, and ecological indicators. Integration of sensor networks with blockchain technology and decentralized data storage solutions offers potential benefits for data security, transparency, and traceability in water quality monitoring initiatives. Blockchain-enabled platforms facilitate secure data sharing, decentralized governance, and enhanced stakeholder engagement, promoting accountability and trust in monitoring outcomes (Li, Z., et al.,2019). Moreover, fostering interdisciplinary collaborations and partnerships between academia, industry, and government agencies is essential for accelerating innovation, knowledge exchange, and technology transfer in sensor development and deployment. Leveraging citizen science initiatives and community-based monitoring programs empowers local stakeholders to participate in data collection efforts, enhance environmental awareness, and support community resilience to water quality challenges. In conclusion, sensor technologies represent a cornerstone of modern water quality monitoring, offering unparalleled opportunities to enhance environmental management, protect public health, and promote sustainable water resource utilization. Addressing technological limitations, improving data reliability, and overcoming implementation barriers will be critical for realizing the full potential of sensors in achieving global water security goals and ensuring clean, safe, and accessible water for all.

AI-Powered Water Quality Prediction and Forecasting Models

This article explores the evolution, applications, capabilities, challenges, and future directions of AI-powered water quality prediction and forecasting models, highlighting their transformative potential in safeguarding water resources and enhancing environmental resilience. The evolution of AI-powered water quality prediction models builds upon traditional statistical methods and hydrological modeling approaches, integrating advanced AI algorithms to enhance predictive accuracy and reliability (Liu, C., et al.,2020). Early applications focused on basic regression models and heuristic approaches to estimate water quality parameters based on linear relationships and historical trends. AI-powered models are applied across diverse water bodies, including lakes, rivers, estuaries, and coastal zones, to predict variations in physical, chemical, and biological parameters critical to water quality assessment. These models simulate water temperature fluctuations, nutrient loading dynamics, dissolved oxygen levels, algal bloom occurrences, and pollutant concentrations, enabling early detection of pollution events and ecosystem disturbances. By assimilating real-time sensor data, satellite imagery, climate forecasts, and hydrological inputs, AI models provide timely insights into spatial and temporal variations in water quality, supporting adaptive management strategies and informed decision-making. AI-powered water quality prediction models offer several key capabilities that enhance their utility in environmental monitoring and management:

Predictive Accuracy: AI algorithms analyze large and heterogeneous datasets to capture complex relationships and non-linear dynamics governing water quality parameters (Mahadev Madgule, et al.,2023). By learning from historical data patterns and environmental inputs, these models improve predictive accuracy and reliability, providing robust forecasts of future water quality conditions.

Real-time Monitoring Integration: Integration with real-time sensor networks, satellite observations, and IoT devices enables continuous data assimilation and model updates. AI models adjust predictions in response to changing environmental conditions, facilitating adaptive management and timely interventions to mitigate pollution impacts.

Scenario Analysis and Decision Support: AI models simulate multiple scenarios and predict the outcomes of management interventions, regulatory measures, or climate change impacts on water quality. Decision-makers utilize scenario analysis to evaluate trade-offs, prioritize actions, and optimize resource allocation strategies for sustainable water resource management.

Early Warning Systems: AI-driven early warning systems detect anomalies and forecast pollution events such as harmful algal blooms, bacterial contamination, or chemical spills. These systems trigger proactive responses, including public advisories, emergency responses, and targeted monitoring efforts to minimize environmental and public health risks.

Spatial and Temporal Resolution: High-resolution modeling capabilities capture fine-scale variations in water quality parameters across spatial gradients and temporal scales. AI models identify localized pollution hotspots, seasonal trends, and long-term trends in water quality dynamics, guiding spatially targeted management interventions and monitoring priorities.

Despite their transformative potential, AI-powered water quality prediction models encounter several challenges that impact their development, implementation, and operational effectiveness:

Data Quality and Availability: Model performance hinges on the quality, quantity, and accessibility of historical data used for training and validation. Incomplete datasets, data gaps, and inconsistencies can introduce biases and uncertainties in model predictions, necessitating robust data preprocessing and quality assurance protocols.

Model Complexity and Interpretability: Complex AI algorithms such as deep learning neural networks offer high predictive power but may lack interpretability, hindering stakeholders' understanding of model outputs and decision-making processes. Enhancing model transparency, explainability, and validation methods is essential for gaining trust and acceptance among users.

Computational Resources and Scalability: AI models require substantial computational resources for training, parameter tuning, and deployment, posing challenges for scalability in resource-constrained environments or regions with limited technical infrastructure. Cloud computing solutions and parallel processing techniques can alleviate computational burdens and enhance model scalability.

Uncertainty Quantification: Predictive uncertainty inherent in AI models due to data variability, model assumptions, and parameter estimation errors must be quantified and communicated effectively to decision-makers. Probabilistic forecasting approaches and ensemble modeling techniques provide insights into uncertainty bounds and confidence intervals, enhancing the reliability of model predictions.

Ethical and Societal Implications: Ethical considerations related to data privacy, bias mitigation, and equitable access to AI-driven technologies require careful attention in model development and deployment. Addressing ethical guidelines, regulatory frameworks, and stakeholder engagement ensures responsible innovation and equitable distribution of benefits from AI-powered water quality prediction models.

Future advancements in AI-powered water quality prediction and forecasting models will focus on overcoming current challenges and expanding application domains:

Integration of Multi-source Data: Enhanced integration of multi-source data, including citizen science observations, social media feeds, and environmental genomics, will enrich model inputs and improve predictive accuracy across diverse spatial and temporal scales.

Adaptive Learning and Transferability: AI models capable of adaptive learning and transferability between different water bodies and geographical regions will enhance scalability, generalizability, and applicability to global water quality monitoring efforts.

Interdisciplinary Collaboration: Collaborative research initiatives integrating AI expertise with domain knowledge in hydrology, ecology, and environmental science will foster interdisciplinary innovations and address complex water quality challenges effectively.

Policy Support and Decision-making Tool: Development of user-friendly decision support tools, policy frameworks, and governance structures will facilitate AI adoption in water resource management, promoting evidence-based policies and stakeholder engagement.

Capacity Building and Knowledge Sharing: Capacity building programs, training workshops, and knowledge-sharing platforms will empower stakeholders to harness AI technologies, interpret model outputs, and implement data-driven solutions for sustainable water management. In conclusion, AI-powered water quality prediction and forecasting models represent a transformative paradigm in environmental monitoring, offering unprecedented opportunities to enhance resilience, mitigate risks, and promote sustainable use of water resources. Addressing technical challenges, enhancing data governance, and fostering collaborative partnerships are essential for realizing the full potential of AI in safeguarding water quality and supporting global water security goals in an era of rapid environmental change and societal challenges. Continued research, innovation, and policy support will drive advancements in AI-driven technologies, ensuring clean, safe, and accessible water for current and future generations.

Machine Learning Algorithms for Water Quality Data Analysis and Pattern Recognition

Traditional statistical methods often struggle with the nonlinear relationships and high dimensionality present in such data. Machine learning addresses these challenges by offering flexible models capable of learning from data without requiring explicit programming of rules. These algorithms can handle diverse types of data, including continuous measurements, categorical variables, and even unstructured data like satellite imagery or sensor readings. Supervised learning techniques, such as regression and classification, are commonly applied to predict water quality parameters or classify water bodies based on their health status (Palaniappan, P., et al.,2020). Regression models can estimate concentrations of pollutants based on correlated variables like temperature, pH, and nutrient levels. For instance, linear regression might predict dissolved oxygen levels based on factors like temperature and water flow rates. Meanwhile, classification algorithms like decision trees or support vector machines classify water bodies into categories such as potable, polluted, or eutrophic based on predefined criteria derived from historical data. Convolutional neural networks (CNNs), originally developed for image processing, can analyze spatial patterns in water quality data derived from satellite imagery or remote sensing platforms. Recurrent neural networks (RNNs) are effective for analyzing temporal trends, such as seasonal variations in water quality parameters. These deep learning techniques require substantial computational resources and large volumes of labeled data but offer unparalleled accuracy in complex pattern recognition tasks (Paramasivam Selvakumar, et al.,2024). The integration of machine learning with geographical information systems (GIS) enhances the spatial analysis of water quality data. GIS platforms provide tools for visualizing and analyzing spatial relationships, such as identifying correlations between water quality parameters and land use patterns. Machine learning algorithms can leverage these spatial insights to predict pollution hotspots, assess the impact of urbanization on water bodies, or optimize the placement of monitoring stations for efficient data collection. Furthermore, ensemble learning techniques combine multiple models to improve prediction accuracy and robustness. Ensemble methods like random forests or gradient boosting machines aggregate predictions from individual models, leveraging their collective intelligence to handle noise and variability in water quality data. These approaches are particularly useful in mitigating overfitting and enhancing generalization capabilities across different environmental conditions and geographical locations.

Real-Time Water Quality Monitoring Systems: Design and Implementation

Central to the design of real-time water quality monitoring systems are the sensors themselves. These sensors must be sensitive, selective, and robust enough to withstand harsh environmental conditions while providing accurate measurements over extended periods. Sensors can be based on various principles, including optical, electrochemical, acoustic, and microbial sensors, each suited for different types of parameters and environments. For instance, optical sensors are commonly used for measuring turbidity and organic matter concentrations, while electrochemical sensors excel in detecting pH, dissolved oxygen, and specific ions (Paramasivam Selvakumar, et al.,2024). The choice of sensor deployment strategy is critical to achieve spatial coverage and optimize data collection. Fixed-point sensors are installed at specific locations, such as near industrial discharge points or in drinking water reservoirs, providing continuous data from predefined monitoring sites. Mobile sensors, on the other hand, can be deployed on autonomous platforms like boats or drones to monitor water quality in remote or dynamic environments, offering flexibility in data collection and spatial coverage. Combining fixed-point and mobile sensors can provide comprehensive spatial and temporal data, essential for understanding water quality dynamics and identifying pollution sources.

Data transmission and communication protocols are another essential aspect of real-time monitoring systems. Sensors typically transmit data either wirelessly (e.g., using WiFi, Bluetooth, or cellular networks) or via wired connections to central data acquisition systems. Wireless communication enables real-time data streaming to centralized databases or cloud platforms, facilitating immediate access to data for analysis and decision-making. Robust communication protocols ensure data integrity and reliability, crucial for transmitting accurate and timely information to stakeholders, including environmental agencies, water utilities, and researchers. Integration with data analytics and visualization tools is imperative for extracting actionable insights from raw sensor data (Ramakrishnan T.,et al.,2022). Advanced analytics techniques, such as machine learning algorithms, statistical modeling, and pattern recognition, process large volumes of real-time data to identify trends, anomalies, and correlations among various water quality parameters. For example, machine learning models can predict pollutant concentrations based on historical data and environmental factors, enabling early detection of contamination events or optimization of treatment processes in wastewater facilities. Visualization tools play a crucial role in presenting monitoring data in an accessible and interpretable manner (Rodriguez, D. J., et al.,2020). Graphical interfaces and dashboards allow stakeholders to monitor trends, view spatial distributions of water quality parameters, and receive real-time alerts for abnormal conditions. These visualizations aid in decision-making processes, enabling quick responses to pollution incidents, evaluating compliance with regulatory standards, and communicating water quality information to the public effectively.

The scalability and interoperability of real-time monitoring systems are essential considerations for their successful implementation across different scales and geographic locations. Scalable systems can accommodate expansions in sensor networks and data volume as monitoring needs evolve or new regulatory requirements emerge. Interoperability ensures compatibility with existing environmental monitoring infrastructure and data management systems, facilitating data sharing and integration with broader environmental information networks (Selvakumar, P., 2024). Furthermore, ensuring the reliability and robustness of real-time monitoring systems requires ongoing maintenance, calibration, and quality assurance procedures. Regular sensor calibration, validation against reference standards, and proactive maintenance minimize measurement errors and ensure the accuracy of monitoring data. Quality assurance protocols, including data validation and quality control checks, verify the integrity and reliability of collected data, enhancing confidence in decision-making based on real-time monitoring results.

AI-Driven Decision Support Systems for Water Quality Management

Unsupervised learning techniques, such as clustering and anomaly detection, play a crucial role in exploring hidden patterns within water quality datasets without predefined labels. Clustering algorithms group similar water samples based on their chemical composition or spatial characteristics, identifying natural clusters that may indicate pollution sources or seasonal variations. Anomaly detection algorithms, on the other hand, flag unusual or unexpected deviations in water quality parameters, alerting stakeholders to potential contamination events or system malfunctions that require immediate attention. Integration with geographic information systems (GIS) enhances the spatial analysis capabilities of AI-driven DSS, enabling stakeholders to visualize and analyze water quality data in relation to geographic features, land use patterns, and human activities. GIS platforms provide tools for spatial interpolation, hotspot identification, and visualization of pollution sources, facilitating targeted interventions and management strategies (Selvakumar, P., et al., 2024). By overlaying water quality data with geographical layers, decision-makers can assess environmental impacts, prioritize mitigation efforts, and optimize the placement of monitoring stations for maximum coverage and effectiveness.

The deployment of AI-driven DSS in water quality management requires robust data infrastructure, including secure data storage, efficient data transmission protocols, and scalable computing resources. Cloud computing platforms offer scalable solutions for storing and processing large volumes of water quality data, enabling seamless integration with AI algorithms and analytics tools (Selvakumar, P., et al.,2024). Data security measures, including encryption and access control, safeguard sensitive information and ensure compliance with data privacy regulations. Furthermore, the effectiveness of AI-driven DSS relies on collaboration and knowledge-sharing among stakeholders, including government agencies, research institutions, water utilities, and community organizations. Collaborative data sharing initiatives facilitate the aggregation of diverse datasets, enhancing the comprehensiveness and accuracy of AI models. Stakeholder engagement and capacity building initiatives are essential to promote understanding and trust in AI-driven DSS capabilities, fostering adoption and implementation across different sectors of water management.

Water Quality Control and Treatment: AI-Optimized Processes and Technologies

Machine learning algorithms analyze historical data to predict water quality variations, optimize chemical dosing schedules, and predict equipment failures or maintenance needs (Sharma, A., et al., 2021). Supervised learning techniques, such as regression and classification models, forecast water demand patterns, anticipate pollutant levels, and optimize treatment plant operations accordingly. These predictive models enable proactive decision-making, minimizing operational costs, reducing energy consumption, and enhancing overall treatment efficiency. Moreover, AI-driven optimization extends to the design and operation of treatment technologies themselves. Computational modeling and simulation tools powered by AI simulate complex hydraulic and chemical processes within treatment systems, aiding in the design of optimized treatment configurations and evaluating their performance under varying conditions. For instance, AI can optimize the placement and sizing of filtration units, chemical dosing pumps, and membrane systems to maximize contaminant removal efficiency while minimizing energy usage and operational footprint.

AI-enhanced control systems enable adaptive management strategies that respond in real time to changing environmental conditions and water quality parameters. Adaptive control algorithms adjust treatment processes dynamically based on incoming data, optimizing operational parameters such as flow rates, chemical dosages, and filtration settings to maintain consistent water quality output (Smith, A. B., et al., 2020). This capability is particularly valuable in managing seasonal variations, weather-induced changes, or unexpected pollution events, ensuring reliable water supply and quality resilience under diverse circumstances. In addition to operational efficiencies, AI supports advanced anomaly detection and fault diagnosis in water treatment infrastructure. AI algorithms analyze sensor data patterns to identify irregularities, diagnose potential equipment malfunctions, and predict maintenance needs before failures occur. This predictive maintenance approach minimizes downtime, extends asset lifespan, and reduces overall maintenance costs, contributing to the sustainability and reliability of water treatment facilities. AI-driven decision support systems (DSS) integrate these capabilities into comprehensive platforms that facilitate holistic water quality management. DSS platforms combine real-time data analytics, predictive modeling, and scenario analysis tools to support informed decision-making by water utility managers, regulators, and policymakers (Thompson, T. M., et al.,2021). These systems provide actionable insights into water quality trends, risk assessments, and compliance monitoring, enabling stakeholders to prioritize resources, implement preventive measures, and optimize long-term planning strategies.

Furthermore, AI enhances environmental sustainability in water quality control by promoting resource efficiency and minimizing environmental impacts. Optimized treatment processes reduce chemical usage, energy consumption, and waste generation, lowering carbon footprints and enhancing overall operational sustainability. AI-driven innovations in water reuse and recycling further support sustainable water management practices by maximizing the recovery of treated water for non-potable applications, agricultural irrigation, or industrial processes, reducing freshwater demand and alleviating pressure on dwindling water resources (Vijayakumar G, et al.,2024). Overall, AI-optimized processes and technologies represent a transformative paradigm in water quality control and treatment, offering unprecedented capabilities in real-time monitoring, predictive modeling, operational optimization, and sustainability. As AI continues to evolve and integrate with advancements in sensor technology, data analytics, and computational modeling, the potential for enhancing water quality management effectiveness and resilience against emerging challenges such as climate change impacts and urbanization pressures grows significantly (Wang, L., et al., 2021). By harnessing AI's capabilities, stakeholders can advance towards more efficient, adaptive, and sustainable water management practices that safeguard public health, protect ecosystems, and ensure equitable access to clean water resources for future generations.

AI-Assisted Water Quality Monitoring for Aquatic Life and Ecosystems

AI-assisted water quality monitoring for aquatic life and ecosystems represents a transformative approach to environmental stewardship, leveraging advanced technologies to enhance understanding, protection, and management of aquatic habitats. By integrating artificial intelligence (AI) with sensor networks, satellite data, and ecological modeling, these systems offer real-time insights into water quality parameters critical for sustaining biodiversity, supporting ecosystem functions, and safeguarding aquatic organisms from anthropogenic impacts and environmental stressors (Wang, Z., et al.,2021). Satellite remote sensing plays a crucial role in AI-assisted monitoring by providing spatially extensive data on water quality across large water bodies. Remote sensing platforms capture multispectral imagery that AI algorithms analyze to derive information on water clarity, algal blooms, sedimentation patterns, and surface temperature variations. These insights enable early detection of environmental stressors such as harmful algal blooms (HABs) or sediment runoff events, facilitating proactive management interventions to mitigate impacts on aquatic ecosystems and public health.

AI-driven ecological modeling frameworks integrate data from water quality sensors, satellite observations, and ecological surveys to simulate ecosystem dynamics and assess ecosystem health. Ecological models simulate interactions between biotic and abiotic factors, predicting ecological responses to disturbances such as climate change, land use changes, or pollution events (Wei, S., et al.,2021). By simulating scenarios and projecting future trends, AI-assisted ecological models inform conservation strategies, ecosystem-based management approaches, and policy decisions aimed at preserving biodiversity and ecosystem resilience. Real-time data processing capabilities of AI-assisted monitoring systems enable timely response to environmental emergencies and emerging threats to aquatic ecosystems. Automated alerts and notifications triggered by AI algorithms notify stakeholders of abnormal water quality conditions, enabling rapid deployment of mitigation measures and emergency response protocols. For instance, AI can detect sudden increases in nutrient concentrations indicative of agricultural runoff, prompting immediate interventions to prevent eutrophication and algal blooms detrimental to aquatic life.

AI-enhanced data integration and visualization tools facilitate accessible and interpretable presentation of monitoring data to stakeholders, including scientists, policymakers, and the general public. Interactive dashboards and spatial mapping applications visualize spatial distributions of water quality parameters, ecological indicators, and habitat characteristics, fostering data-driven decision-making and community engagement in environmental conservation efforts (Wu, Q., et al.,2021). These tools enhance transparency, accountability, and public awareness of water quality issues, promoting collective efforts towards sustainable management of aquatic resources. Furthermore, AI-driven monitoring systems support biodiversity conservation efforts by identifying priority areas for habitat restoration, establishing ecological corridors, and enhancing connectivity between aquatic ecosystems. By analyzing spatial patterns and ecological dependencies, AI assists in identifying critical habitats for endangered species, optimizing conservation strategies, and mitigating habitat fragmentation and degradation caused by human activities (Wu, X., et al.,2021). These conservation initiatives are crucial for maintaining ecosystem services, supporting fisheries, and preserving cultural and recreational values associated with aquatic environments. The scalability and interoperability of AI-assisted monitoring systems facilitate their implementation across diverse geographic scales, from local watersheds to transboundary river basins and coastal marine environments. Scalable sensor networks and cloud-based data platforms accommodate varying data volumes and spatial resolutions, ensuring comprehensive coverage and robust data integration.

Cybersecurity and Data Privacy in AI-Powered Water Quality Monitoring Systems

Regular audits and monitoring of access logs further enhance transparency and accountability, enabling timely detection and response to suspicious activities or security breaches. Incorporating robust cybersecurity protocols into the design and development phases of AI-powered monitoring systems strengthens resilience against cyber threats and vulnerabilities. Cloud-based infrastructure and data storage solutions offer scalability and flexibility in managing large volumes of water quality data but require robust cybersecurity measures to protect against cloud-specific risks (Zhang, L., et al.,2019). Implementing data encryption, secure APIs (Application Programming Interfaces), and data tokenization techniques safeguards data integrity and confidentiality in cloud environments, ensuring compliance with industry standards and regulatory requirements. Contractual agreements with cloud service providers (CSPs) outline data protection responsibilities, service-level agreements (SLAs), and incident response protocols to mitigate risks associated with cloud-based data storage and processing.

Sharing best practices, threat intelligence, and lessons learned from cybersecurity incidents enhance sector-wide preparedness and resilience against evolving cyber threats. Public awareness campaigns and stakeholder engagement initiatives foster understanding of cybersecurity risks and responsibilities, promoting a culture of vigilance and accountability in safeguarding water quality data and infrastructure. In conclusion, cybersecurity and data privacy considerations are integral to the deployment, operation, and sustainability of AI-powered water quality monitoring systems. By implementing robust cybersecurity measures, adhering to data protection regulations, and fostering collaboration among stakeholders, these systems can effectively mitigate cyber threats, protect sensitive data, and ensure the reliability and integrity of environmental monitoring efforts (Zhang, Y., et al.,2022). Continued innovation in cybersecurity technologies and proactive risk management strategies will further enhance the resilience of AI-powered monitoring systems, supporting sustainable water management practices and safeguarding water resources for future generations.

Case Studies: AI-Driven Water Quality Monitoring and Control in Various Applications

One prominent application of AI-driven water quality monitoring is in urban water supply systems. Cities worldwide face challenges in maintaining safe drinking water supplies amidst increasing population growth, aging infrastructure, and environmental pressures. AI-powered systems integrate real-time sensor data from water treatment plants and distribution networks to monitor key parameters such as pH, turbidity, chlorine levels, and microbial contaminants. For instance, in Singapore, PUB, Singapore's National Water Agency, employs AI algorithms to predict water demand patterns, optimize water treatment processes, and detect anomalies in water quality that could indicate potential contamination events. These systems enable proactive management strategies, ensuring continuous supply of safe drinking water to urban residents while minimizing operational costs and resource wastage (Zhou, Z., et al.,2021). In agricultural settings, AI-driven water quality monitoring plays a crucial role in managing irrigation efficiency and nutrient management. Precision agriculture techniques utilize AI algorithms to analyze soil moisture levels, nutrient concentrations, and crop growth indicators, guiding farmers in optimizing irrigation schedules and fertilizer applications. For example, in California's Central Valley, AI-powered sensors monitor water quality in irrigation channels and groundwater aquifers, providing real-time data on salinity levels, nutrient runoff, and water availability. This information enables farmers to make data-driven decisions, conserve water resources, and mitigate environmental impacts associated with agricultural practices.

AI-powered monitoring systems also contribute to environmental conservation efforts in natural water bodies, such as lakes, rivers, and coastal ecosystems. These systems integrate satellite remote sensing data with in-situ sensor measurements to monitor water quality indicators, detect algal blooms, and assess habitat conditions critical for aquatic biodiversity. In Lake Erie, for instance, AI algorithms analyze satellite imagery to detect harmful algal blooms (HABs) and predict their growth patterns based on environmental factors such as water temperature and nutrient concentrations. Early detection of HABs enables authorities to implement timely mitigation measures, protect aquatic ecosystems, and safeguard public health from toxins associated with algal blooms. In industrial applications, AI-driven water quality monitoring enhances compliance with regulatory standards and operational efficiency in manufacturing processes. Industries such as pharmaceuticals, chemicals, and electronics rely on AI-powered sensors and real-time analytics to monitor wastewater discharge, detect pollutants, and optimize treatment processes. For example, in the semiconductor manufacturing industry, AI algorithms analyze sensor data from water treatment facilities to predict chemical contamination events and adjust filtration systems accordingly. These predictive capabilities ensure adherence to environmental regulations, reduce environmental impact, and support sustainable industrial practices.

Furthermore, AI-driven monitoring systems support ecosystem-based management approaches in marine environments and coastal zones. These systems integrate data from buoy-mounted sensors, underwater drones, and satellite observations to monitor oceanographic parameters, marine biodiversity, and habitat conditions. For example, AI algorithms analyze acoustic data to detect marine mammals, monitor their behavior, and mitigate potential impacts of human activities such as shipping or offshore energy development. These applications contribute to sustainable marine conservation efforts, protect endangered species, and inform marine spatial planning initiatives aimed at balancing conservation goals with economic activities. In conclusion, AI-driven water quality monitoring and control systems demonstrate versatile applications across diverse environmental and operational contexts, from urban water supply systems and agricultural irrigation to natural water bodies, industrial facilities, disaster response scenarios, and marine ecosystems. By harnessing AI's predictive capabilities, real-time monitoring functionalities, and data-driven insights, these systems enhance environmental stewardship, promote resource efficiency, and support adaptive management strategies in safeguarding water resources and ecosystems.

Future of Water Quality Monitoring and Control: Emerging Trends and Directions

Advanced sensor technologies are evolving to provide higher sensitivity, selectivity, and durability in measuring diverse water quality parameters. Innovations in sensor design, such as miniaturization, wearable sensors, and nanotechnology-based sensors, enable continuous monitoring at high temporal and spatial resolutions. These sensors detect contaminants, nutrients, pathogens, and emerging pollutants with greater precision, facilitating early detection of water quality issues and prompt response strategies to mitigate environmental impacts. IoT-enabled sensors and data loggers transmit real-time measurements to centralized databases or cloud platforms, enabling remote monitoring, data visualization, and decision support functionalities. Wireless communication protocols, including LoRaWAN, NB-IoT, and 5G networks, support scalable deployment of IoT devices in diverse environmental settings, enhancing coverage and accessibility of water quality data for stakeholders. Decentralized monitoring solutions empower communities, citizen scientists, and local authorities to participate actively in water quality monitoring efforts. Low-cost sensor technologies, mobile applications, and community-based monitoring programs engage stakeholders in data collection, environmental education, and advocacy initiatives. Decentralized networks complement traditional monitoring approaches by enhancing spatial coverage, facilitating early warning systems for pollution events, and promoting public awareness of water quality issues.

Remote sensing technologies continue to evolve as powerful tools for monitoring water quality dynamics at regional and global scales. Satellite-based sensors capture multispectral and hyperspectral imagery to assess water clarity, chlorophyll-a concentrations, and sediment transport in inland water bodies, coastal zones, and oceans. Advances in remote sensing algorithms and data fusion techniques integrate satellite observations with ground-based measurements to enhance spatial resolution and temporal coverage, supporting comprehensive monitoring of environmental changes and ecosystem health. Blockchain technology emerges as a potential solution to enhance transparency, data integrity, and traceability in water quality monitoring and management. Blockchain-enabled platforms facilitate secure data sharing, immutable record-keeping of monitoring results, and smart contracts for decentralized governance of water resources. By establishing trust among stakeholders and ensuring data authenticity, blockchain enhances collaboration among water utilities, regulatory agencies, and research institutions in managing shared water resources and addressing transboundary water quality challenges. Incorporating bioinformatics and biotechnology approaches offers innovative solutions for biomonitoring and bioassessments of water quality. Bioanalytical tools, such as biosensors and microbial assays, detect biological responses to environmental stressors and pollutants, providing early warning indicators of ecological health. Genomics, transcriptomics, and metabolomics techniques analyze microbial communities and biomarker responses to assess water quality impacts on aquatic organisms and ecosystem functions, informing ecosystem-based management strategies and biodiversity conservation efforts. The convergence of these emerging trends and technological innovations signifies a paradigm shift towards holistic, data-driven approaches to water quality monitoring and control. By leveraging advancements in sensor technologies, AI algorithms, IoT connectivity, decentralized monitoring solutions, remote sensing capabilities, blockchain applications, and bioinformatics, stakeholders can address complex water quality challenges effectively.

AI and IoT Integration for Smart Water Quality Monitoring and Management

CNNs analyze spatial patterns and detect environmental features in satellite images, while RNNs model temporal dependencies to forecast water quality variations over time. These AI-driven insights inform adaptive management strategies, optimize resource allocation, and improve operational efficiencies in water treatment processes and environmental monitoring initiatives. The integration of AI and IoT facilitates real-time monitoring and predictive analytics capabilities essential for addressing water quality challenges in urban, industrial, agricultural, and natural environments. In urban water supply systems, AI-powered predictive models optimize water distribution, predict demand patterns, and detect anomalies in water quality that could indicate contamination events or infrastructure failures. For instance, cities like Barcelona and Singapore utilize AI-driven systems to manage water resources efficiently, ensuring reliable supply and quality compliance through data-driven decision-making and proactive maintenance strategies.

In industrial applications, AI and IoT integration enhances regulatory compliance and operational efficiency in managing wastewater treatment processes. Industrial facilities deploy IoT sensors to monitor effluent quality parameters and operational metrics, transmitting real-time data to AI algorithms that optimize chemical dosing, filtration systems, and energy consumption. AI-driven predictive maintenance models identify equipment failures before they occur, minimizing downtime and reducing maintenance costs while ensuring continuous compliance with environmental regulations. Agricultural practices benefit from AI and IoT-enabled precision agriculture techniques that optimize irrigation efficiency, nutrient management, and soil health. IoT sensors deployed in fields monitor soil moisture levels, nutrient concentrations, and crop growth indicators, transmitting data to AI platforms that recommend precise irrigation schedules and fertilizer applications. These data-driven insights enable farmers to conserve water resources, minimize nutrient runoff, and enhance crop yields while mitigating environmental impacts associated with agricultural activities. In natural water bodies, AI and IoT integration supports ecosystem-based management approaches by monitoring water quality parameters, detecting harmful algal blooms (HABs), and assessing habitat conditions critical for aquatic biodiversity. Remote sensing technologies capture multispectral imagery to analyze water clarity, chlorophyll-a concentrations, and sediment dynamics, providing early warning systems for environmental disturbances and facilitating timely intervention strategies. AI-driven ecological models simulate ecosystem dynamics, predict species distributions, and inform conservation efforts aimed at preserving aquatic habitats and ecosystem resilience.

The scalability and interoperability of AI and IoT technologies enable their deployment across diverse geographic scales, from local watersheds to transboundary river basins and coastal marine environments. Cloud-based data platforms facilitate data integration, storage, and analysis, supporting collaborative efforts among stakeholders, including government agencies, research institutions, water utilities, and community organizations. Interoperable IoT networks and standardized data protocols ensure compatibility and seamless data exchange, promoting transparency, accountability, and knowledge sharing in water quality monitoring and management initiatives. In conclusion, AI and IoT integration for smart water quality monitoring and management represents a paradigm shift towards data-driven, adaptive approaches to safeguarding water resources and supporting sustainable development goals. By harnessing the synergies between AI's analytical capabilities and IoT's real-time monitoring functionalities, stakeholders can address complex water quality challenges effectively, enhance resilience to environmental changes, and ensure equitable access to clean water for present and future generations. Continued innovation, investment, and collaboration in AI and IoT technologies will drive the evolution of smart water management practices, fostering environmental stewardship and enhancing water security worldwide.

Ethical and Social Implications of AI-Powered Water Quality Monitoring and Control

Ethical and social implications surrounding AI-powered water quality monitoring and control systems are increasingly pertinent as these technologies evolve and integrate into environmental management practices. While AI offers significant benefits in enhancing efficiency, accuracy, and sustainability in water resource management, several ethical considerations and societal impacts must be addressed to ensure responsible deployment and equitable outcomes. AI algorithms used in water quality monitoring and control should be transparently developed, validated, and audited to mitigate biases and ensure equitable treatment across diverse communities and environmental contexts. Fairness metrics and explainable AI techniques enable stakeholders to understand how decisions are made, identify potential biases in data inputs or algorithmic outputs, and uphold principles of fairness, justice, and nondiscrimination in environmental decision-making. Equitable access to AI-powered technologies and benefits is critical to avoid exacerbating digital divides and disparities in water resource management. Ensuring that underserved communities, marginalized populations, and developing regions have access to affordable, user-friendly AI tools and technologies promotes inclusivity and empowers stakeholders to participate in environmental monitoring, decision-making processes, and advocacy efforts.

Social acceptance and public trust in AI-powered water quality monitoring systems hinge on effective stakeholder engagement, public consultation, and participatory governance frameworks. Engaging diverse stakeholders, including local communities, indigenous groups, environmental organizations, and regulatory agencies, fosters dialogue, transparency, and shared decision-making in designing and implementing AI-driven solutions. Incorporating stakeholder perspectives, cultural values, and traditional knowledge systems enhances the relevance, effectiveness, and legitimacy of water quality monitoring initiatives, promoting inclusive governance and social acceptance of AI technologies in environmental management. Ethical considerations also extend to the broader implications of AI on human rights, environmental justice, and sustainable development goals. AI-powered water quality monitoring systems should uphold principles of human rights, including the rights to clean water, health, and a healthy environment, as enshrined in international agreements and conventions. Promoting environmental justice entails addressing inequities in access to clean water resources and mitigating disproportionate environmental burdens borne by vulnerable populations, underserved communities, and marginalized groups affected by water pollution or resource exploitation. Moreover, AI's role in shaping environmental policies, regulatory frameworks, and governance structures necessitates ethical guidelines, standards, and accountability mechanisms to guide responsible innovation and mitigate potential risks. Establishing ethical guidelines for AI development and deployment in water quality monitoring encourages ethical decision-making, risk assessment, and mitigation strategies that prioritize environmental sustainability, social equity, and public health outcomes. Regulatory oversight, independent audits, and ethical impact assessments facilitate compliance with ethical standards, foster responsible AI adoption, and safeguard against unintended consequences or misuse of AI technologies in environmental governance.

CONCLUSION

In conclusion, the integration of artificial intelligence (AI) into water quality monitoring and control represents a transformative advancement with profound implications for environmental stewardship, public health protection, and sustainable development. AI-powered systems leverage advanced technologies, including IoT sensors, machine learning algorithms, remote sensing, and cloud computing, to revolutionize how we monitor, analyze, and manage water resources across diverse environmental settings. For instance, AI algorithms can analyze historical data and current sensor readings to forecast algal blooms in lakes or detect changes in pollutant concentrations in rivers, enabling water utilities and environmental agencies to implement preventive measures and safeguard public health. Furthermore, AI-driven predictive modeling supports adaptive management strategies by simulating scenarios and predicting the impacts of climate change, land use changes, and anthropogenic activities on water quality. These predictive capabilities inform policy decisions, resource allocation, and conservation efforts aimed at preserving aquatic ecosystems, mitigating pollution sources, and promoting sustainable water use practices. By integrating AI into ecological modeling frameworks, stakeholders can assess ecosystem health, biodiversity trends, and habitat suitability, supporting ecosystem-based approaches to water resource management and biodiversity conservation. The scalability and interoperability of AI technologies enable their deployment across various scales, from local watersheds to regional river basins and global marine environments. Cloud-based platforms facilitate real-time data integration, storage, and analysis, supporting collaborative efforts among stakeholders, including government agencies, research institutions, water utilities, and community organizations. Interoperable IoT networks and standardized data protocols ensure seamless data exchange and interoperability, promoting transparency, accountability, and knowledge sharing in water quality monitoring initiatives. By implementing robust data protection measures, transparent AI algorithms, and inclusive governance frameworks, stakeholders can uphold principles of fairness, accountability, and environmental justice in managing water resources and protecting vulnerable communities from water-related risks. In conclusion, AI-powered water quality monitoring and control systems represent a paradigm shift towards data-driven, adaptive approaches to safeguarding water resources and supporting sustainable development goals.

REFERENCES

Abhilasha, A., Sreenivasulu, A., & Manimozhi, T. Satheesh kumar P., Selvakumar P. and Singh P.,(2022). "The Model of Smart Sensing Device For Sensitive Nano-clusters Modification in Sensing Properties," 2022 2nd International Conference on Advance Computing and Innovative Technologies in Engineering (ICACITE), Greater Noida, India, 2022, pp. 1043-1047, DOI: 10.1109/ICACITE53722.2022.9823846

Brown, E. C., & Horton, A. A. (Eds.). (2018). *Advances in water quality monitoring and assessment*. Academic Press.

Chen, Y., & Liu, C. (2021). Deep learning applications in water quality monitoring: A review. *The Science of the Total Environment*, 761, 144306. DOI: 10.1016/j.scitotenv.2020.144306

Cui, T., & Zhang, H. (2020). Artificial intelligence in water quality monitoring and control: Applications, challenges, and perspectives. *Water Supply*, 20(5), 1837–1852. DOI: 10.2166/ws.2020.224

Garcia-Cuerva, L., & Minguez, R. (2023). Artificial intelligence techniques for smart water quality monitoring. *Sensors (Basel)*, 23(1), 118. DOI: 10.3390/s23010118 PMID: 38202979

Gupta, S. K., & Deshmukh, R. R. (Eds.). (2019). *Emerging trends in water quality monitoring and assessment*. Springer.

Hu, H., Li, S., & Wen, Y. (2020). Review of artificial intelligence technologies in water quality management. *Water Environment Research*, 92(12), 2041–2052. DOI: 10.1002/wer.1361 PMID: 32449943

Huang, G., & Wang, J. (2022). Intelligent water quality monitoring based on machine learning and big data analysis. *Environmental Science and Pollution Research International*, 29(10), 15742–15755. DOI: 10.1007/s11356-022-18789-6

Johnson, L. E., & Smith, R. D. (2017). Remote sensing technologies for water quality monitoring. *Environmental Monitoring and Assessment*, 189(8), 407. DOI: 10.1007/s10661-017-6099-4 PMID: 28730581

Kumar, A., & Kaur, M. (2021). Applications of artificial intelligence in water quality management. *Environmental Technology & Innovation*, 21, 101313. DOI: 10.1016/j.eti.2020.101313

Lee, J., Kim, H., & Lee, K. (2019). IoT-based water quality monitoring system using machine learning techniques. *Sustainability*, 11(15), 4135. DOI: 10.3390/su11154135

Li, J., & Li, Z. (2022). Application of artificial intelligence in water quality monitoring based on multi-source data fusion. *Environmental Monitoring and Assessment*, 194(7), 445. DOI: 10.1007/s10661-022-10190-1

Li, W., Zhang, H., Liu, L., & Ye, J. (2022). A review on AI technologies for water quality monitoring. *Frontiers in Environmental Science*, 10, 780412. DOI: 10.3389/fenvs.2022.780412

Li, Z., Ma, J., & Wang, P. (2019). Artificial intelligence for smart water quality monitoring and control. *Journal of Hydroinformatics*, 21(1), 171–188. DOI: 10.2166/hydro.2018.169

Liu, C., & Liu, H. (2020). Water quality monitoring and management based on internet of things and artificial intelligence. *IEEE Internet of Things Journal*, 8(3), 1702–1712. DOI: 10.1109/JIOT.2020.2961604

Mahadev Madgule, N. (2023). Vinayaka, Yeshwant M. Sonkhaskar, Dhiren Ramanbhai Patel, R. Karthikeyan, P. Selvakumar,(2023). Mechanical properties and microstructure of activated TIG welded similar joints of Inconel alloys by desirability approaches. *Materials Today: Proceedings*, 77, 528–533. DOI: 10.1016/j.matpr.2022.12.250

Palaniappan, P., & Choo, K. K. R. (2020). Applications of artificial intelligence in water quality monitoring and control. *Journal of Computational Science*, 41, 101092. DOI: 10.1016/j.jocs.2019.101092

Ramakrishnan, T., Mohan Gift, M. D., Chitradevi, S., Jegan, R., & Subha Hency Jose, P. Nagaraja, Rajneesh Sharma H.N., Selvakumar P., Sintayehu Mekuria Hailegiorgis, (2022). "Study of Numerous Resins Used in Polymer Matrix Composite Materials", Advances in Materials Science and Engineering, vol. 2022, Article ID 1088926, 8 pages, 2022. DOI: 10.1155/2022/1088926

Rodriguez, D. J., & Singh, J. (2020). Machine learning and AI for water quality monitoring and control. *Journal of Water Process Engineering*, 36, 101309. DOI: 10.1016/j.jwpe.2020.101309

Selvakumar, P. (2024). Emerging Engineering Technologies: Design, Development, and Industrial Applications of IoT-Based Aquaculture Monitoring System. In El Kacimi, Y., & Alaoui, K. (Eds.), *Emerging Engineering Technologies and Industrial Applications* (pp. 332–341). IGI Global., DOI: 10.4018/979-8-3693-1335-0.ch016

Selvakumar, P.. (2024). Prediction of Earthquakes, Volcanic Eruptions, Tornadoes, Wildfires, and Droughts Through Artificial Intelligence. In Satishkumar, D., Sivaraja, M., & Global, I. G. I. (Eds.), *Utilizing AI and Machine Learning for Natural Disaster Management* (pp. 133–145)., DOI: 10.4018/979-8-3693-3362-4.ch008

Selvakumar, P., Mariappan, P., Shunmugam, S., Paul, R., & Jabez, S. (2024). Removal of chemicals from effluent using photobioreactor technology to improve environmental and health impacts. *Zeitschrift für Physikalische Chemie*, 2024(7), 1295–1302. Advance online publication. DOI: 10.1515/zpch-2023-0450

Selvakumar, P., Muthusamy, S., Satishkumar, D., Vigneshkumar, P., Selvamurugan, C., & Satheesh Kumar, P. (2024). AI-Powered Tools. In Satishkumar, D., & Sivaraja, M. (Eds.), *Using Real-Time Data and AI for Thrust Manufacturing* (pp. 20–42). IGI Global., DOI: 10.4018/979-8-3693-2615-2.ch002

Selvakumar, P., Palanisamy, S. K., Cinthaikinian, S., Palanisamy, V., Mariappan, R., & Selvakumar, P. K. (2024). Biosensors and its diverse applications in healthcare systems. *Zeitschrift für Physikalische Chemie*, 2024. Advance online publication. DOI: 10.1515/zpch-2023-0406

Sharma, A., & Sivapragasam, M. (2021). Application of machine learning techniques in water quality assessment: A review. *Environmental Progress & Sustainable Energy*, 40(5), e13313. DOI: 10.1002/ep.13313

Smith, A. B., & Johnson, C. D. (2020). Advances in water quality monitoring systems. *Environmental Science and Pollution Research International*, 27(4), 4193–4206. DOI: 10.1007/s11356-019-07059-5

Thompson, T. M., & Kim, J. (2021). Artificial intelligence applications in water quality monitoring: A review. *Water Research*, 196, 116979. DOI: 10.1016/j.watres.2021.116979

Vijayakumar, G., Rajkumar, M., Rajiv Chandar, N., Selvakumar, P., & Duraisamy, R. (2024). Muniyandi Rajkumar, Rajiv Chandar N, Selvakumar P and Ramesh Duraisamy, (2024). Environmentally friendly TDS removal from waste water by electrochemical ion exchange batch-type recirculation (EIR) technique. *Environmental Science. Water Research & Technology*, 10(4), 826–835. DOI: 10.1039/D3EW00793F

Wang, L., & Chen, C. (2021). Advances in artificial intelligence techniques for water quality prediction and control. *Journal of Hydrology and Hydromechanics*, 69(1), 29–40. DOI: 10.2478/johh-2021-0003

Wang, Z., Zhang, Y., & Ma, J. (2021). Application of machine learning techniques in water quality management: A review. *The Science of the Total Environment*, 758, 143607. DOI: 10.1016/j.scitotenv.2020.143607

Wei, S., & Huang, G. (2021). Review on machine learning applications in water quality monitoring and control. *Water Science and Technology*, 84(7), 1522–1537. DOI: 10.2166/wst.2021.358 PMID: 33843740

Wu, Q., & Liu, Y. (2021). Machine learning in water quality monitoring and control: A review. *Journal of Environmental Management*, 297, 113334. DOI: 10.1016/j.jenvman.2021.113334

Wu, X., & Li, X. (2021). Application of artificial intelligence in water quality assessment and prediction based on multiple data sources. *Water (Basel)*, 13(21), 3048. DOI: 10.3390/w13213048

Zhang, L., & Zou, Y. (2019). A survey on artificial intelligence in water quality monitoring and management. *Water Resources Management*, 33(2), 527–544. DOI: 10.1007/s11269-018-2127-5

Zhang, Y., & Wang, H. (2022). A review on artificial intelligence in water quality prediction and optimization. *Water Resources and Economics*, 38, 100184. DOI: 10.1016/j.wre.2021.100184

Zhou, Z., & Huang, G. (2021). Machine learning for water quality prediction: Challenges, strategies, and applications. *Science China. Technological Sciences*, 64(6), 1044–1060. DOI: 10.1007/s11431-020-1685-7

Chapter 4
Real-Time Water Quality Monitoring:
Use of AI Sensors and Algorithms for Real-Time Monitoring of Water

Francisco R. Trejo-Macotela
https://orcid.org/0000-0003-2133-3456
Universidad Politécnica de Pachuca, Mexico

Daniel Robles-Camarillo
https://orcid.org/0009-0002-4990-8522
Universidad Politécnica de Pachuca, Mexico

Uriel A. Ramírez-Hernández
Universidad Politécnica de Pachuca, Mexico

ABSTRACT

This chapter discusses the significance of real-time water quality monitoring utilizing AI sensors and algorithms, emphasizing water as a renewable yet finite resource. With freshwater representing only 2.5% of Earth's water supply, its uneven distribution among continents necessitates efficient management. The increasing global population exacerbates pressure on these limited resources, particularly through leaks that contribute significantly to water loss during transportation. Advanced monitoring systems employing various sensor technologies—such as differential pressure, ultrasonic, and electromagnetic sensors—facilitate continuous tracking of essential water quality parameters. Algorithms, particularly neural networks and decision trees, enhance predictive modeling, enabling timely detection of contaminants. These innovations not only optimize resource management but also promote sustainability, ultimately protecting public health and the environment by facilitating

DOI: 10.4018/979-8-3693-8074-1.ch004

informed decision-making in water treatment and agricultural practices.

INTRODUCTION

Water is one of the vital elements on our planet, considered a renewable but finite resource. Covering over 70% of the Earth's surface, water is found in oceans, lakes, rivers, in the air, and in the ground. However, it's known that oceans account for nearly 97.5% of the planet's water, with only 2.5% being freshwater. Glaciers, snow, and ice caps from the polar regions represent almost 80% of freshwater, while groundwater represents 19%, and surface water, which is quickly accessible, only accounts for 1%. This small amount of easily accessible surface water is mainly found in lakes (52%), rivers (10%) and wetlands (38%), (Cirelli, 2012) as illustrated in Figure 1.

Figure 1. Distribution of water on earth.

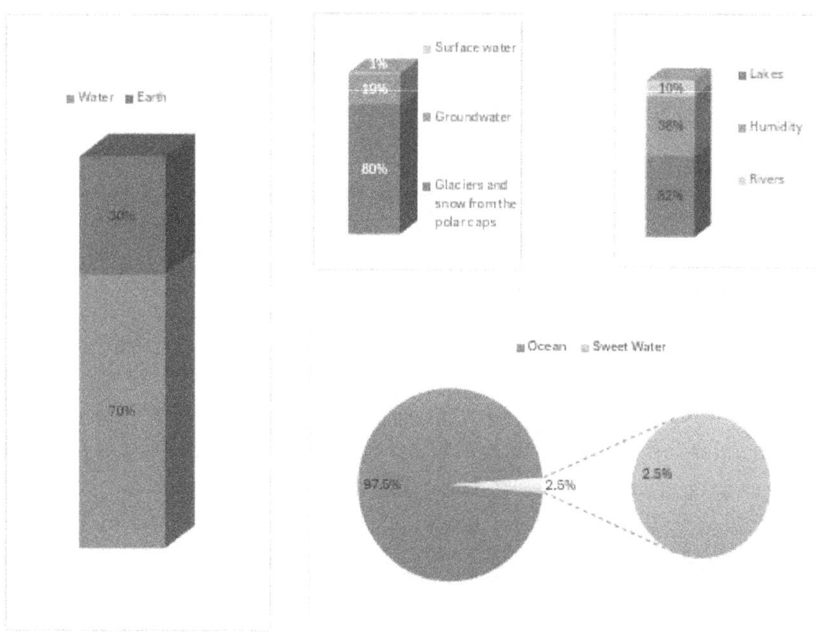

An analysis conducted in 1999 regarding water resources in the world's population estimated the percentages by continent and population as follows (Figure 2): Asia with 60% of the population and only 36% of water, Europe with 13% of the population and 8% of water, in Africa 13% of humanity and only 11% of water, while in North and Central America 8% of the population with 15% of water, and finally, South America with 6% of the population, but with 26% of the water resource (Fernández-Jáuregui, 1999).

Figure 2. Percentages of population and water resources by continent.

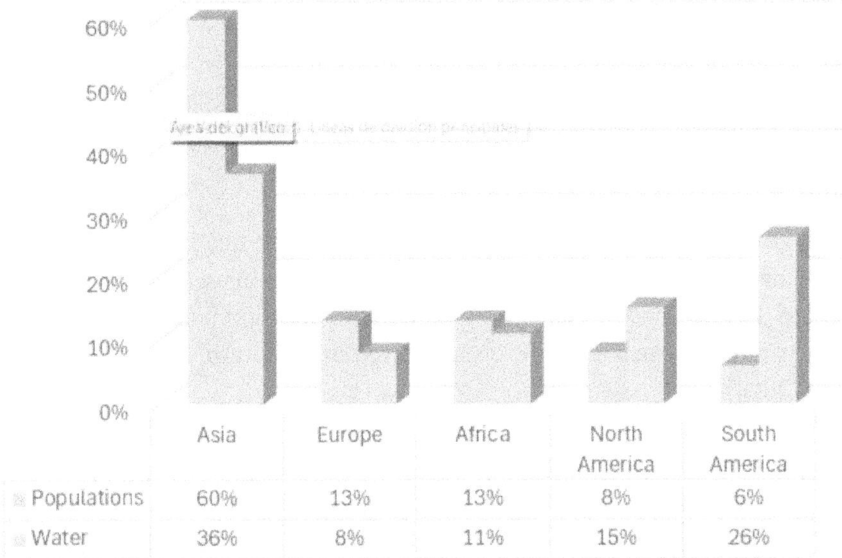

These analyses are fundamental for understanding that the use of potable water in each part of the planet is necessary for sustaining civilizations. However, despite freshwater being essential for human existence, the quantity available on our planet does not guarantee that all humans will always have access to this resource. As observed in Figure 1, freshwater is the resource humanity needs, but its quantity does not ensure universal access.

Figure 2 concludes that for 100% of the population, only 96% of water is available, and this allocation is not proportionate across continents. In other words, the amount of water accessible to continents is not proportional to their population.

From 1999 to the present day, the population has grown significantly, indicating that access to water is decreasing in relation to population growth and the use that civilization gives it through domestic consumption, agriculture, and industry (Escolero et al., 2016).

This chapter focuses on the analysis using sensors to act in controlling and monitoring water flow in cities. Through the information obtained and the applied artificial intelligence (AI), insight is expected to be gained, leading to an alternative solution to the current problem of water loss due to pipe leaks when supplying cities. Leaks constitute the largest water losses recorded in 2021, accounting for 40%, although currently, it is estimated to be approximately 60% during transportation (Izazola, 2001).

(Asociación de Tecnología Alternativa, 2007) assumes that a dripping tap at a rate of one drop per second will waste more than 12,000 liters of water per year. That is why civilizations opt for the use of artificial intelligence applied to the monitoring and control of water flow, as the assumption estimated is very minimal considering that usually the flow of leaks can be more continuous.

REAL-TIME WATER QUALITY MONITORING

One of the most significant advancements in water quality monitoring is the development of real-time monitoring systems. These innovative systems utilize advanced sensors and data analysis techniques to continuously monitor water quality parameters such as pH levels, dissolved oxygen, turbidity, and temperature (Faster, 2024).

IoT devices can be deployed in bodies of water to collect real-time data and wirelessly transmit it to a central server for analysis (Figure 3).

To ensure accuracy and reliability in flow measurement, there are various flow meter technologies available that cater to the requirements of different sectors, such as the oil and gas industry, water and wastewater treatment, chemical and petrochemical industries, among others (Bid, 2023). Some sensors are described below.

Figure 3. Water monitoring with the Real-time Water Quality Monitoring sensor, and IoT (Shenbaga Lakshmi et al., 2019).

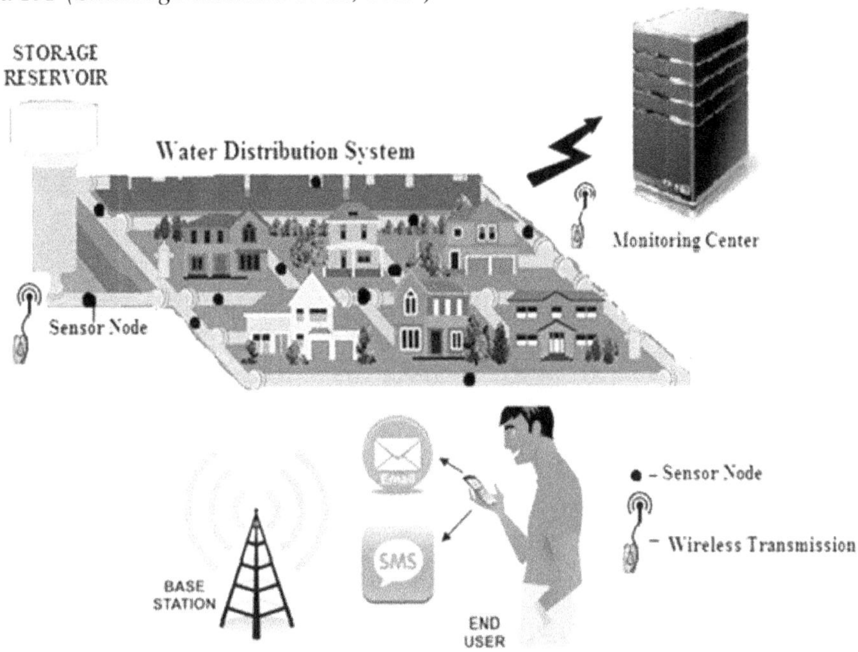

Differential Pressure (DP) Sensors

Differential pressure meters measure the flow of liquids in pipes by creating a constriction that causes a pressure drop. Pressure sensors measure the pressure before and after the constriction, and the difference between them correlates with the liquid flow rate (Figure 4).

> Advantages: its measurements are wide-ranging, they can be safely isolated from the process; Measurements can be digitally networked for remote computer access
> Disadvantages: they require power and active electronics, they require two penetrations in the vessel, they depend on the density and temperature of the liquid (Nikolov & Nikolova, 2008).

Figure 4. Differential pressure transmitter explained (Gregory, 30-Dec-19).

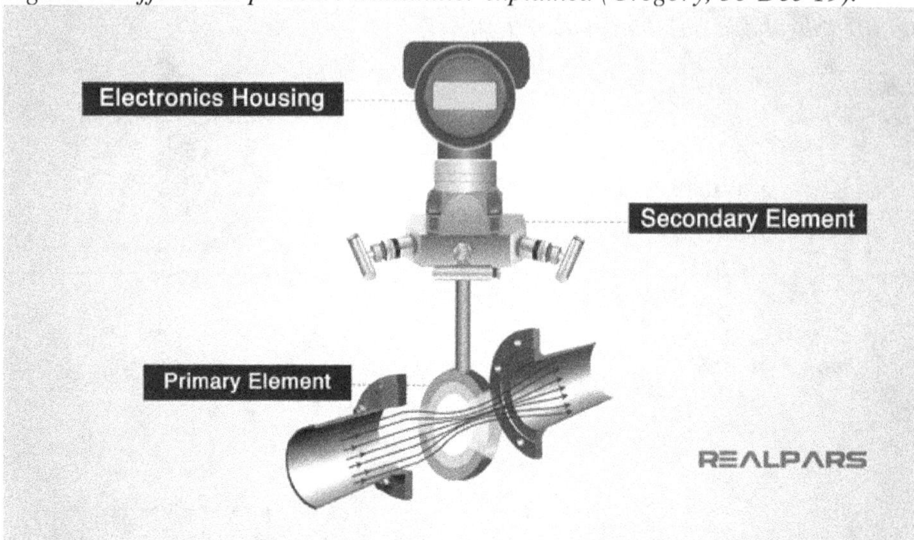

Positive Displacement (PD) Sensors

Positive displacement meters measure the volumetric flow rate of liquids or gasses by repeatedly trapping them with rotating parts that measure volume (Figure 5). There are various types of positive displacement meters, such as oscillating piston, nutating disk, industrial gear, and others. These meters are often chosen for their high precision, excellent repeatability, and large reduction ratios.

> Advantages: Good accuracy and wide measuring range, suitable for fluids with very high viscosity, allowing for large changes in viscosity, Direct local reading in volume units. The reading can be transmitted through a pulse system, No electrical power required, Wide flow range.
> Disadvantages: It is expensive, especially in large sizes, can be damaged by overspeed, not suitable for abrasive or dirty fluids (Villajulca, 2009).

Figure 5. Sensor performance characterization (Omega Engineering, 2024).

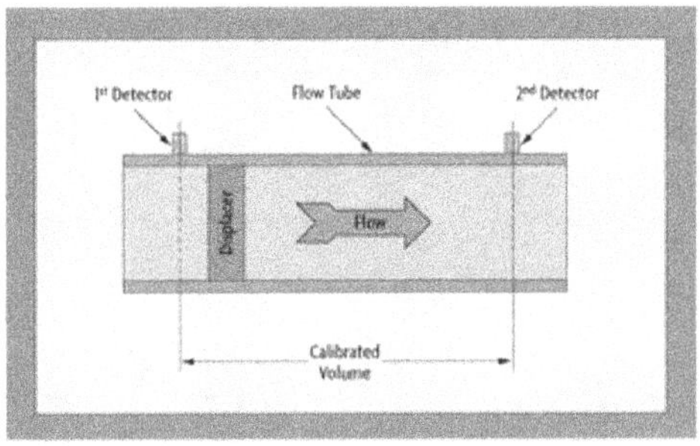

Ultrasonics Sensors

This technique uses two sensors placed at opposing positions at each end of the measuring tube. Each sensor alternately emits and receives ultrasonic signals while measuring the transit time of these signals.

When a fluid flows through the measuring tube, signals traveling in the same direction as the flow accelerate, while those traveling in the opposite direction decelerate. The differential between the transit times measured by both sensors is directly proportional to the flow rate (Figure 6).

Advantages: Lower wiring cost, activated input connection for each proximity switch quick response, as all proximity switches are permanently activated
Disadvantages: The object cannot be assigned to a specific proximity switch (Pepperl & Fuchs, 2020).

Figure 6. Functionality of the ultrasonic sensor (Dubey, 2020).

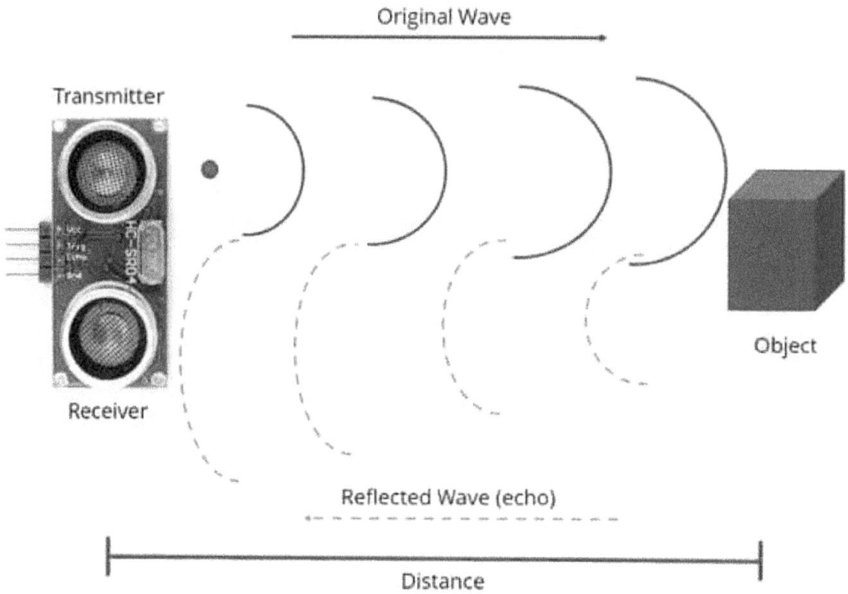

Coriolis Sensors

Each Coriolis flow meter consists of one or more measuring tubes that are artificially oscillated by an exciter. When a fluid passes through the measuring tube, its inertia causes a torsional overlap in these oscillations. Two sensors detect this change in tube oscillation in time and space, recorded as "phase shift." This phase shift is a direct measure of the mass flow rate of the fluid (Figure 7).

Advantages: Due to having no moving parts, they are easy to maintain and their accuracy is not affected by erosion, corrosion, or sensor coating. They are easy to install and purge. They have high accuracy in measurement.
Disadvantages: Their use in gas flow measurement is limited. Their use is limited in flow measurement applications in pipes with diameters larger than 6 inches due to the size of the sensors. They are expensive, so they are not recommended for simple applications where accuracy is not required and where another meter can be used satisfactorily (Ihuhwa Marta Tau, 2014).

Figure 7. 7-A and Figure 7-B: Two-Tube and Straight-Tube Coriolis Meter Operation (Omega Engineering, 2023).

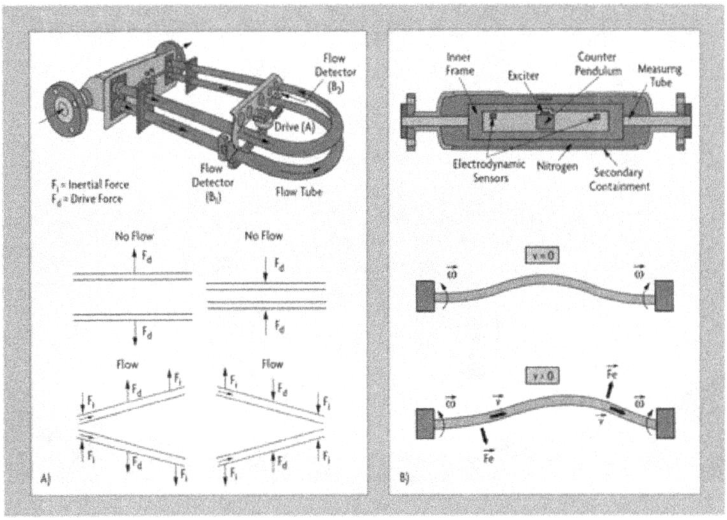

Electromagnetic Sensors

Faraday's law of induction states that when a conductor moves through a magnetic field, an electric voltage is induced in the conductor. This principle is applied in electromagnetic flow meters, where electrically charged particles passing through the magnetic field generated by two coils induce an electric voltage that is directly proportional to the flow velocity and, therefore, the volumetric flow rate (Figure 8).

Pulsating direct current of alternating polarity is responsible for generating the magnetic field. This approach ensures zero point stability and ensures that the measurement is insensitive to non-homogeneous or multi-phase liquids, and can be performed in low-conductivity liquids.

Advantages: Its installation is very easy since it is mounted on the pipe, and the maintenance it requires is minimal.
Disadvantages: It is highly susceptible to electromagnetic interference, and the precision is not very adequate (Madrigal et al., 2010).

Figure 8. Electromagnetic sensor system (Bob, 2022).

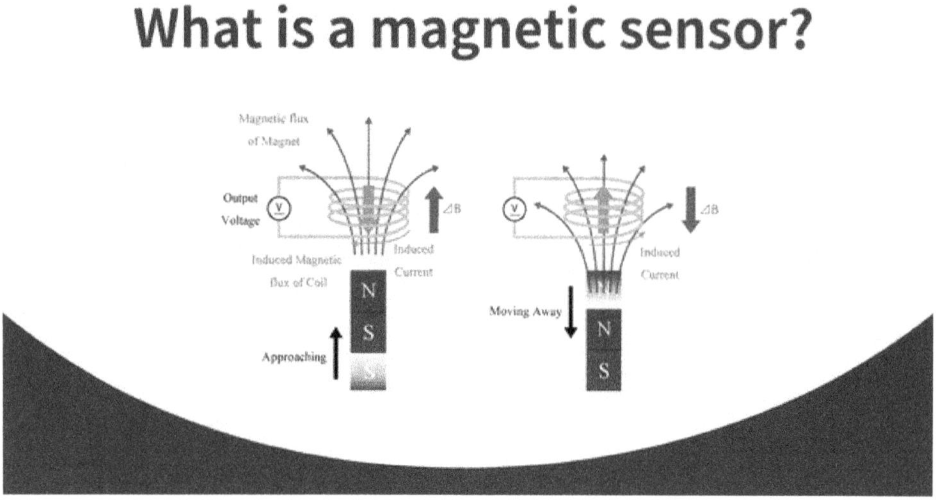

Thermal Dispersion Sensors

Thermal dispersion is a flow measurement principle based on the cooling experienced by a hotter body when traversed by a fluid flow.

One sensor sets the reference temperature by measuring the effective temperature of the fluid, while the other maintains a constant temperature difference with respect to the first under zero flow conditions.

As the fluid circulates through the measuring tube, the hotter sensor is cooled by the fluid passing alongside it, with this cooling being proportional to the fluid velocity. The electric current needed to maintain a constant temperature difference between the sensors is used to calculate the mass flow rate (Figure 9).

> Advantages: It has a high precision in water flow measurement, its installation is external on the pipe, and it is cost-effective.
> Disadvantages: It can be damaged by high temperatures, and it requires a stabilization time before providing readings (Varona et al., 2013).

Figure 9. The operation of the sensor in a pipeline (Melina, 2018).

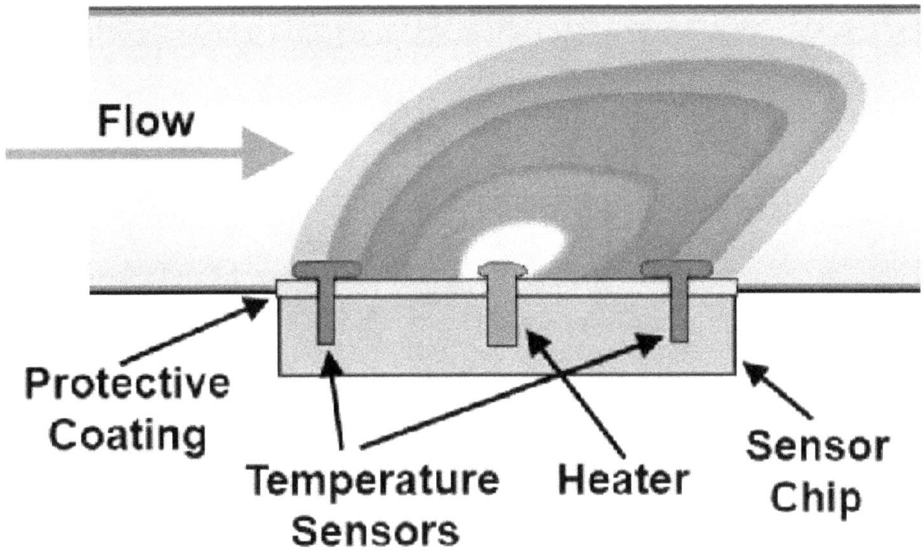

Turbine-Type Sensors

These flow meters harness the mechanical energy of liquids to rotate a rotor within the fluid flow. The rotation speed of the rotor is proportional to the fluid velocity passing through the meter (Figure 10).

These meters reliably measure the velocity of liquids, gasses, and vapors and are often chosen for their durable measurement structure, high accuracy range at low cost, and remarkable repeatability over a wide range of temperatures and pressures.

Advantages: Its use is suitable for obtaining readings at different speeds; it is economical.

Disadvantages: Its motor tends to wear out; installation is complex; it can be damaged if it interacts with solid material (Penereiro & Araujo, 2018).

Figure 10. Sensor operating in a physical environment (RealPars, 2023).

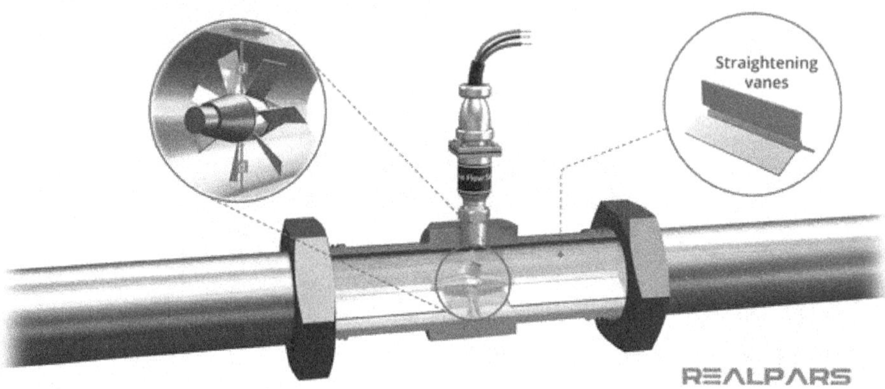

Vortex Sensors

These flow sensors use the von Kármán effect for the measurement of liquids, gasses, and vapors. To measure, an obstruction, also known as a bluff body, is placed in the flow path, generating alternating differential pressure vortices.

These vortices cause a small sensor device to oscillate with a frequency directly proportional to the velocity of the moving fluid. Subsequently, the sensor element converts the oscillation frequency into an electrical signal, which is transformed into a quantifiable reading of the velocity (Figure 11).

Vortex meters are highly valued for their wide range capability, repeatability, and precision in measuring liquids, gasses, and saturated vapor (Bino, 2023).

> Advantages: They have high durability in high temperatures, good sensitivity to measure different speeds, are not expensive, and maintenance is very easy.
> Disadvantages: They are very sensitive to vibration, which affects reading accuracy, sensitive to magnetic interference, and installation is complex (González et al., 2017).

Figure 11. Characterization of the sensor mechanism with flow readings in a pipeline (Smart Sensor, 2024).

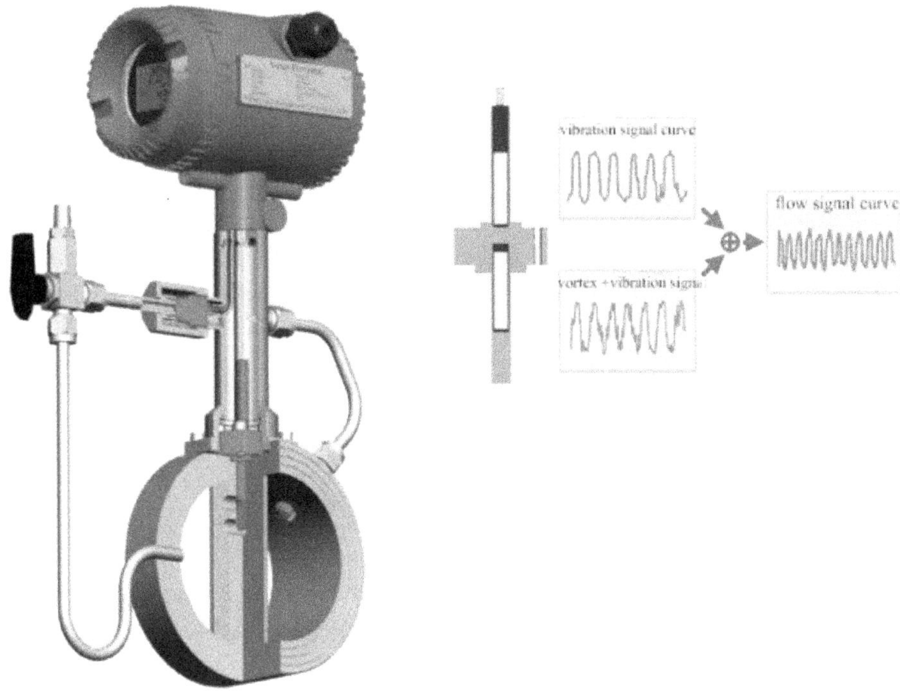

USE OF ALGORITHMS TO MONITOR WATER QUALITY AND CHARACTERISTICS

Currently, all sets of tasks and jobs are carried out through protocols or defined instructions to execute an action in an orderly manner to obtain a benefit. The computer and mechanical industries work essentially with algorithms as they implement these techniques with ordered instructions and rules to solve problems, whether complex or simple. They have no limitations because it is a sequence of structured and well-defined steps to achieve the expected outcome.

(Apaza et al., 2022) implemented a decision tree with predictive modeling for the detection, control, and quality of drinking water. This autonomous learning tool is of paramount importance for solving complex problems related to water quality, offering a predictive approach that benefits people's health and community supply control.

(Gutiérrez et al., 2006) used neural networks with activation neurons as a modeling tool for predicting seawater quality, focusing on four quality indicators, namely water temperature, pH, dissolved oxygen content, and turbidity.

(García et ai, 2010) employed Artificial Neural Networks (ANNs) to predict and forecast quantitative characteristics of water bodies in a coastal area of Singapore, facilitating the rapid assessment and forecast of certain water quality variables such as salinity, temperature, dissolved oxygen, and chlorophyll-a. They also implement an ANN methodology as a tool for estimating water quality in the upper and middle basin of the Bogotá River.

(Pelletier, 2005) implemented QUAL2Kw, a tool for simulating water quality in streams and rivers; it includes a genetic algorithm to facilitate model calibration for application to particular water bodies. The genetic algorithm is used to find the combination of kinetic rate parameters and constants, resulting in a better fit for a model application compared to observed data.

(Romero & Shan, 2005) developed a software tool based on neural networks to predict the discharge water temperature from the channel in a coal-fired power plant. They include plant operating parameters and local weather conditions, as well as tidal information.

(López et al., 2016) developed an ANN model to predict contaminant transport in groundwater; for this, researchers use backpropagation in a 3-layer structure with training data.

Technological Implementations for Water Control and Management

It is known that Artificial Intelligence (AI) is essentially the evolution of algorithms, as they are used to train AI models through data analysis and pattern recognition to solve tasks that would normally require human intelligence, allowing them to be done efficiently and effectively.

Another great benefit of working with artificial intelligence is that, in addition to automating processes, it can be monitored and adapted to any type of problem remotely through the Internet of Things (IoT), providing a wide range of possibilities for control, analysis, operation, quality, and optimization.

Using this technology offers a great opportunity to implement a variety of techniques for controlling, saving, and dispersing the flow of water, adapting water usage to different environments as mentioned below.

In Ecuador, (León et al., 2024) utilized the ESP3246 module and IoT, a flow sensor, and a YF-S201 shut-off valve to propose the monitoring and remote control of potable water, aiming for a more accurate and reliable reading. This benefits both the economy of the distributing company and the consumer. They also developed a

Telegram Bot to directly check consumption at any time of the day, and it will send a notification every 24 hours with daily or monthly consumption details.

(Yacelga et al., 2023) implemented a smart water meter that allows unlimited reading data collection. They also developed an application that lets consumers and service personnel visualize this information. This project was developed in Ecuador for use in its communities.

(Paredes Paucar, 2022) implemented the ESP8266 module by adapting the smart node in meters. This sensor captures 10 liters/impulse of water consumption and uses GPS to capture the current positioning of this tool. It also performs billing based on the accumulated consumption at the end of each month and the contracted category, allowing visualization through a web application.

(Kleiman et al., 2016) describes the support obtained by applying AI in Valencia, Spain. Their company focuses on implementing smart water meters to send daily or hourly information on water usage to the utility, using this detailed data to identify water leaks and frauds, as well as providing information to users about their water consumption. Additionally, utility employees no longer need to go house by house to read the water meter. They implement the use of artificial intelligence for plant irrigation using a microcontroller ATmega328, moisture sensor, temperature sensor, servo motor, pump driver LM298 for water pump control, and LED indicator. This was a functional scale project carried out at Amrita School of Engineering, India. Although the irrigation project was focused on watering a single plant, it is not limited to adapting this project for large-scale irrigation control for agricultural purposes.

CONCLUSION

Real-time monitoring of water quality is crucial to ensure public health and environmental sustainability. This approach allows rapid detection of contaminants and variations in key parameters such as pH, temperature, turbidity, dissolved oxygen and the presence of toxic substances. The ability to obtain instant data makes it easier to identify and mitigate problems before they become serious threats to aquatic ecosystems and human health. Additionally, continuous monitoring provides valuable information for water resources management, helping authorities and companies make informed decisions about water treatment, waste management and emergency response. In industries such as agriculture, real-time monitoring can optimize the use of fertilizers and pesticides, reducing runoff pollution and protecting water sources. Advanced technology, such as IoT sensors, AI and data analytics platforms, has revolutionized water quality monitoring, making it more accessible and accurate. This not only improves the capacity to respond to contamination, but also promotes transparency and public trust in water management. In summary, real-time monitoring is an essential tool to preserve water quality and protect health and the environment.

REFERENCES

Apaza, A. A. Z., Gárate, S. S. O., Cuadros, A. E. J. C., & Ccasa, P. M. C. (2022). Modelo predictivo de la potabilidad del agua mediante un árbol de decisión en Inteligencia Artificial. *Innovación y Software*, 3(2), 121–131. DOI: 10.48168/innosoft.s9.a72

Asociación de Tecnología, A. (2007). Smart water solutions. *Sanctuary: Modern Green Homes*, 1, 34–35.

Bid, I. (2023, noviembre 28). *Cómo las nuevas tecnologías están transformando el abastecimiento de agua y saneamiento en Latinoamérica y el Caribe*. https://idbinvest.org/es/publicaciones/como-las-nuevas-tecnologias-estan-transformando-el-abastecimiento-de-agua-y

Bino, ovo. (2023, abril 13). *Sensores de flujo para medir la velocidad de flujo, volumen...* Binoovo. https://binoovo.com/blog/sensores-de-flujo-medir-flujo-volumen-masa-liquidos-gases

Bob, D. (2022, junio 8). *What is a magnetic sensor? What are the categories?*https://www.linkedin.com/pulse/what-magnetic-sensor-categories-bob-dong

Cirelli, A. F. (2012). El agua: Un recurso esencial. *Química Viva*, 11(3), 147–170.

Dubey, A. (2020, agosto 4). Working of an Ultrasonic Sensor. *NerdyElectronics*. https://nerdyelectronics.com/working-of-an-ultrasonic-sensor/

Engineering, O. (2023). *What is a Coriolis Flow Meter and How Does it Work*

Engineering, O. (2024). *Lessons about Positive Displacement Flow Meters*. https://www.omega.com/en-us/resources/positive-displacement-flow-meter

Escolero, O., Kralisch, S., Martínez, S. E., & Perevochtchikova, M. (2016). Diagnóstico y análisis de los factores que influyen en la vulnerabilidad de las fuentes de abastecimiento de agua potable a la Ciudad de México, México. *Boletín de la Sociedad Geológica Mexicana*, 68(3), 409–427. DOI: 10.18268/BSGM2016v68n3a3

Faster, C. (2024). *Monitoreo De Calidad Del Agua*. FasterCapital. https://fastercapital.com/keyword/monitoreo-de-calidad-del-agua.html

Fernández-Jáuregui, C. A. (1999). El agua como fuente de conflictos: Repaso de los focos de conflictos en el mundo. *Revista CIDOB d'Afers Internacionals*, 45/46, 179–194.

García et ai. (2010). *Transporte de Contaminantes en Aguas Subterráneas mediante Redes Neuronales Artificiales*. https://www.scielo.cl/scielo.php?script=sci_arttext&pid=S0718-07642010000500011

González, L. F. D., Casierra, G. A. T., & Moreno, E. del R. (2017). Confinated horizontal vortex generator. *Ingeniería y Desarrollo*, 35(1), 38–57.

Gregory, D. (30-Dec-19). *Differential Pressure Transmitter Explained—RealPars*. https://www.realpars.com/blog/differential-pressure-transmitter

Gutiérrez, J. D., Riss, W., & Ospina, R. (2006). Bioindicación De La Calidad Del Agua En La Sabana De Bogota - Colombia, Mediante La Utilización De La Lógica Difusa Neuro-Adaptativa Como Herramienta. *Caldasia*, 28(1), 45–56.

Ihuhwa Marta Tau. (2014). *Efecto Coriolis | PDF | Física | Física Aplicada e Interdisciplinaria*. Scribd. https://es.scribd.com/presentation/199377222/Efecto-Coriolis

Izazola, H. (2001). Agua y sustentabilidad en la Ciudad de México. *Estudios Demográficos y Urbanos*, 16(2 (47)), 285–320. DOI: 10.24201/edu.v16i2.1121

Kleiman, N., Hillard, T., Forman, A., Gold, A., Weng, D., Adibhatla, V., Carillo, S., Coffy, A., Gichinga, R., Kleinfeld, Z., Prakash, M., Riaz, S., & Sokolsky, J. (2016). *Smart Water Management* (Innovation and the City, pp. 26-27). Center for an Urban Future. https://www.jstor.org/stable/resrep14850.15

León, L., & Rolando Israel, M. S. Juan, & Gabriel. (2024). *Construcción de un medidor de agua inteligente mediante Arduino para reducir las pérdidas en la distribución de agua*. https://dspace.ucacue.edu.ec/items/94539eb1-9644-437a-9aef-e00acf619889

López, I. D., Figueroa, A., & Corrales, J. C. (2016). Un mapeo sistemático sobre predicción de calidad del agua mediante técnicas de inteligencia computacional. *Revista Ingenierías Universidad de Medellín*, 15(28), 35–51. DOI: 10.22395/rium.v15n28a2

Madrigal, L. P., Meraz, J. G., Wiegand, C. L., Reyes, J. I., & Delgado, J. M. (2010). Monitoreo De La Salinidad Mediante Sensores Remotos. *Terra Latinoamericana*, 28(1), 15–26.

Melina. (2018, September). *Fig. 2. Principle of operation for thermal mass flow sensor. Extracted...* ResearchGate. https://www.researchgate.net/figure/Principle-of-operation-for-thermal-mass-flow-sensor-Extracted-from-18_fig1_342260098

Nikolov, G., & Nikolova, B. (2008). *Virtual Techniques For Liquid Level Monitoring Using Differential Pressure Sensors. 9*(2).

Paredes Paucar, H. A. (2022). *Diseño de un prototipo de nodo inteligente integrable a una red inalámbrica para telemedición del consumo de agua potable como propuesta para el GADM de Guano.* http://dspace.espoch.edu.ec/handle/123456789/20774

Pelletier, G. J., Chapra, S. C., & Tao, H. (2005). *QUAL2Kw—A framework for modeling water quality in streams and rivers using a genetic algorithm for calibration.* https://apps.ecology.wa.gov/publications/summarypages/0503044.html

Penereiro, J. C., & Araujo, G. de S. (2018). Proposal of using sensor networks for measurements of the efficiencies of a low cost vertical wind turbine. *Acta Scientiarum. Technology*, 40(1), 35490. https://www.redalyc.org/journal/3032/303258327039/. DOI: 10.4025/actascitechnol.v40i1.35490

Pepperl & Fuchs. (2020, octubre 14). *Ultrasonic Sensors Knowledge (Part 6) – Synchronizing Ultrasonic Sensors.* Pepperl+Fuchs. https://www.pepperl-fuchs.com/mexico/es/27967.htm

RealPars. (2023). *Turbine Flow Meter Explained | Operation and Calibration—RealPars.* https://www.realpars.com/blog/turbine-flow-meter

Romero, C., & Shan, J. (2005). Development of an artificial neural network-based software for prediction of power plant canal water discharge temperature. *Expert Systems with Applications*, 29(4), 831–838. DOI: 10.1016/j.eswa.2005.06.009

Shenbaga Lakshmi, G., Revathi T., & Kavi Priya Shanthosam. (2019, diciembre). *Fig. 6. System Architecture.* ResearchGate. https://www.researchgate.net/figure/System-Architecture_fig6_337733237

Smart Sensor, C. (2024). *Comate meter | Flow measurement for compressed air, gases, steam.* Comate Flow Meter. https://www.comateflowmeter.com/

Varona, J., Tecpoyotl-Torres, M., & Velázquez, R. (2013). Micro sensor-actuador térmico sin baterías para aplicaciones en microelectrónica de ultra-bajo consumo de potencia. *Revista Mexicana de Física*, 59(1), 26–38.

Villajulca, J. C. (2009, septiembre 27). Caudal - Flujo: Medidores por desplazamiento positivo. *Instrumentacion y Automatizacion Industrial.* https://instrumentacionycontrol.net/caudal-flujo-medidores-por-desplazamiento-positivo/

Yacelga, A. R. L., Vásquez, R. A. D., Espinoza, J. L. A., & Cabrera, M. A. C. (2023). Implementación de prototipo de medición del consumo de agua potable mediante utilización de tecnologías inteligentes en la parroquia Julio Andrad, Ecuador. *Universidad y Sociedad*, 15(S2), S2.

Chapter 5
The Critical Role of Water Quality:
Health Impacts, Contaminants, and Sustainable Solutions for Environmental and Human Well-being

Shalom Akhai
https://orcid.org/0000-0002-7533-457X
MMEC MM(DU), Mullana, India

Tanu Taneja
https://orcid.org/0009-0004-3183-2819
Department of Civil Engineering, RIMT University, India

ABSTRACT

Clean water is essential for human health and environmental sustainability. This paper explores the intricate link between water quality, public health, and ecosystem stability, highlighting the dangers posed by contaminants such as heavy metals, pathogens, and chemicals. It examines the health risks associated with polluted water and emphasizes the urgent need for sustainable solutions to mitigate these threats. By addressing water contamination through advanced treatment technologies, policy reforms, and community-driven initiatives, this work underscores the critical role of clean water in fostering a healthier, more sustainable future for all.

DOI: 10.4018/979-8-3693-8074-1.ch005

1. INTRODUCTION

Water is a vital resource for life, ecosystems, agricultural productivity, and human well-being. It plays a crucial role in agriculture by transforming sunlight and soil nutrients into nourishment that sustains human existence. However, our water sources are increasingly vulnerable to contaminants such as bacteria, viruses, toxins, and heavy metals, which affect their overall quality. The advancement of technology has resulted in the buildup of chemical pollutants, while the rapid expansion of urban areas has strained sewage systems, leading to the presence of hazardous microbes. The unintended consequences of intensive farming efforts include the contamination of rivers and aquifers due to the discharge of excessive fertilisers and pesticides.

Water's most detrimental property is its capacity to create a discrepancy between appearances and reality, making it difficult to perceive the danger presented by contaminated water. Preserving the purity of drinking water is crucial, as it directly impacts our well-being and general quality of life. Access to clean water is not only essential for maintaining good health but also considered a fundamental human right (Barzilay, Weinberg, & Eley, 1999). Extended exposure to contaminants can lead to subtle but broad effects, such as slight cognitive impairments, minor hormone abnormalities, or low-grade inflammation. Contaminants present in seemingly innocuous water sources may potentially cause significant health hazards, such as gastrointestinal illnesses and long-lasting chronic conditions (Bozorg-Haddad, Delpasand, & Loáiciga, 2021). The scope of these health hazards is extensive, including organisms such as E. coli or Giardia that may lead to severe diarrhea, rapid dehydration, and potentially life-threatening outcomes in children. Lead exposure may result in developmental delays, learning difficulties, and behavioral problems. Per- and polyfluoroalkyl substances (PFAS) are often known as "forever chemicals" because they can induce liver damage, thyroid disease, and certain forms of cancer. Water testing is an essential technique for identifying water contamination, enabling proactive measures to avert possible hazards and allowing for focused and cost-effective interventions (Zhang et al., 2023; Kumar & Puri, 2012). The complexity of water contamination arises from the many interconnections between rivers, aquifers, soil, air, and living organisms. Chemical pollutants from industrial output, agricultural practices, and domestic waste pose additional health risks, including acute toxicity and long-term diseases like cancer. Heavy metals, such as lead, mercury, and arsenic, can cause severe harm to human health, including neurotoxicity, cardiovascular problems, and increased susceptibility to cancer. Thorough water testing is crucial for safeguarding the health and welfare of people and maintaining the integrity of the environment. Interdisciplinary collaboration, including public and private sectors, and international cooperation can contribute to the preservation of water as an essential resource for life and community well-being (Awuchi, Igwe, & Amagwula, 2020).

2. THE CRUCIAL ROLE OF CLEAN DRINKING WATER IN HUMAN HEALTH

Clean drinking water is essential for human health, playing a critical role in digestion, waste disposal, temperature control, and joint lubrication (Kumar & Puri, 2012; Awuchi et al., 2020; Munteanu et al., 2021). Water's solvent properties aid in nutrient absorption, electrolyte balance, and detoxification. Even mild dehydration can cause digestive issues, fatigue, and joint pain (Ji et al., 2020). Water contamination can lead to malabsorption, vitamin deficiencies, and increased risk of kidney disease (Saravanan et al., 2021; Samanta, 2019). Furthermore, clean water is necessary for thermoregulation, with dehydration impairing the body's ability to cool itself (Hassan & Suhail, 2023; Ugran et al., 2017).

2.1 The Impact of Water Contamination

Water pollution can have severe consequences, including:

- Digestive issues and malabsorption
- Kidney damage and disease
- Joint pain and inflammation
- Impaired thermoregulation
- Nutritional deficiencies

2.2 The Importance of Safe Drinking Water

Understanding the value of safe drinking water is crucial for protecting public health. Clean water is essential for maintaining cellular and physiological processes, and its importance cannot be overstated (Barzilay et al., 1999). With water shortages and pollution becoming increasingly significant global issues, ensuring access to clean drinking water is a critical concern for community resilience and public health.

Table 1. Importance of clean water for human body functions

Key Aspect	Importance of Clean Water	Consequences of Contaminated Water	References
Digestion	Crucial for breaking down food particles and facilitating nutrient absorption, essential for energy, cellular function, and biological activities	Disruption of the digestive process, leading to malabsorption and nutrient deficiencies	Kumar & Puri, 2012
Waste Disposal	Aids kidneys and urinary tract in eliminating toxins and metabolic byproducts	Accumulation of harmful compounds, compromising detoxification capabilities	Awuchi et al., 2020; Saravanan et al., 2021; Samanta et al., 2019
Temperature Regulation	Maintains body temperature, cooling through sweating	Disruption of body's cooling mechanisms, impairing temperature regulation	Munteanu et al., 2021; Akhai et al., 2016a; Akhai et al., 2016b; Saghiv et al., 2020; Lim, 2020; Hassan & Suhail, 2023; Ugran et al., 2017
Joint Lubrication	Lubricates joints, reducing friction and allowing painless movement	Trigger inflammation and joint issues, limiting mobility	Ji et al., 2020

Table 2 highlights the impact of drinking water quality on health and wellbeing, including digestion, waste elimination, temperature regulation, and disease prevention.

Table 2. Drinking water quality: Impact on health and wellbeing

Function	Importance	Consequences of compromised quality
Digestion	Breaks down food particles, facilitates nutrient absorption	Malabsorption, nutritional deficiencies
Waste elimination	Enables efficient kidney and urinary system function	Toxin accumulation, compromised detoxification
Temperature regulation	Facilitates sweating and heat dissipation	Overheating, impaired cooling ability
Musculoskeletal system	Lubricates joints for smooth movement	Inflammation, joint problems
Disease prevention	Prevents waterborne diseases and long-term health issues	Gastrointestinal distress, dehydration, chronic health conditions

3. COMMON WATER CONTAMINANTS

Pollutants found in water include bacteria, viruses, and parasites, all of which have the potential to cause a wide variety of illnesses. While viruses and parasites are capable of causing cholera, giardiasis, and cryptosporidiosis, these bacteria have the potential to cause devastating infections that affect the gastrointestinal tract. It is necessary for these bacteria to go via water sources in order to undergo stringent testing and purification (Ali et al., 2021). The presence of these microorganisms in drinking water poses a significant threat to public health, particularly in regions with inadequate water treatment facilities. In developing countries, where access to clean water is often limited, waterborne diseases remain a leading cause of morbidity and mortality, especially among children under five years old.

There are significant health risks associated with the introduction of many types of chemicals into water sources. Streams and rivers are contaminated with heavy metals and synthetic chemicals as a result of industrial runoff. Agriculture, including the use of pesticides, fertilisers, and home chemical waste, makes the problem much worse. In addition to causing acute and chronic ailments, chemical pollution may also generate neurotoxic chemicals, which can increase the likelihood of developing cancer (Sarma, 2022). The pervasive nature of these chemical contaminants is particularly alarming. Even in developed nations with advanced water treatment technologies, trace amounts of pharmaceuticals, endocrine disruptors, and microplastics are increasingly being detected in drinking water supplies. These emerging contaminants pose new challenges for water treatment professionals and raise concerns about their long-term effects on human health.

According to Pagano et al. (2023) and Egbuna et al. (2021), the fight against chemical contamination requires the implementation of regulatory measures, the promotion of sustainable agriculture, and the raising of public awareness. This multifaceted approach recognizes that water pollution is not merely a technological problem but also a socio-economic and political issue. Stricter regulations can compel industries to adopt cleaner production methods and properly treat their wastewater. Simultaneously, promoting sustainable agricultural practices, such as organic farming and precision fertilizer application, can reduce the runoff of agrochemicals into water bodies. Public awareness campaigns play a crucial role in this battle, educating consumers about the environmental impact of their choices and encouraging them to support eco-friendly products and practices.

It is very hazardous for lead, mercury, and arsenic to make their way into water systems. Activities that are caused by humans have the potential to introduce these elements into water sources, particularly in vulnerable populations such as pregnant women and newborns. In addition to neurological and cardiovascular disorders, chronic exposure to heavy metals has been linked to the development of malignant diseases (Balali-Mood et al., 2021; Zhang et al., 2024; Koyama et al., 2024). The insidious nature of heavy metal contamination lies in its bioaccumulative properties. Unlike some pollutants that break down over time, heavy metals persist in the environment, accumulating in aquatic organisms and moving up the food chain. When humans consume contaminated fish or use polluted water for irrigation, these toxic elements can enter their bodies. Over time, even low-level exposures can lead to serious health issues, as heavy metals tend to accumulate in specific organs like the brain, liver, and kidneys.

The presence of heavy metals in water sources necessitates the implementation of stringent monitoring and mitigation strategies in order to protect both the environment and human health. This involves not only treating contaminated water but also preventing pollution at its source. Old lead pipes in aging infrastructure, for example, continue to leach this toxic metal into drinking water in many cities. Replacing these pipes is a massive undertaking but essential for public health. Similarly, better regulation and cleanup of mining sites, which are major sources of mercury and arsenic pollution, are critical steps. In some cases, phytoremediation—using plants that absorb heavy metals from soil and water—offers a promising, eco-friendly solution for cleaning up contaminated sites.

Graphic 1 depicts bacteria in water as seen through a microscope, offering a stark visual reminder that even clear water can harbor invisible threats. This microscopic view underscores the importance of rigorous water testing, as many dangerous microorganisms are undetectable to the naked eye. **Graphic 2** depicts a river that is polluted by industrial waste, showcasing the stark reality of human-induced water contamination. The image likely shows discolored water, perhaps with an oily sheen or floating debris, highlighting how industrial activities can dramatically alter aquatic ecosystems. **Graphic 3** depicts contaminated water flowing from a tap, bringing the issue directly into people's homes. This powerful image serves as a wake-up call, demonstrating that water pollution is not just an abstract environmental problem but a direct threat to the safety of our household water supply.

Table 3 provides an overview of some common water pollutants and the health hazards that are connected with them. This table likely categorizes various contaminants—microbiological, chemical, and heavy metals—and lists their specific health impacts. For example, it might show that E. coli can cause severe gastrointestinal illness, nitrates from fertilizers can lead to "blue baby syndrome" in infants, and lead exposure can result in developmental delays. Such a comprehensive table serves as a valuable educational tool, helping the public understand the diverse range of water pollutants and their corresponding health risks. This knowledge empowers individuals to take proactive steps, whether it's using water filters, supporting local water quality initiatives, or advocating for stronger environmental policies.

Table 3. Common water contaminants and health risks

Category	Contaminant Examples	Potential Health Effects	Image
Microorganisms	- Bacteria (E. coli, Legionella) - Viruses (Hepatitis A, Rotavirus) - Parasites (Giardia, - Cryptosporidium)	Gastrointestinal illness, vomiting, diarrhea, fever, dehydration	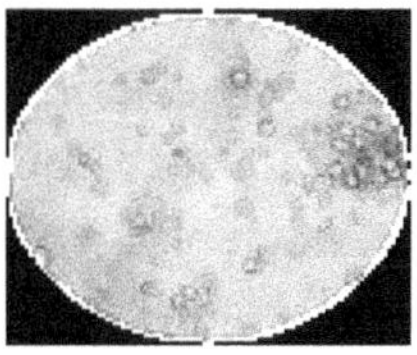
Chemicals	- Pesticides (Atrazine, Glyphosate) - Industrial chemicals (PCBs, PFAS) - Household chemicals (Chlorine byproducts)	Cancer, organ damage, endocrine disruption, developmental problems	
Heavy Metals	- Lead - Mercury - Arsenic - Cadmium	Neurological damage, kidney problems, developmental delays, cancer	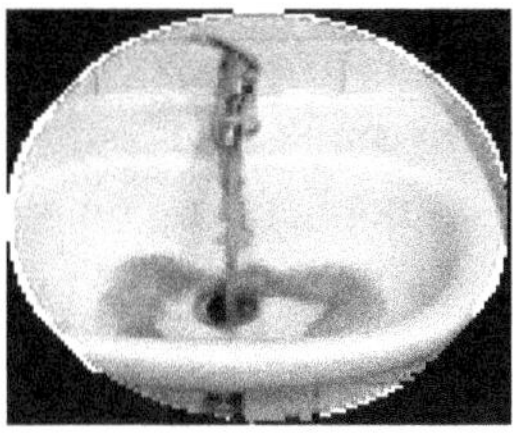

4. CHALLENGES AND SOLUTIONS: PROTECTING OUR PRECIOUS WATER

The world's water future is at risk due to community disengagement, aging infrastructure, and increasing contaminants (Richter et al., 2013). These interconnected issues threaten our relationship with water, highlighting the need for urgent attention.

4.1 Challenges

- **Aging Infrastructure:** Defective pipes, inadequate treatment systems, and old equipment lead to water loss, pollution, and increased operating costs (Spellman, 2013).
- **Emerging Contaminants:** Pharmaceuticals and microplastics pose uncertain health risks, requiring ongoing surveillance and adjustment of treatment methods (Pashaei et al., 2022).
- **Community Disengagement:** Insufficient understanding of water quality and lack of information about conservation hinder community engagement.

4.2 Solutions

- **Infrastructure Improvement:** Implementing long-term maintenance strategies, fostering public-private collaborations, and incorporating intelligent technology (Fältström & Anderberg, 2020).
- **Addressing Emerging Contaminants:** Enhanced scientific investigation, adaptable treatment methods, and global collaboration (Pashaei et al., 2022).
- **Community Engagement:** Implementing public education programs, establishing community-based monitoring initiatives, and providing water conservation incentives.

4.3 Ensuring a Sustainable Water Future

Tackling these challenges requires a comprehensive and future-oriented strategy, involving technical innovation, scientific rigor, and cultural rejuvenation. By addressing these issues, we can transform society's connection to water, ensuring a sustainable future for generations to come (Srinivas et al., 2020; Tinoco et al., 2022).

Table 4 examines water safety issues and remedies, highlighting long-term neglect's impact on water quality, supply disruptions, and expenditures. Smart water management and emerging pollutants research are key tools, while "Community Engagement" encourages education and community-based efforts to address water quality's psychological impact. Technical innovation and cultural regeneration also improve water society.

Table 4. Addressing key challenges and finding solutions for safeguarding water.

Challenge	Description	Potential Impacts	Solutions
Aging Infrastructure	Leaky pipes, outdated treatment plants, and inadequate maintenance increase the risk of contamination and water loss.	Compromised water quality, disruptions in supply, increased costs.	Investments in: - Infrastructure renewal and maintenance: Replacing old pipes, upgrading treatment facilities, and implementing leak detection technologies. - Smart water management: Utilizing sensors and data analytics to optimize resource usage and reduce waste. - Public-private partnerships: Fostering collaboration to secure funding and expertise for infrastructure improvements.
Emerging Contaminants	Pharmaceuticals, personal care products, and other new pollutants pose challenges in detection and treatment due to their evolving nature and unknown long-term effects.	Threats to human health and aquatic ecosystems, increased treatment costs.	Focus on: - Research and development: Studying emerging contaminants, their sources, and effective treatment methods. - Improved monitoring: Implementing advanced testing technologies and expanding the range of monitored contaminants. - Regulatory frameworks: Establishing guidelines and standards for managing emerging contaminants in water sources and wastewater.
Community Engagement	Lack of awareness about water quality issues and limited participation in monitoring and conservation efforts can hinder progress.	Inadequate appreciation for water's value, ineffective resource management, increased vulnerability to contamination.	Prioritize: - Education and outreach: Raising awareness about water quality, its importance for health and the environment, and individual roles in protecting it. - Community-based initiatives: Engaging citizens in water quality monitoring, conservation projects, and decision-making processes. - Empowerment and capacity building: Providing communities with resources and training to actively participate in water management.

5. SUSTAINABLE WATER MANAGEMENT PRACTICES

The article recommends sustainable water management strategies, including water conservation, water reuse and recycling programs, rainwater harvesting systems, integrated water resource management, and green infrastructure investments. These measures aim to reduce water usage, promote environmental stewardship, and mitigate the impacts of urbanization on water resources. Implementing these strategies can help reduce water usage, improve water quality, and foster collaboration among stakeholders. By incorporating these practices, communities can safeguard public health and promote environmental stewardship in the face of growing water challenges.

Recommendations for Sustainable Water Management:

- **Water Conservation Strategies:** Implement water conservation measures such as promoting water-efficient appliances, fixing leaks, and implementing water-saving technologies in industrial, commercial, and residential sectors. Encourage behavioral changes through educational campaigns to reduce unnecessary water usage.
- **Water Reuse and Recycling Programs:** Develop and implement water reuse and recycling programs to treat and repurpose wastewater for non-potable applications such as irrigation, industrial processes, and toilet flushing. Invest in advanced treatment technologies to ensure the safety and quality of reclaimed water.
- **Rainwater Harvesting Systems:** Encourage the adoption of rainwater harvesting systems to capture and store rainwater for various uses, including landscape irrigation, toilet flushing, and groundwater recharge. Provide incentives and subsidies to incentivize the installation of rainwater harvesting infrastructure in both urban and rural areas.
- **Integrated Water Resource Management:** Promote integrated water resource management approaches that consider the interconnectedness of water systems, ecosystems, and human activities. Foster collaboration among stakeholders from different sectors to develop comprehensive water management plans that prioritize sustainability, resilience, and equity.
- **Green Infrastructure Investments:** Invest in green infrastructure projects such as constructed wetlands, green roofs, and permeable pavements to enhance natural water filtration, reduce stormwater runoff, and improve water quality. Integrate nature-based solutions into urban planning and development to mitigate the impacts of urbanization on water resources.

By incorporating sustainable water management practices into water quality initiatives, communities can not only safeguard public health but also promote environmental stewardship and resilience in the face of growing water challenges.

6. RECOMMENDATIONS

The report recommends comprehensive water testing protocols, continuous monitoring systems, enhanced public awareness campaigns, infrastructure modernization, research and innovation funding, regulatory framework strengthening, community engagement and participation in water quality monitoring, and capacity building and training programs. These recommendations aim to ensure accurate detection and analysis of contaminants, improve public health, and preserve clean drinking water for future generations. They also suggest investing in infrastructure modernization projects, fostering interdisciplinary collaboration, strengthening regulatory frameworks, and promoting community engagement in water quality monitoring and conservation efforts. These recommendations aim to protect public health and preserve clean drinking water for future generations:

- **Comprehensive Water Testing Protocols:** Implement thorough water testing protocols that encompass microbiological, chemical, and heavy metal contaminants. Utilize advanced testing methodologies to ensure accurate detection and analysis of water quality parameters.
- **Continuous Monitoring Systems:** Deploy continuous water quality monitoring systems in key water sources to enable real-time detection of contaminants. These systems can provide early warnings of water quality issues, allowing for prompt intervention and mitigation measures.
- **Enhanced Public Awareness Campaigns:** Launch robust public awareness campaigns to educate communities about the importance of clean drinking water and the risks associated with water contamination. Empower individuals to take proactive steps to safeguard their water sources and advocate for water quality improvements.
- **Infrastructure Modernization Initiatives:** Invest in infrastructure modernization projects to address aging pipelines, upgrade treatment facilities, and implement leak detection technologies. Collaborate with public and private stakeholders to secure funding and expertise for infrastructure improvements.
- **Research and Innovation Funding:** Allocate resources for research and innovation initiatives aimed at addressing emerging contaminants and developing effective treatment methods. Foster interdisciplinary collaboration

to advance scientific understanding and technological innovations in water quality management.

- **Regulatory Framework Strengthening:** Enhance regulatory frameworks governing water quality standards and pollution control measures. Establish stringent guidelines for managing emerging contaminants and enforcing compliance to ensure the protection of human health and the environment.
- **Community Engagement and Participation:** Promote community engagement and participation in water quality monitoring and conservation efforts. Foster partnerships between local communities, government agencies, and environmental organizations to create a shared responsibility for protecting water resources.
- **Capacity Building and Training Programs:** Offer capacity building and training programs to empower communities with the knowledge and skills needed to actively participate in water management initiatives. Provide resources and support to facilitate community-led conservation projects and decision-making processes.

By implementing these recommendations, stakeholders can work together to address water quality challenges, protect public health, and preserve clean drinking water as a vital resource for present and future generations.

7. CONCLUSION

- The ecosystem and human health depend on clean drinking water for processes including digestion, waste elimination, temperature regulation, and joint lubrication. Toxins may be safely transported out of the body by the hydration shells that water's molecular makeup creates around them. Through sweating, it also contributes significantly to temperature regulation by absorbing a large amount of heat.
- Health problems include malabsorption, malnutrition, waterborne infections, chronic diseases, and developmental difficulties may be brought on by low water quality. Lead and mercury are examples of heavy metals that may tamper with the function of water in gene expression and cell signalling, possibly changing a developing child's genetic makeup. These are violations of water's basic covenant with life, not only health problems.
- Modern infrastructure, innovative technology, and public-private partnerships are required to ensure water quality. Water is pleading silently for preservation, and renewable technologies are humanity's intellectual answer to it. Infrastructure is the physical manifestation of our social compact.

Partnerships between the public and corporate sectors recognise that water flows across sectoral lines and that no one organisation can adequately manage water quality on its own.

- New contaminants need for ongoing research, improved surveillance, and adaptable laws. Conventional water testing often searches for known toxins at predefined intervals, but in an era of fast technological advancement, real-time, broad-spectrum monitoring is required. Restoring the relationship between humanity and their water sources and establishing a shared responsibility for water resources are made possible by flexible rules that quickly incorporate new information and harmonise human law with natural law.
- Educating people about water quality is a way to restore culture by reawakening long-dormant senses that used to tell us whether the water was safe to drink. This reconnection is furthered by including communities in monitoring programmes, which foster environmental communion and the restoration of a feeling of custodianship. As shown by many indigenous cultures, where water is seen as a living thing with rights and demands of its own, the shared responsibility for water resources is an old understanding.

REFERENCES

Akhai, S., Singh, V. P., & John, S. (2016). Investigating Indoor Air Quality for the Split-Type Air Conditioners in an Office Environment and Its Effect on Human Performance. *Journal of Mechanical Civil Engineering*, 13(6), 113–118.

Akhai, S., Singh, V. P., & John, S. (2016). Human performance in industrial design centers with small unit air conditioning systems. *Journal of Advanced Research in Production Industrial Engineering*, 3(2), 5–11.

Ali, S., Amir, S., Ali, S., Rehman, M. U., Majid, S., &Yatoo, A. M. (2021). Water pollution: Diseases and health impacts. *Freshwater Pollution and Aquatic Ecosystems: Environmental Impact and Sustainable Management*, 1.

Awuchi, C. G., Igwe, V. S., & Amagwula, I. O. (2020). Nutritional diseases and nutrient toxicities: A systematic review of the diets and nutrition for prevention and treatment. *International Journal of Advanced Academic Research*, 6(1), 1–46. DOI: 10.46654/ij.24889849.e61112

Balali-Mood, M., Naseri, K., Tahergorabi, Z., Khazdair, M. R., & Sadeghi, M. (2021). Toxic mechanisms of five heavy metals: Mercury, lead, chromium, cadmium, and arsenic. *Frontiers in Pharmacology*, 12, 227. DOI: 10.3389/fphar.2021.643972 PMID: 33927623

Barzilay, J. I., Weinberg, W. G., & Eley, J. W. (1999). *The water we drink: Water quality and its effects on health*. Rutgers University Press.

Bozorg-Haddad, O., Delpasand, M., & Loáiciga, H. A. (2021). Water quality, hygiene, and health. In *Economical, Political, and Social Issues in Water Resources* (pp. 217-257). Elsevier. DOI: 10.1016/B978-0-323-90567-1.00008-5

Egbuna, C., Amadi, C. N., Patrick-Iwuanyanwu, K. C., Ezzat, S. M., Awuchi, C. G., Ugonwa, P. O., & Orisakwe, O. E. (2021). Emerging pollutants in Nigeria: A systematic review. *Environmental Toxicology and Pharmacology*, 85, 103638. DOI: 10.1016/j.etap.2021.103638 PMID: 33757839

Fältström, E., & Anderberg, S. (2020). Towards control strategies for microplastics in urban water. *Environmental Science and Pollution Research International*, 27(32), 40421–40433. DOI: 10.1007/s11356-020-10064-z PMID: 32666462

Hassan, T. A., & Suhail, Z. A. (2023). The Efficacy of Ozonized Water Versus Ringer Lactate Arthrocentesis for the Treatment of Temporomandibular Joint Internal Derangement. *The Journal of Craniofacial Surgery*, 34(3), e238–e241. DOI: 10.1097/SCS.0000000000009085 PMID: 36730935

Ji, Y., Wu, J., Wang, Y., Elumalai, V., & Subramani, T. (2020). Seasonal variation of drinking water quality and human health risk assessment in Hancheng City of Guanzhong Plain, China. *Exposure and Health*, 12(3), 469–485. DOI: 10.1007/s12403-020-00357-6

Koyama, H., Kamogashira, T., & Yamasoba, T. (2024). Heavy Metal Exposure: Molecular Pathways, Clinical Implications, and Protective Strategies. *Antioxidants*, 13(1), 76. DOI: 10.3390/antiox13010076 PMID: 38247500

Kumar, M., & Puri, A. (2012). A review of permissible limits of drinking water. *Indian Journal of Occupational and Environmental Medicine*, 16(1), 40. DOI: 10.4103/0019-5278.99696 PMID: 23112507

Lim, C. L. (2020). Fundamental concepts of human thermoregulation and adaptation to heat: A review in the context of global warming. *International Journal of Environmental Research and Public Health*, 17(21), 7795. DOI: 10.3390/ijerph17217795 PMID: 33114437

Munteanu, C., Teoibas-Serban, D., Iordache, L., Balaurea, M., & Blendea, C. D. (2021). Water intake meets the Water from inside the human body—Physiological, cultural, and health perspectives-Synthetic and Systematic literature review. *Balneo and PRM Research Journal*, 12(3), 196–209. DOI: 10.12680/balneo.2021.439

Pagano, C., Navarra, G., Coppola, L., Savarese, B., Avilia, G., Giarra, A., Pagano, G., Marano, A., Trifuoggi, M., Bifulco, M., & Laezza, C. (2023). Impacts of Environmental Pollution on Brain Tumorigenesis. *International Journal of Molecular Sciences*, 24(5), 5045. DOI: 10.3390/ijms24055045 PMID: 36902485

Pashaei, R., Dzingelevičienė, R., Abbasi, S., Szultka-Młyńska, M., & Buszewski, B. (2022). Determination of the pharmaceuticals–nano/microplastics in aquatic systems by analytical and instrumental methods. *Environmental Monitoring and Assessment*, 194(2), 1–33. DOI: 10.1007/s10661-022-09751-w

Richter, B. D., Abell, D., Bacha, E., Brauman, K., Calos, S., Cohn, A., Disla, C., O'Brien, S. F., Hodges, D., Kaiser, S., Loughran, M., Mestre, C., Reardon, M., & Siegfried, E. (2013). Tapped out: How can cities secure their water future? *Water Policy*, 15(3), 335–363. DOI: 10.2166/wp.2013.105

Saghiv, M. S., Sagiv, M. S., Saghiv, M. S., & Sagiv, M. S. (2020). Thermoregulation. *Basic Exercise Physiology: Clinical and Laboratory Perspectives*, 437-463.

Samanta, P., Desai, A. V., Let, S., & Ghosh, S. K. (2019). Advanced porous materials for sensing, capture and detoxification of organic pollutants toward water remediation. *ACS Sustainable Chemistry & Engineering*, 7(8), 7456–7478. DOI: 10.1021/acssuschemeng.9b00155

Saravanan, A., Kumar, P. S., Jeevanantham, S., Karishma, S., Tajsabreen, B., Yaashikaa, P. R., & Reshma, B. (2021). Effective water/wastewater treatment methodologies for toxic pollutants removal: Processes and applications towards sustainable development. *Chemosphere*, 280, 130595. DOI: 10.1016/j.chemosphere.2021.130595 PMID: 33940449

Sarma, H. (2022). Understanding emerging contaminants in soil and water: current perspectives on integrated remediation approaches. *Emerging Contaminants in the Environment*, 1-38.

Spellman, F. R. (2013). *Water & wastewater infrastructure: Energy efficiency and sustainability*. Crc Press. DOI: 10.1201/b13968

Srinivas, R., Singh, A. P., & Shankar, D. (2020). Understanding the threats and challenges concerning Ganges River basin for effective policy recommendations towards sustainable development. *Environment, Development and Sustainability*, 22(4), 3655–3690. DOI: 10.1007/s10668-019-00361-0

Tinoco, C., Julio, N., Meirelles, B., Pineda, R., Figueroa, R., Urrutia, R., & Parra, Ó. (2022). Water Resources Management in Mexico, Chile and Brazil: Comparative Analysis of Their Progress on SDG 6.5. 1 and the Role of Governance. *Sustainability (Basel)*, 14(10), 5814. DOI: 10.3390/su14105814

Ugran, V., Desai, N. N., Chakraborti, D., Masali, K. A., Mantur, P., Kulkarni, S., Deshmukh, N., Chadchan, K. S., Das, S. N., Tanksali, A. S., Arwikar, A. S., Guggarigoudar, S. P., Vallabha, T., Patil, S. S., & Das, K. K. (2017). Groundwater fluoride contamination and its possible health implications in Indi taluk of Vijayapura District (Karnataka State), India. *Environmental Geochemistry and Health*, 39(5), 1017–1029. DOI: 10.1007/s10653-016-9869-2 PMID: 27591763

Zhang, P., Yang, M., Lan, J., Huang, Y., Zhang, J., Huang, S., Yang, Y., & Ru, J. (2023). Water Quality Degradation Due to Heavy Metal Contamination: Health Impacts and Eco-Friendly Approaches for Heavy Metal Remediation. *Toxics*, 11(10), 828. DOI: 10.3390/toxics11100828 PMID: 37888679

Zhang, Y., Gao, Y., Liu, Q. S., Zhou, Q., & Jiang, G. (2024). Chemical Contaminants in Blood and Their Implications in Chronic Diseases. *Journal of Hazardous Materials*, 466, 133511. DOI: 10.1016/j.jhazmat.2024.133511 PMID: 38262316

Chapter 6
Applying the Water Sensitive City Index in a House of Quality Framework to Enhance Urban Water Governance and Sustainability:
Evaluating Smart Water Cities

Ocotlan Diaz-Parra
Universidad Politécnica de Pachuca, Mexico

Jorge A. Ruiz-Vanoye
https://orcid.org/0000-0003-4928-5716
Universidad Politécnica de Pachuca, Mexico

Eric Simancas-Acevedo
https://orcid.org/0000-0001-7823-709X
Universidad Politécnica de Pachuca, Mexico

Julio C. Ramos-Fernández
https://orcid.org/0000-0002-9997-6550
Universidad Politécnica de Pachuca, Mexico

Juan M. Xicotencatl-Pérez
https://orcid.org/0009-0008-8554-4192
Universidad Politécnica de Pachuca, Mexico

Francisco Marroquín-Gutierrez
Universidad Politécnica de Pachuca, Mexico

Julio C. Salgado-Ramírez
https://orcid.org/0000-0003-1666-9924
Universidad Politécnica de Pachuca, Mexico

Yaneth Reyes-Hernández
Universidad Politécnica de Pachuca, Mexico

DOI: 10.4018/979-8-3693-8074-1.ch006

ABSTRACT

This article presents a combination of tools to determine if a city could be considered Smart Water City. Smart Water cities are those that use advanced technologies and innovative solutions to improve water management and ensure the long-term sustainability of water supplies. This includes optimizing water use, reducing water losses, treating, and reusing wastewater, and conserving water resources. This paper proposes the use of the evaluation of the water sensitive city index applied in the house of quality, to identify the fulfillment of the seven goals that the water sensitive city index manages within the scheme of the house of quality and thus precisely observe the areas of opportunity for decision making.

INTRODUCTION

The concept of Smart Water Cities integrates advanced technologies and sustainable practices to manage urban water resources effectively. This chapter explores the application of the Water Sensitive City Index (WSCI) within a House of Quality (HoQ) framework to enhance urban water governance and sustainability. By synthesizing relevant literature, we identify key indicators and governance challenges that can inform the development of resilient urban water systems.

Urban water management is increasingly challenged by rapid urbanization, climate change, and resource scarcity. The Water Sensitive City Index (WSCI) provides a comprehensive framework for assessing the sustainability of urban water systems by evaluating various indicators related to water governance, quality, and infrastructure. Integrating the WSCI with a House of Quality (HoQ) framework can facilitate a structured approach to enhance urban water governance and sustainability. This paper reviews existing literature to identify the critical components of this integrated framework.

There are different types of water cities, some of which are mentioned below (Brown et al., 2009):

- **Water Supply City:** To be a drinking water supply city is one that enables water to move from the point of catchment to the point of consumption in a fit condition. Suitable means not only in terms of sanitary conditions of quality, but also in terms of quantity. The source of water that gives rise to the system can be from: Natural spring water, seawater, which is desalinated before entering the supply network, surface water, such as from lakes, rivers, reservoirs or streams, groundwater, captured through abstractions and other, such as rainwater stored in cisterns.

- **Sewered City:** The final layout of a city's sewerage network will be influenced by several factors, including the topography and geology of the terrain and the urban characteristics of the area. Provided that the topography permits, it is a priority that all or most of the network allows gravity flow of water, otherwise pumping would be necessary, which is technically difficult and costly. Another important factor in the construction of the network will be the type of soil in the area, which will affect both the layout and the construction methods used. The occurrence of hard soil that is difficult to excavate can considerably increase the cost. Finally, the sewerage network must coincide with the city's roads and spaces, so that land occupation, urban planning classification and possible future developments will have to be considered in its development.
- **Drained City:** Flow of water never stops, but its speed is controllable. The deep drainage system was originally designed to absorb this counter-slope and to carry the water out naturally within the "natural" ecological discharge system, from a main drainage channel for flood control and other secondary channels for river transport and irrigation activities.
- **Waterways City:** The term fluvial is used in geography and earth sciences to refer to the processes associated with rivers, streams, and the deposits and reliefs created by them, which enable trade, mobility, and economics.
- **Water Cycle City:** Simplified, the urban water cycle is divided into supply (clean water or drinking water), consumption (use) and sanitation (dirty water or wastewater). In short: After abstraction, water is treated to make it drinkable. This is done in the Drinking Water Treatment Plants (DWTP's).

Water Sensitive City is a concept of Australian origin that is now considered the ideal state of the city in terms of liveability, productivity, sustainability, and resilience in terms of flooding and sustainable access to water. Its strategies, such as stormwater reuse and water-efficient landscaping, can increase sustainability while making communities better places to live and work. As urbanization and climate change continue, WSUD is likely to become increasingly important.

SMART WATER CITIES

Smart Water Cities are those that use advanced technologies and innovative solutions to improve water management and ensure the long-term sustainability of water supply. This includes optimizing water use, reducing water losses, treating, and reusing wastewater and conserving water resources. Some of the common characteristics of Smart Water Cities include:

- **Water Supply Monitoring and Control:** Use of monitoring and control technologies to optimize water use and reduce water losses.
- **Wastewater Treatment and Reuse:** Use of advanced water treatment technologies to transform wastewater into drinking water or water that can be used for other purposes, such as irrigating gardens or cleaning buildings.
- **Public Awareness and Participation:** Promoting public participation in water-related decision-making and education on sustainable water use.
- **Integrated Water Cycle Management:** Integration of water management in the context of other urban resources and services, such as energy and transport.

Smart Water City has three general kinds of problems and ways to resolve them

- **Water-Hard Problems:** The water-hard problems can be optimized (from Combinatorial Optimization: maximization or minimization) of the water's resources. A Combinatorial Optimization problem *w* of the Water problems is either a minimization problem or a maximization problem and consists of the following three parts: 1) a set *Dw* of instances of the Water Problem, 2) for each instance $I \in Dw$, a finite set *Sw(I)* of candidate solutions for *I*; and 3) a function m_W that assigns to each instance $I \in Dw$ and each candidate solution $\sigma \epsilon S_w(I)$ a positive rational number $m_w(I, \sigma)$ called the solution value for σ.
- **Water-Soft Problems:** The water-soft problems are a fuzzy concept used to drive specific security government agendas and others. The water-Soft Problems are the water problems that handle information inaccurate or incomplete, with uncertainty and ambiguity, volatile, poorly understood and dynamic (water policy, administration, decision-making, water government reforms, and others).
- **Water-Wicked Problems:** The complex, intractable, open-ended and wicked water problems are a problem that is difficult or impossible to solve because it presents incomplete, contradictory and changing requirements that are generally difficult to recognize. This problem requires deliberation and debate.

There is an index to determine whether it is a water city. Water Sensitive Cities (WSC) Index is a benchmarking tool created by Cooperative Research Centre for Water Sensitive Cities (Rogers et al., 2020), for evaluating the city sensibility of water viewed in different levels of urbanization. Its 34 indicators are organized into seven goals: 1.- Ensure good water sensitive governance, 2.- Increase community capital, 3.- Achieve equity of essential services, 4.- Improve productivity and resource efficiency, 5.-Improve ecological health, 6.-Ensure quality urban spaces, 7.-Promote adaptive infrastructure.

The WSC Index is used as a diagnostic and planning tool to help cities identify areas for improvement and establish strategies for more effective water management. It helps city authorities and planners to make informed decisions and implement policies and actions that promote water sustainability in cities.

The WSCI serves as a tool for municipalities to assess their performance in managing water resources sustainably. According to Koop et al. (Hananda et al., 2023), the WSCI encompasses various dimensions, including water quality, governance capacity, and community engagement. By evaluating these dimensions, cities can identify strengths and weaknesses in their water management practices, leading to more informed decision-making. Li et al. Bennehalli (2024) emphasize the importance of comprehensive evaluation systems in water management, highlighting the need for indicators that reflect the relationship between urbanization and environmental sustainability. The WSCI aligns with this perspective by providing a holistic view of urban water systems, enabling cities to address governance gaps and improve overall performance.

To determine that a city is a Water Smart City it must meet the requirements or elements contained in the water sensitive index. In order to have a tool for decision making related to the management of water-related aspects we use the House of Quality.

The Quality Function Deployment concept appeared in Japan in the 60s decade and quickly spread due to several factors. The construction of the QFD instrument makes use of a graphical method that determine the relations between the demands of customers and the characteristics of the design, so in the matrix structure, emptied the expectations of customers (attributes) and is it contrasts with the characteristics of design (Sullivan, 1986; Hauser and Clausing, 1988; Cohen, 1995).

Framework Applied to Decision-Making in Water Cities

The House of Quality (HoQ) framework, derived from Quality Function Deployment (QFD), is a structured approach to translating customer requirements into specific design factors. In the context of urban water governance, the HoQ framework can be employed to prioritize key indicators identified through the WSCI. While Chang and Kim Kumarappa (2024) discuss the application of the HoQ framework in different contexts, its specific application to urban water governance requires further exploration. By utilizing the HoQ framework, urban planners and policymakers can systematically assess the importance of various water governance indicators, such as institutional capacity, stakeholder engagement, and infrastructure resilience. This structured approach facilitates the identification of critical areas for improvement and resource allocation.

The House of Quality applied to Water Smart City is a framework to convert the city's needs into a set of specifications to improve the Smart City. The House of Quality for the Smart Water City (Figure 1) includes:

- **Requirements of the Smart Water City:** It is one of the House of Quality inputs, and it defines the analysis, interviews in the Smart Water City.
- **The Relationship Matrix:** This is the dimension where requirements correspond or match with the characteristics or specifications of the management of the water.
- **City Characteristics:** Product features or specifications to improve the management of the water.
- **Correlation Matrix:** This stage is classified as strongly positive, positive, negative, strongly negative and none.
- **Competitive Benchmark:** The result of the relationship matrix is compared with the security products available on the market. And is used to enhance weaknesses identified in the comparison.
- **Technical Details:** In this section, the matrix relationship begins to analyze and measure the city's needs related to the times, costs and difficulties.
- **Technical Benchmark:** This is the assessment of the improvements to the needs of the Cities and the specifications or City characteristics.
- **Goals:** Herein determining the goals that must be obtained to improve the needs of the city.

Figure 1. Framework to decision-making of smart water cities

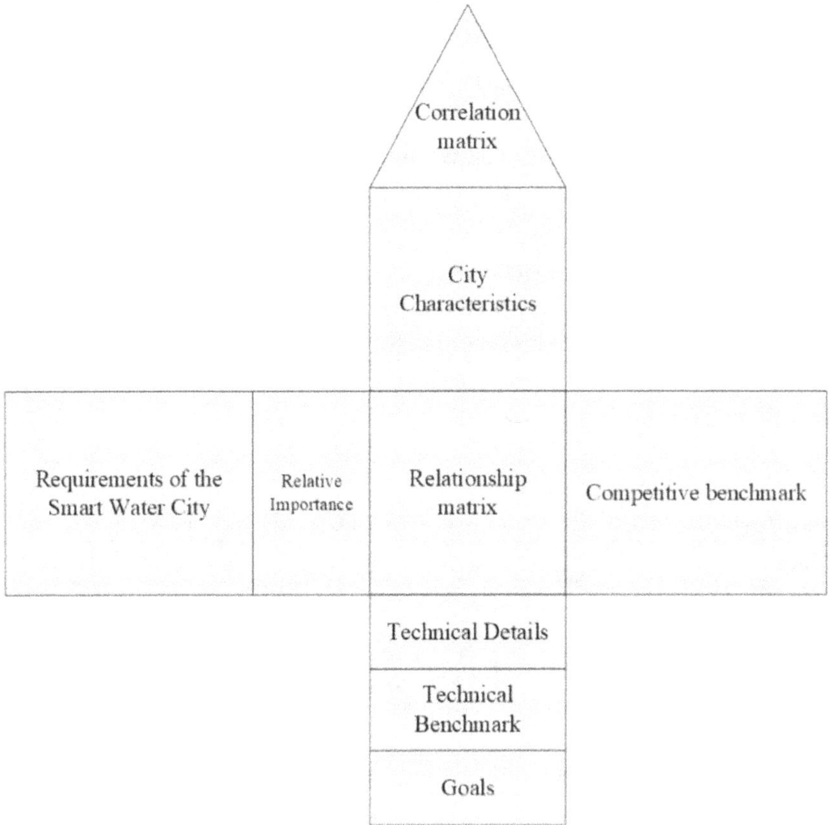

To determine that a city is a Smart Water City, it must meet the requirements or elements contained in the water sensitive index, determine the City Characteristics (CC) that focus on how you should implement the aspects in the Smart Water City, determines the time and costs of each solution, and the difficulty required to implement the business plan according to the Smart Water City and determine the goals to be fulfilled at the time of improving the Smart Water City.

Effective urban water governance is essential for addressing the challenges posed by climate change and urbanization. Hossain and Haque Zhu et al. (2019) highlight the role of responsible institutions in implementing water management plans and infrastructure development. Their findings underscore the necessity of prioritizing investments in water and sanitation infrastructure to enhance governance capacity. Koop et al. Hananda et al. (2023) further emphasize the need for integrated water resources management (IWRM) to address governance gaps. By applying the WSCI within the HoQ framework, cities can develop targeted strategies to enhance governance capacity, focusing on key challenges such as water scarcity, flood risk, and wastewater treatment.

Experimentation and Results

In this section, the House of Quality applies to analyses strategies to improve o determine a Smart Water City. The House of Quality is applied to the city of Cuernavaca, México. The process for making the House of Quality for Smart Water City consists of the following steps.

Make a list or select 10 needs of the Smart Water City (from Table 1). And the ranks assigned section for the needs of the Smart City (Table 2). It also determines the relative importance (RI) of each of the needs with values between 0 (unimportant) to 10 (very important).

Table 1. Requirements of the smart water city (Rogers et al., 2020)

WSC Index	
Goal 1: Ensure good water sensitive governance	-Knowledge, skills, and organizational capacity. -Water is key element in city planning and Design. -Cross-sector institutional arrangements and Processes. -Public engagement, participation and Transparency. -Leadership, long-term vision and commitment. -Water resourcing and funding to deliver broad societal value. -Equitable representation of perspectives
Goal 2: Increase community capital	-Water literacy. -Connection with water. -Shared ownership, management & Responsibility. -Community preparedness and response to extreme events. -Indigenous involvement in water planning.
Goal 3: Achieve equity of essential services	-Equitable access to safe and secure water supply. -Equitable access to safe and reliable sanitation. -Equitable access to flood protection. -Equitable and affordable access to amenity values of water-related assets.
Goal 4: Improve productivity & resource efficiency	-Optimized resource recovery. -Low GHG emission in water sector. -Water-related business opportunities. -Low end-user potable water demand. -Broad community benefits from water services.
Goal 5: Improve ecological health	-Healthy and biodiverse habitat. -Surface water quality and flows. -Groundwater quality and replenishment. -Protect existing areas of high ecological value

continued on following page

Table 1. Continued

WSC Index	
Goal 6: Ensure quality urban space	-Activating connected green / blue space. -Urban elements functioning as part of the urban water system. -Vegetation coverage.
Goal 7: Promote adaptive infrastructure	-Diverse fit-for-purpose water supply. -Multi-functional water infrastructure. -Integration and intelligent control. -Robust infrastructure. -Infrastructure and ownership at multiple scales. -Adequate maintenance.

Table 2. City characteristics or needs of the smart water city (Cuernavaca)

#		Requirements of the Smart City
R1	9	Water efficiency: Implement measures and technologies that reduce water consumption in the city, including smart irrigation systems, water conservation technologies in buildings and awareness-raising on responsible water use.
R2	7	Flood management: Develop flood management strategies that include green infrastructure, such as water retention parks, permeable areas, and sustainable drainage systems.
R3	7	Water reuse: Promote the reuse of treated water in different sectors, such as irrigation of green areas, industrial use and aquifer recharge.
R4	7	Water conservation: Encourage water conservation practices at the individual and community level, through educational campaigns, tiered tariff policies and the installation of efficient devices such as low flow taps and showers.
R5	8	Community participation: Actively involve the community in water management and decision-making through environmental education programmes, participatory workshops, and citizen forums.
R6	8	Sustainable urban planning: Integrate water considerations into urban planning, ensuring conservation and restoration of aquatic ecosystems, protection of floodplains and promotion of water systems connectivity.
R7	9	Smart water monitoring and management: Implement smart water monitoring and management systems, using technologies such as sensors, smart meters and data analytics to optimize water use and distribution.
R8	8	Education and awareness: Develop educational programmes and awareness campaigns on the importance of water, its sustainable management, and the impact of individual actions on the quality and availability of water resources.
R9	9	Protection of water sources: Safeguard and protect local water sources, such as rivers, lakes, and aquifers, by conserving catchment areas, implementing protection zones and controlling polluting activities.
R10	8	Climate change resilience: Adopt measures to cope with the impacts of climate change on water availability and quality, such as the implementation of adaptation infrastructure, integrated drought management and the promotion of sustainable agricultural practices.

Determine 10 City Characteristics (CC) that focus on how you should implement the city's aspects. And assigns in the columns for the features section of the city (Table 3).

Table 3. City characteristics

#	Characteristics
CC1	Green infrastructure: Implementation of sustainable drainage systems, green roofs, rain gardens and permeable areas to reduce runoff and improve rainwater infiltration.
CC2	Water harvesting and storage technologies: Use of technologies such as rainwater harvesting, installation of cisterns or storage tanks to collect and store water for non-potable uses.
CC3	Water treatment and reuse: Establishment of wastewater treatment systems that allow for reuse in activities such as irrigation of green areas, street washing or industrial use.
CC4	Smart irrigation management: Implementation of smart irrigation systems based on control and monitoring technologies, such as soil moisture sensors or automated scheduling, to optimize water use in green area irrigation.
CC5	Loss and leakage reduction: Implementation of water loss management programmes, such as detection and repair of leaks in the distribution network, to reduce losses and improve the efficiency of the water system.
CC6	Integrated watershed management: Development of plans and strategies for integrated watershed management, addressing the protection of aquatic ecosystems, conservation of ecological flows and coordination between different stakeholders.
CC7	Information and communication technologies (ICTs): Use of ICTs to monitor and control water consumption, share information on water quality, provide flood warnings and promote citizen participation in water management.
CC8	Water-oriented urban planning and design: Integration of water management criteria into urban planning and design, considering the location of green areas, the protection of flood zones and the connection of river corridors.
CC9	Drought and water stress management: Development of drought management plans that promote water conservation, efficient use of water resources and promotion of sustainable agricultural practices in times of scarcity.
CC10	Water cooperation and governance: Encouraging cooperation between different actors, such as local governments, institutions, businesses and civil society, for participatory and efficient water management, promoting transparency and collaborative decision-making.

Develops the evaluation corresponding to the matrix of relationship. Assigns Θ to the strong relationship (9), O moderate relationship (3), or ? weak relationship (1) where the needs corresponding or match the City characteristics (Table 4).

Table 4. The matrix of relationship

	CC 1	CC 2	CC 3	CC 4	CC 5	CC 6	CC 7	CC 8	CC 9	CC 10
R1	Θ		Θ	Θ		?	?	?		Θ
R2			?						O	?
R3			?			O			?	
R4		Θ		Θ		O			O	
R5			?			O				
R6							Θ			

continued on following page

Table 4. Continued

	CC 1	CC 2	CC 3	CC 4	CC 5	CC 6	CC 7	CC 8	CC 9	CC 10
R7		O			Θ	?				
R8	O				?			O	O	
R9			Θ				O	O		
R10	?	Θ					Θ	?		

Assigns 5-10 to City products on the market at competitive benchmark section (Figure 2). City products will serve to improve the weaknesses identified in the city. Rate 0 to 5, where 0 is the worst rating of the product that solves the city's weaknesses and 5 is the best.

Figure 2. Competitive benchmark

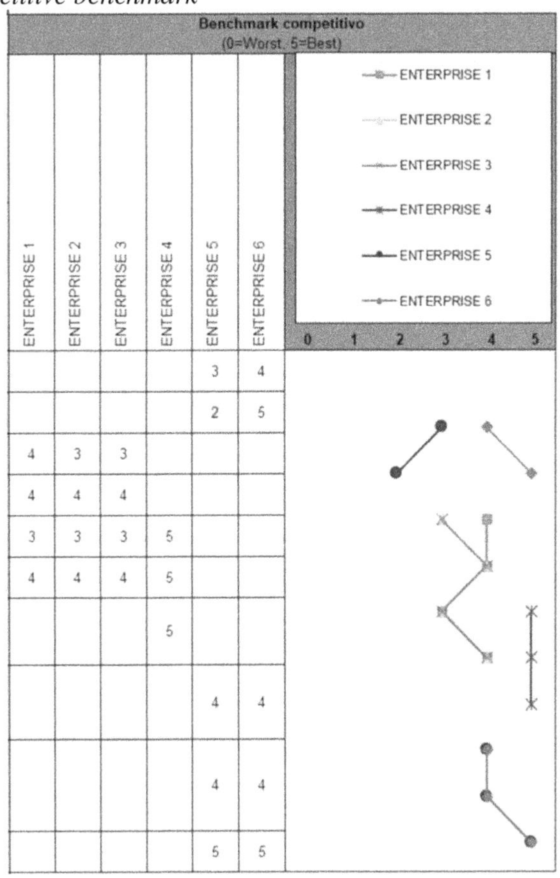

ENTERPRISE 1	ENTERPRISE 2	ENTERPRISE 3	ENTERPRISE 4	ENTERPRISE 5	ENTERPRISE 6
				3	4
				2	5
4	3	3			
4	4	4			
3	3	3	5		
4	4	4	5		
			5		
				4	4
				4	4
				5	5

Determines the time and computational costs of each solution, and the difficulty required to implement the business plan according to the city in the technical details section. Assign between 0 to implementation that is easy to perform and 10 if it is extremely difficult (Table 5).

Calculate the importance of improving the aspects of the city and allocate in the section technical benchmark from 0 (unimportant) to 10 (very important), (Table 5).

Determine the goals or targets to be fulfilled at the time of improving the Smart City and assigned them to the goals section (Table 5).

Table 5. Technical details, technical benchmark and goals

Target 1:	**Reduce drinking water consumption by 20% over the next 5 years.**
Timeframe:	5 years.
Performance indicator:	Percentage reduction in drinking water consumption.
Metric:	Comparison of current drinking water consumption with target consumption.
Target 2:	Implement green infrastructure in 30% of the city's urban areas in the next 3 years.
Timeframe:	3 years.
Performance indicator:	Percentage of urban areas with implemented green infrastructure.
Metric:	Comparison of the number of urban areas with implemented green infrastructure with the set target.
Target 3:	Achieve a 50% treated water reuse rate over the next 5 years.
Timeframe:	5 years.
Performance indicator:	Percentage of treated water reused.
Metric:	Comparison of volume of treated water reused with total volume of treated water.
Target 4:	Reduce water losses in the distribution network by 30% over the next 4 years.
Timeframe:	4 years.
Performance indicator:	Percentage reduction in water losses.
Metric:	Comparison of current volume of water lost with target volume of water lost.
Target 5:	Increase community participation in water management by 40% over the next 2 years.
Timeframe:	2 years.

continued on following page

Table 5. Continued

Target 1:	**Reduce drinking water consumption by 20% over the next 5 years.**
Performance indicator:	Percentage increase in community participation.
Metric:	Comparison of the number of participants in current community water management activities with the set target.

Rate the correlation between each of the City features and assign ++ to the strong positive correlation, + the positive correlation, ▬ the negative correlation, ▼ if the objective is to minimize, ▲if the objective is to maximize, *x* if the goal is only the fulfilment of the activity (Figure 3).

Figure 3. Correlations

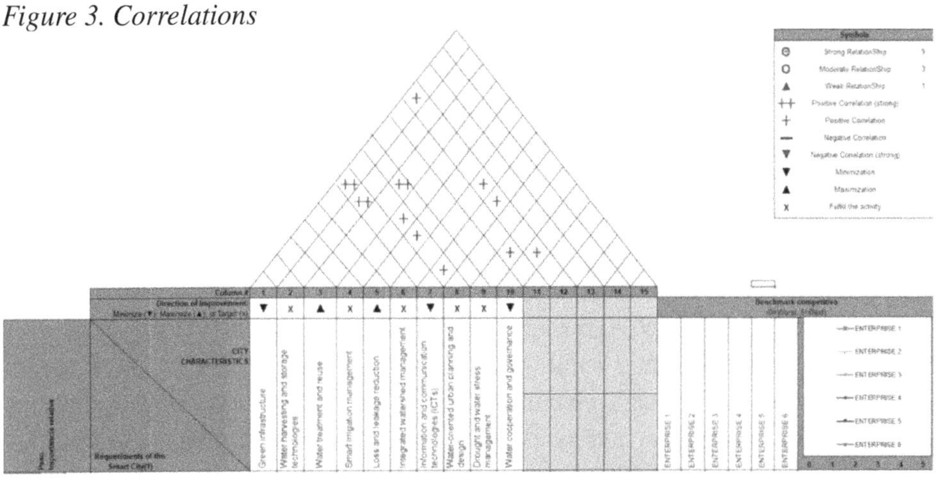

CONCLUSION

The integration of the Water Sensitive City Index with the House of Quality framework offers a robust approach to enhancing urban water governance and sustainability. By systematically evaluating key indicators and prioritizing governance challenges, cities can develop targeted strategies to improve their water management practices. Continued research and collaboration among stakeholders are essential for advancing the implementation of this integrated framework in urban water governance.

The importance of good management of a basic natural resource for human beings is indisputable, which is why any initiative to improve it will always be of high value to humanity and what better than decision makers such as governments have tools to manage not only a part of the context but more broadly the various aspects involved in water management for strategic planning and therefore the definition of actions that have a projection of results in the short, medium and long term. The Water Sensitive City Index is a valuable tool to assess the current position of cities regarding water design and management. By implementing the analysis of the objectives established by the index within a schematic tool such as the house of quality, the interaction of the elements and their level of performance can be observed, therefore the areas of opportunity are exposed, as well as the necessary requirements for the achievement of the objectives, which allows decisions to be made more quickly and accurately to improve conditions. The progress of humanity involves carrying out actions more and more conscientiously and knowing that any action requires an investment of resources, having a clear guide to the objective that is pursued, gives greater certainty in decision-making and its impact. in the future.

In addition to the House of Quality, there are other frameworks and approaches you can consider for city water management. Here are some examples:

Water Resilient Cities Framework: This framework focuses on building cities that can withstand, absorb and adapt to the impacts of climate change and water-related events. It provides guidelines for strengthening urban resilience, including integrated water resources management, green infrastructure, ecosystem-based planning and community participation.

Blue City Approach: This approach is based on the integration of natural systems and the water cycle into urban planning and design. It promotes sustainable water management, restoration of aquatic ecosystems, rainwater harvesting and reuse, and the creation of public spaces with aquatic features.

Water Circular City Strategy: This approach focuses on closing water cycles in cities, maximising the reuse and recovery of water resources. It includes the implementation of advanced wastewater treatment technologies, the reuse of treated water for various uses and the efficient management of water flows in the city.

Smart Water Cities Approach: This approach combines information and communication technologies (ICT) with water management to improve the efficiency, sustainability and resilience of cities. It includes the use of sensors, smart meters, data analytics and geographic information systems to optimise water resources management and improve decision-making.

Circular Water City Framework: This framework focuses on circular water management, resource conservation and waste minimisation in cities. It promotes the implementation of water treatment and reuse technologies, integrated water resources management, citizen education and participation, and collaboration between different stakeholders to achieve sustainable water management.

REFERENCES

Australian Government. Water sensitive urban desing. Available online: https://cdn.environment.sa.gov.au/environment/docs/water-sensitive-urban-design-policy-gen.pdf (accessed on 21 March 2023).

Bennehalli, B. (2024). Machine learning approach to battery management and balancing techniques for enhancing electric vehicle battery performance. *Journal of Electrical Systems*, 20(2s), 885–892. DOI: 10.52783/jes.1685

Brown, R. R., Keath, N., & Wong, T. H. (2009). Urban water management in cities: Historical, current and future regimes. *Water Science and Technology*, 59(5), 847–855. DOI: 10.2166/wst.2009.029 PMID: 19273883

Cartegraph for Asset Management Software for Water Treatment Plants. Retrieved Available online: https://www.cartegraph.com/software-for/water-treatment-plants (accessed on 20 Jan, 2023).

Cohen, L. Quality function deployment: how to make QFD work for you. Prentice Hall, Hoboken, New, Jersey, U.S.A, pp 211-231, 1995.

EPA. EPANET: Application for Modeling Drinking Water Distribution Systems. Retrieved. Available online: https://www.epa.gov/water-research/epanet (accessed on 22 Jan, 2023).

Hananda, N., Kamul, A., Harito, C., Djuana, E., Elwirehardja, G. N., Pardamean, B., … & Pasang, T. (2023). Battery optimization by machine learning algorithms: research gap via bibliometric analysis. E3S Web of Conferences, 388, 01020. DOI: 10.1051/e3sconf/202338801020

Hananda, N., Kamul, A., Harito, C., Djuana, E., Elwirehardja, G. N., Pardamean, B., … & Pasang, T. (2023). Battery optimization by machine learning algorithms: research gap via bibliometric analysis. E3S Web of Conferences, 388, 01020. DOI: 10.1051/e3sconf/202338801020

Hauser, J. R., & Clausing, D. (1988). *The house of quality*. Harvard Business Review.

HidraSoftware. Available online: https://www.hidrasoftware.com/ (accessed on 22 Jan, 2023)

Hoban, A. Chapter 2 Water Sensitive Urban Design Approaches and Their Description. In Book Approaches to Water Sensitive Urban Design A. K. Sharma, T. Gardner, & D. Begbie., Woodhead Publishing, Sawston, Cambridge, England, pp. 25-47, 2019

Intergovernmental Agreement On A National Water Initiative. Available online: https://www.pc.gov.au/inquiries/completed/water-reform/national-water-initiative-agreement-2004.pdf (accessed on 15 April 2023).

Kumarappa, S., & Manjunatha, H. M. (2024). Machine learning-based prediction of lithium-ion battery life cycle for capacity degradation modelling. *World Journal of Advanced Research and Reviews*, 21(2), 1299–1309. DOI: 10.30574/wjarr.2024.21.2.0553

Pathak, S., Shukla, A. K., Ojha, C. S. P., & Garg, R. D. Identifying hot spots for stormwater harvesting and reuse in urban areas applying remote sensing, GIS and system optimization techniques. *IEEE International Geoscience and Remote Sensing Symposium (IGARSS)*, pp 967-970, 2016. DOI: 10.1109/IGARSS.2016.7729245

Quorum Business Solutions. (2023). Quorum EnergySuite. Available online: https://www.quorumsoftware.com/solutions/ (accessed on 23 Jan, 2023)

Rogers, B. C., Dunn, G., Hammer, K., & Novalia, W. DE Haan FJ., Brown L., Brown R.R., Lloyd S., Urich C., Wong T.H.F., Chesterfield, C. Water Sensitive Cities Index: A diagnostic tool to assess water sensitivity and guide management actions. In Book Water research, Eberhard Morgenroth, Guang-Hao Chen, Ana Deletic, David Dixon, Xiaodi Hao, Peter Jarvis, Gregory Korshin, Hui Lu, Sun Yat-Sen, Yoshihiko Matsui, Eberhard Morgenroth, Ameet J. Pinto, Maria Reis, Hang-Sik Shin, Thomas Ternes, Stefan Wuertz, Zhiguao Yuan, Yan Zhou, Julie Zilles., Publisher: ELSEVIER Publisher location, Country, Amsterdam, Netherlands, Volume **186**, pp. 116411, **2020**.

Ruiz-Vanoye, J. A., Díaz-Parra, O., Ramos-Fernández, J. C., Trejo-Macotela, F. R., & Xicoténcatl-Pérez, J. M. (2023). Editorial: A Brief Panorama of Water Smart Cities. *International Journal of Combinatorial Optimization Problems and Informatics*, 14(1), 1. https://www.ijcopi.org/ojs/article/view/33. DOI: 10.61467/2007.1558.2023.v14i1.332

Sullivan, L. P. (1986). *Quality Function Deployment. Quality Progress*. Springer.

Tecnologías para el uso sostenible del agua: una contribución a la seguridad alimentaria y la adaptación al cambio climático. Available online: https://www.fao.org/3/i3442s/i3442s.pdf (accessed on 21 Jan 2023).

Water Sensitive Cities Australia. Available online: http://URL https://wscaustralia.org.au/2022/08/09/water-sensitive-cities-australia-our-first-12-months/ (accessed on 11 May 2023).

Zhu, S., Zhao, N., & Sha, J. (2019). Predicting battery life with early cyclic data by machine learning. *Energy Storage*, 1(6), e98. Advance online publication. DOI: 10.1002/est2.98

Chapter 7
Water Demand Forecast and Efficient Supply Management

Jaime Aguilar Ortiz
Universidad Politécnica de Pachuca, Mexico

ABSTRACT

This chapter explores the use of artificial intelligence (AI) models for forecasting water demand and managing water supply systems efficiently. It highlights the need for accurate prediction models in urban areas, where fluctuating water demand poses significant challenges. Various AI techniques, including Artificial Neural Networks (ANN), Convolutional Neural Networks (CNN), and Long Short-Term Memory (LSTM) networks, are discussed for their ability to handle complex consumption data. The methodology for building these models is also addressed, covering data collection, preparation, and model training and validation. Optimization algorithms, such as Particle Swarm Optimization (PSO) and Genetic Algorithms (GA), are emphasized to enhance accuracy and robustness. Finally, a practical example illustrates the application of these AI methods in water management, highlighting AI's potential to improve the sustainability of water supply systems and identifying areas for future research to refine these models across various urban settings.

DOI: 10.4018/979-8-3693-8074-1.ch007

1. INTRODUCTION

1.1. Context and Relevance of Water Management

Efficient water supply management is essential to ensure the availability and quality of water resources in urban areas. The use of artificial intelligence (AI) models for water demand forecasting has been highlighted as an innovative and effective solution. These models allow water consumption to be predicted with high accuracy, which is essential for the planning and operation of water distribution systems. The adoption of AI in this context not only improves the accuracy of predictions, but also facilitates more dynamic and adaptable management to changes in demand and environmental conditions (Herrera et al., 2010).

Traditional forecasting methods, such as time series analysis and regression, have been complemented and, in many cases, surpassed by AI techniques such as neural networks and expert systems. Research has shown that AI models, including artificial neural networks (ANN) and rule-based inference systems, offer better results in predicting short-term demand compared to conventional methods. These models are able to handle non-linear data and adapt quickly to new consumption patterns, which is crucial for effective water resource management in changing urban environments (Jain & Ormsbee, 2002).

Furthermore, the use of adaptive and continuously learning models, such as Evolutionary Artificial Neural Networks (EANN), has been shown to be particularly effective for real-time prediction of water demand. These models not only predict with high accuracy, but also continuously adjust to variations in consumption patterns, providing a robust tool for the daily management of water distribution systems. This adaptive capacity is crucial to meet the challenges associated with variability in demand and climatic conditions, allowing proactive and efficient management of water supply (Romano & Kapelan, 2014).

1.2. Problems of Water Demand in Urban Systems

Water demand in urban environments is subject to various challenges, particularly due to increasing populations and the unpredictability of climate and human behavior. As water usage fluctuates, it can create pressure inconsistencies within distribution systems, heightening the risk of contamination through infiltration of external substances. Traditional models, which typically fail to account for the dynamic variability of these factors, often struggle to provide accurate predictions, making them less effective for modern urban water management. The inaccuracies can lead to inefficient water supply operations, potential resource wastage, and failure to meet the evolving demands of urban populations.

In this context, AI models offer significant advantages by improving the accuracy of water demand forecasts and enabling more efficient management of water distribution systems. These models, such as neural networks and other AI-driven approaches, are better equipped to handle the non-linear and complex nature of water consumption data. AI models, as described by Campisi-Pinto et al. (2012), also provide superior forecasting capabilities by incorporating external variables like climate and behavioral data, which allow for dynamic adjustments to varying conditions. This adaptability is crucial in reducing the risk of contamination, optimizing system performance, and ensuring a sustainable and safe water supply for urban populations.

Hybrid models that combine Convolutional Neural Networks (CNN) and Long Short-Term Memory (Bi-LSTM) networks offer a powerful solution to the complexities inherent in forecasting urban water demand. These models stand out due to their ability to process both the spatial and temporal patterns of water consumption simultaneously. By leveraging CNNs, which are highly effective in analyzing spatial data, and Bi-LSTMs, which excel in handling temporal sequences, these hybrid models provide a comprehensive understanding of water usage trends.

Hybrid models improve water demand forecasting by incorporating historical consumption patterns and their evolution over time, leading to precise predictions crucial for efficient water distribution in dynamic urban environments. These models adapt to real-time conditions, such as climate variations and events like holidays, enhancing prediction stability and reducing risks of shortages or inefficiencies. By dynamically integrating external factors, they ensure a balanced supply-demand management. As highlighted by Hu et al. (2019), these models not only boost accuracy but also strengthen the sustainability and resilience of urban water systems.

Distributed modeling of urban water demand enhances water supply and drainage management by accounting for land use patterns and their impact on water consumption (see Figure 1). This approach generates localized predictions for residential, commercial, industrial, and recreational zones, enabling efficient infrastructure development tailored to varying demands. Its flexibility allows for parameter calibration specific to each land use type, ensuring accurate projections of future needs. By adapting to unique characteristics, such as peak usage in commercial or residential areas, distributed models support sustainable and resilient urban water systems, crucial for growing cities (Luan & Liu, 2017).

Figure 1. Water demand challengers.

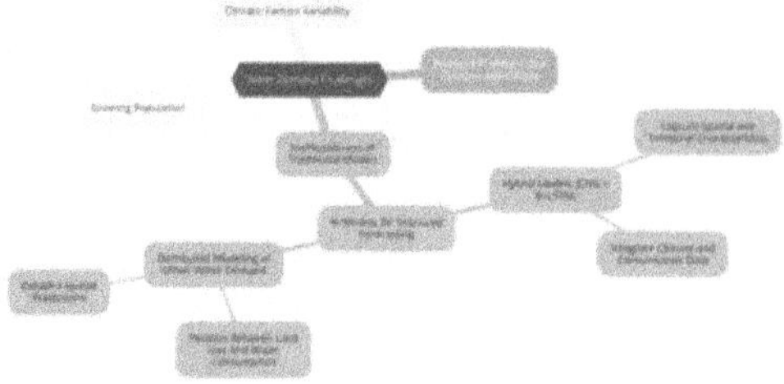

1.3. Objectives of the Study

Develop an artificial intelligence (AI) model for accurate water demand forecasting in urban systems, using a combination of advanced techniques such as Artificial Neural Networks (ANN), Convolutional Neural Networks (CNN), and Long Short-Term Memory (LSTM) networks. This objective is essential for improving the planning and management of water supply systems, allowing for the anticipation of needs and the adjustment of operations accordingly.

Evaluate and compare the effectiveness of different AI modeling techniques (such as ANN, Adaptive Neuro-Fuzzy Inference Systems, and hybrid models) in predicting water demand. This evaluation will include the implementation of optimization algorithms, such as Particle Swarm Optimization (PSO) and Genetic Algorithms (GA), to identify the most robust and accurate models capable of handling nonlinear and variable data common in urban consumption patterns.

Develop a water demand forecasting system that is adaptive and capable of operating in real-time, using Evolutionary Artificial Neural Networks (EANN) models that continuously adjust to changing water demand patterns. The implementation of a real-time forecasting system will enable water resource managers to respond more quickly and effectively to consumption variations, improving the resilience and sustainability of the water supply system (See Figure 2).

2. LITERATURE REVIEW

2.1. AI Applications in Water Demand Forecasting

Artificial intelligence (AI) has revolutionized water demand forecasting by improving accuracy and efficiency in resource management. Unlike traditional methods like regression or time series, AI techniques, such as Artificial Neural Networks (ANNs), excel in handling the complexities of non-linear relationships and sudden demand shifts in urban water systems. ANNs are particularly effective in identifying hidden patterns in large, irregular datasets, continuously improving with more data. This capability enhances short-term demand forecasting, as demonstrated by studies like Jain & Ormsbee (2002), enabling precise resource allocation, reducing waste, and supporting the sustainability of urban water systems.

Figure 2. AI models for water demand forecasting.

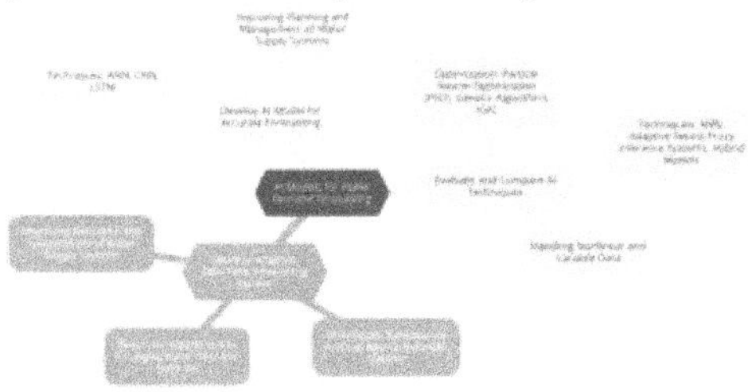

A significant advancement in AI-based water demand forecasting is the use of neuro-genetic networks, which combine neural networks with genetic algorithms to optimize forecasting accuracy. Neural networks excel at identifying complex, non-linear patterns in data, making them ideal for predicting water consumption influenced by variables like weather, crop types, and soil moisture. However, optimizing the architecture of these networks is challenging, which is where genetic algorithms come in. These algorithms use evolutionary techniques to iteratively improve network configurations by selecting the best-performing models. In the Bembézar Irrigation District in southern Spain, this approach achieved a remarkable 93% accuracy in daily water demand prediction. This high level of precision is especially important for irrigation systems, where accurate forecasts reduce water waste and optimize irrigation timing and volume, directly benefiting crop yields and resource management. By supplying water precisely when and where it's needed, neuro-genetic networks enhance both water efficiency and the sustainability of agricultural practices. This approach illustrates the potential of AI to transform water resource management, aligning with the broader goal of improving accuracy, efficiency, and sustainability across different sectors (Perea et al., 2015).

Deep learning techniques, such as convolutional neural networks (CNNs) and long-short-term memory (LSTM) neural networks, have significantly advanced water demand forecasting by addressing spatial and temporal complexities. CNNs excel at capturing geographic and climatic spatial features, while LSTMs handle time-dependent data like daily and seasonal consumption fluctuations. Their combined use improves short-term forecasting accuracy, enabling real-time adjustments in water distribution, reducing shortages, and minimizing resource overuse, thus promoting sustainability in urban water management systems (Rajakumar et al., 2021).

Adaptive Neuro-Fuzzy Inference Systems (ANFIS) further enhance forecasting by integrating neural networks and fuzzy logic to manage uncertainty and imprecision in water consumption data. ANFIS adapts to complex and variable patterns, effectively handling incomplete or noisy data, making it highly suitable for unpredictable consumption scenarios. Its proven success in groundwater quality prediction highlights its potential for broader water resource management, aligning with goals of improving accuracy, adaptability, and resilience in water systems. By leveraging ANFIS, water managers can ensure sustainable and efficient resource use while supporting long-term environmental and urban planning objectives (Nordin et al., 2021).

Wavelet decomposition combined with neural networks enhances water demand forecasting by isolating and reducing noise in data while leveraging neural networks' ability to model non-linear relationships. This hybrid approach, proven effective in studies like those conducted in Syracuse, Italy, surpasses traditional methods, such as multiple linear regression, by delivering more precise predictions. Its ability to address seasonal, human, and climate variability supports sustainable and efficient urban water management systems, aligning with the broader goal of integrating adaptive AI techniques for improved reliability and sustainability (Campisi-Pinto et al., 2012).

Hybrid AI models, integrating neural networks and support vector machines (SVM), further advance water demand forecasting. Neural networks handle non-linear data patterns, while SVM excels in precise classification and regression tasks. Optimization methods like differential evolution enhance these models by fine-tuning parameters for superior accuracy, even with noisy or complex datasets. These hybrid models outperform single AI techniques, supporting intelligent, adaptive urban water systems and ensuring supply meets the dynamic demands of modern cities (Oyebode & Ighravwe, 2019).

Additionally, AI-driven predictive maintenance in water infrastructure improves failure detection and planning, boosting the sustainability and efficiency of water distribution systems (Mehmood et al., 2020).

AI-driven monitoring and prediction systems have significantly advanced water quality management and disaster response, promoting sustainability and resilience in water systems. AI models enable real-time analysis of vast data from sensors, accurately forecasting critical parameters like dissolved oxygen and biochemical oxygen demand (BOD). Unlike traditional, time-intensive laboratory methods, AI delivers continuous monitoring, allowing for rapid responses to contamination or quality changes, particularly in urban systems prone to industrial and climatic shifts.

This proactive approach safeguards public health and enhances resource management efficiency (Najah et al., 2011).

In disaster management, AI techniques improve the prediction of water-related events, such as floods, by analyzing environmental data like rainfall patterns and river levels in real-time. AI systems detect early warning signs, providing accurate forecasts on the likelihood, timing, and severity of disasters. This enables timely evacuations, infrastructure adjustments, and resource deployment, enhancing both short-term response and long-term preparedness. By mitigating risks and improving urban planning, AI-driven disaster forecasting aligns with the goal of creating resilient and sustainable water management systems (Mehmood et al., 2020).

Evolutionary Artificial Neural Network (EANN) models combine neural networks with evolutionary algorithms to optimize real-time water demand prediction. These models continuously adapt to changing conditions, such as seasonal shifts or sudden usage variations, enhancing predictive accuracy and operational efficiency. EANNs allow for dynamic adjustments in water distribution, minimizing waste, preventing over- or under-supply, and ensuring resources are allocated where needed. This approach is particularly effective in urban areas with fluctuating demand and aligns with the document's goal of improving the adaptability, sustainability, and efficiency of water systems through advanced AI techniques (Romano & Kapelan, 2014).

AI-based Intelligent Decision Support Systems (IDSS) optimize real-time water management by integrating machine learning and advanced optimization algorithms. These systems analyze data from various sources—like consumption patterns and weather forecasts—to predict demand with high accuracy and provide actionable insights. IDSS enhances resource allocation, minimizes wastage, reduces operational costs, and optimizes infrastructure performance, such as pumps and reservoirs. By improving forecasting and resource management, IDSS contributes to sustainable, efficient urban water systems, supporting the document's vision for smarter, adaptive water management solutions (Ponte et al., 2016).

Figure 3. AI applications in water demand forecasting.

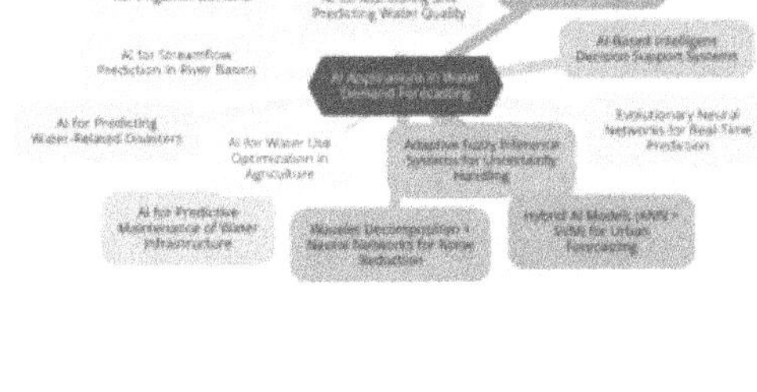

AI has revolutionized water use in agriculture by optimizing irrigation efficiency and reducing water consumption. Neural network models analyze variables like soil moisture, weather, and crop type to accurately predict irrigation needs, ensuring precise water application and preventing over- or under-irrigation. Optimization techniques, such as genetic algorithms and particle swarm optimization, further enhance irrigation strategies by fine-tuning schedules and resource allocation. These AI-driven solutions conserve water resources, improve agricultural sustainability, and reduce environmental impact, aligning with the document's goal of advancing efficient and adaptive water management across sectors (Perea et al., 2015).

In river basin management, AI techniques have proven invaluable for streamflow prediction, addressing the complexities of hydrological data that traditional methods struggle to manage. Models like artificial neural networks and support vector machines process large datasets to identify patterns in variables such as precipitation, evaporation, and soil saturation. These models deliver accurate streamflow forecasts, critical for managing water supply, flood prevention, and ecosystem health. By enabling precise planning for water storage and distribution, AI supports sustainable resource management and mitigates risks associated with extreme events. This application highlights AI's role in advancing the sustainability and adaptability of water management systems, in line with the document's objectives (Yaseen et al., 2015).

2.2. AI Strategies in Efficient Water Supply Management

Efficient water supply management with artificial intelligence (AI) leverages advanced strategies to optimize resource use, reduce costs, and enhance sustainability. One effective approach is pump scheduling optimization using genetic algorithms, which can improve energy efficiency by an average of 15%. These algorithms, inspired by natural selection, analyze factors such as water demand, energy prices, and system constraints to determine the most efficient pump operating schedules. This not only lowers operational costs but also reduces energy consumption, extending equipment lifespan and decreasing the carbon footprint of water systems. This aligns with the goal of creating adaptive and sustainable water management systems (Luna et al., 2019).

Another impactful strategy is the use of Intelligent Decision Support Systems (IDSS) with multi-agent methodologies and artificial neural networks (ANNs) for real-time water demand management. IDSS integrates components like supply sources and distribution networks into a unified framework, utilizing ANNs to forecast demand accurately, even amid fluctuating patterns. By operating in real time, IDSS ensures precise water delivery, reducing wastage and minimizing energy use. This approach lowers operational costs and enhances system efficiency, supporting the document's objective of advancing sustainable and adaptive water management systems through AI technologies (Ponte et al., 2016).

Adaptive Intelligent Water Resource Planning (AIDWRP) utilizes Markov decision processes (MDPs) to dynamically optimize urban water systems, offering a flexible and efficient approach to managing complex water resource challenges. MDPs model various future states of water systems, including changes in demand, supply, and environmental conditions, to determine optimal actions for long-term efficiency. By combining numerical AI tools, such as predictive models and optimization algorithms, with human expertise, AIDWRP enables real-time strategy adjustments based on current data and forecasts. This approach enhances the system's adaptability to events like droughts or demand surges, ensuring sustainable and resilient urban water management. It aligns with the document's goal of promoting smart, adaptive decision-making for long-term water sustainability (Xiang et al., 2021).

Hybrid neural network models and decision trees have been applied to monitor critical parameters like pressure and flow in water supply networks, improving operational decision-making. Neural networks excel at identifying complex patterns in large datasets, while decision trees offer clear, interpretable decision-making structures. These models accurately predict network loads, such as pressure fluctuations and flow rates, enabling water managers to anticipate high-demand periods and adjust infrastructure operations proactively. This precision reduces energy consumption, operational costs, and risks of service disruptions, particularly in urban settings with variable water usage. The integration of these hybrid models supports the document's objective of leveraging AI technologies to develop efficient, adaptive, and sustainable water management systems for modern urban environments (Rojek, 2014).

The integration of Internet of Things (IoT)-based intelligent systems for pressure monitoring in water distribution networks has significantly improved efficiency and sustainability. These systems use real-time data from sensors to monitor pressure, flow, and water levels, enabling early detection of leaks, faults, or imbalances. This proactive approach allows for timely interventions, reducing water loss, operational costs, and downtime while improving service reliability. IoT-enabled solutions align with smart city initiatives by optimizing resource use and sustainability, supporting the document's aim of smarter, adaptive water management systems for urban areas (Pérez-Padillo et al., 2020).

Artificial intelligence (AI) applications in drinking water treatment have enhanced process efficiency and water quality by optimizing coagulation and disinfection processes. AI models analyze operational data to determine optimal dosages and timings for chemical treatments, improving water safety while reducing waste, costs, and energy consumption. Similarly, AI-driven irrigation optimization in agriculture predicts water needs based on soil moisture, weather, and crop type, ensuring efficient water use without excess. These AI applications in urban water treatment and agriculture align with the document's goal of promoting efficient, adaptive, and sustainable water management practices, contributing to resource conservation and long-term environmental sustainability (Li et al., 2021; Perea et al., 2015).

The use of smart meters and water end-use data has transformed urban water supply management by providing real-time, detailed insights into consumption patterns. These devices enable utilities to tailor supply strategies to specific areas or users, optimize resource distribution, and make data-driven decisions about infrastructure investments. Smart meter data also supports proactive water conservation by detecting anomalies such as leaks for timely intervention. This approach aligns with the document's aim of smarter, adaptive, and sustainable water management systems (Gurung et al., 2015).

AI has revolutionized desalination processes by improving efficiency and reducing costs. Neural networks enhance control over variables like pressure and temperature, while genetic algorithms optimize system parameters, minimizing resource use and energy consumption. These AI-driven advancements support global efforts to address water scarcity and align with the document's goals of sustainable water management. AI-powered desalination offers scalable solutions for arid regions, ensuring efficient potable water production with reduced environmental impact (Mahadeva et al., 2023).

The integration of AI and IoT in water supply management has created efficient systems for smart cities. By collecting and analyzing real-time data from sensors, these systems optimize water distribution, detect inefficiencies, and adapt to dynamic demand patterns. IoT devices monitor performance and send alerts for immediate issue resolution, reducing resource waste and energy consumption. This innovative approach supports the document's objective of sustainable, adaptive urban water infrastructure to meet the demands of interconnected urban environments (Gonçalves et al., 2020).

In recent years, the introduction of artificial intelligence (AI) and machine learning (ML) into drinking water treatment has transformed the way we manage this essential resource. Imagine a bustling city where the demand for clean water never slows down. Hidden behind the scenes, AI and ML quietly analyze massive streams of data from the water treatment process, much like an orchestra conductor guiding musicians. But instead of music, these technologies are fine-tuning complex, nonlinear relationships within the water treatment systems—factors like water quality, chemical dosing, and pressure control that are constantly fluctuating.

Traditionally, managing these variables was challenging, often leading to inefficiencies or inconsistencies in water quality. However, AI and ML have revolutionized this process. These technologies now handle the complexity by predicting and adjusting chemical dosing levels in real time, ensuring that the water meets safety standards with minimal resource waste. Through advanced algorithms, AI models recognize patterns in the data that humans might miss, allowing treatment plants to make precise adjustments, lowering costs and improving water quality.

This narrative of transformation aligns perfectly with the document's broader goal: to leverage AI technologies for smarter, more adaptive water management. By using these cutting-edge tools, cities can provide cleaner, safer water to their citizens more efficiently than ever before, contributing to the long-term sustainability of urban water supplies (Li et al., 2021).

Intelligent pressure monitoring systems have revolutionized water distribution efficiency, especially in arid regions where water scarcity is critical. By using real-time data from strategically placed sensors, these systems detect anomalies like pressure drops, enabling proactive responses to leaks or inefficiencies. This approach prevents water loss, ensures optimal network performance, and supports sustainable water management. Such technology aligns with the document's focus on smarter, adaptive solutions for resource conservation and infrastructure development in water-stressed environments (Pérez-Padillo et al., 2020).

Deep learning algorithms have transformed water resource management by improving the accuracy of demand predictions. These models analyze historical and real-time data to identify patterns influenced by factors like population growth and weather, enabling precise forecasts. Enhanced accuracy allows water managers to optimize supply systems, reducing resource waste, energy consumption, and costs. By aligning distribution with actual demand, deep learning supports the document's goal of sustainable and adaptive water management, contributing to the resilience of urban and rural infrastructures (Krishnan et al., 2022).

The use of AI-based models for the analysis and optimization of desalination and water treatment processes has proven to be highly effective in enhancing both operational efficiency and sustainability. Desalination and water treatment are energy-intensive processes, often requiring precise control over multiple variables such as temperature, pressure, and chemical dosing to ensure the production of safe, high-quality water. AI-based models, including machine learning algorithms and neural networks, are capable of analyzing vast amounts of operational data in real time, allowing for the continuous adjustment and fine-tuning of these processes. By optimizing parameters that affect energy use and resource consumption, AI helps to minimize waste, reduce costs, and extend the lifespan of treatment infrastructure. Moreover, AI's predictive capabilities allow managers to anticipate maintenance needs, avoiding downtime and improving system reliability. This leads to more sustainable operations by ensuring that water is treated efficiently and with minimal environmental impact. The application of AI in desalination and water treatment aligns closely with the document's objective of promoting smarter, more adaptive, and sustainable water management systems. Through the use of AI, these processes can be streamlined, making them more resilient and capable of meeting the growing global demand for clean water while conserving resources (Aani et al., 2019). (See Figure 4).

3. METHODOLOGY

3.1. AI Models to be Applied

3.1.1. Neural Networks for Water Demand Forecasting

Artificial Neural Networks (ANN) have emerged as a powerful and highly adaptable technique for water demand forecasting, primarily because of their ability to model complex, non-linear relationships between multiple input and output variables. These networks excel at learning from large datasets, making them particularly effective in capturing the intricate factors that influence water consumption. For example, in one study, ANNs were applied to forecast daily water demand in a city by incorporating a wide range of variables, including maximum and minimum temperatures, precipitation, day of the week, and historical water usage data. The flexibility of ANNs allowed the model to account for these diverse factors and adapt to changing conditions, resulting in highly accurate predictions. The study demonstrated that the prediction errors were minimal and well within acceptable limits for managing the water supply system efficiently. This precise forecasting ability is crucial for urban water management, where anticipating fluctuations in demand is essential to prevent shortages or over-distribution, both of which can lead to inefficiencies and resource wastage. The use of ANNs directly supports the document's objective of promoting smarter, more data-driven, and adaptive water management systems that can respond to the evolving needs of urban populations while optimizing resource use (Msiza, Nelwamondo & Marwala, 2008).

The use of artificial neural networks (ANNs) in urban water demand forecasting has been thoroughly explored in case studies utilizing MATLAB, showcasing the impressive versatility and effectiveness of these models across various timeframes—short-, medium-, and long-term. What makes these ANN models particularly powerful is their ability to generate highly accurate predictions without the need to explicitly include complex variables such as climatic factors or the exact number of consumers. By learning from historical consumption data and recognizing patterns in water usage, ANNs are able to account for underlying trends and shifts in demand, making them highly adaptable to different urban contexts. This ability to predict water demand with high accuracy, even when simplifying certain inputs, is crucial for urban water management systems that need to be both flexible and efficient. In rapidly growing urban areas, the demand for water can fluctuate significantly, and the capacity to predict these changes without relying on overly detailed inputs simplifies the forecasting process while maintaining reliability. The use of ANNs in these case studies directly aligns with the document's goal of promoting advanced AI-based solutions that improve the accuracy, efficiency, and sustainability of water resource management, particularly in densely populated urban environments (Lorente-Leyva et al., 2019).

Figure 4. AI strategies in efficient water supply management.

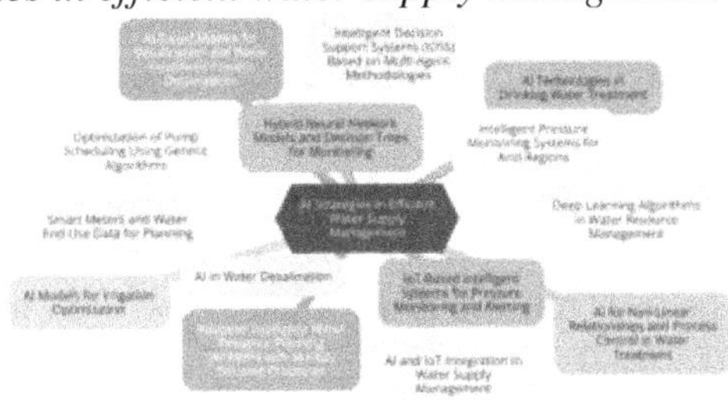

In addition to traditional artificial neural networks (ANNs), dynamic neural network (DAN2) models have been developed specifically for urban water demand forecasting, offering an even more sophisticated approach to handling time-series data. These models are particularly notable for their ability to provide highly accurate predictions, consistently achieving over 99% accuracy in monthly, weekly, and daily water demand forecasts. One of the key strengths of DAN2 models is their ability to model complex temporal relationships in water usage patterns without the need to explicitly include external factors such as climate data. By learning directly from historical consumption data, DAN2 models adapt to the inherent fluctuations in water demand, making them highly reliable across different time frames. This high level of precision is critical for urban water management, where accurate demand forecasting ensures that water supply systems operate efficiently, reducing both resource wastage and operational costs. Moreover, the use of DAN2 models aligns perfectly with the document's broader goal of promoting the use of advanced AI-driven techniques to enhance the accuracy, sustainability, and responsiveness of water management systems. By offering exceptional predictive capabilities with minimal input requirements, these models contribute to the development of smarter, more adaptive water supply infrastructures capable of meeting the evolving demands of urban populations (Ghiassi et al., 2008).

Another innovative approach in water demand forecasting is the use of neural networks with intervention series, specifically designed for large supply and distribution systems. This method introduces units with variable gain in the transfer function, which allows the model to adjust its sensitivity to different input signals based on changing conditions within the water distribution network. By treating the forecasting process as a standard optimization problem, these models are able to streamline the training process, improving both the efficiency and accuracy of the forecasts. The ability to adjust dynamically through intervention series makes these neural networks particularly effective in handling the complexities of large-scale systems, where factors such as sudden shifts in demand or disruptions in supply can have significant impacts. This approach not only enhances the precision of water demand predictions but also optimizes the performance of the entire distribution system by allowing more accurate adjustments to be made in real-time. By increasing forecast accuracy and operational efficiency, these models align with the document's goal of promoting more sustainable and adaptive water management practices. Neural networks with intervention series contribute to the development of smarter, more resilient infrastructure, ensuring that water supply systems can meet the challenges of growing urban populations while minimizing waste and inefficiencies (Cubero, 1991).

Figure 5. AI models and optimization algorithms in water management.

The use of radial basis function (RBF) neural networks for water demand forecasting has proven to be highly effective in identifying and modeling complex, non-linear relationships within water consumption data. RBF networks are particularly well-suited for this task due to their ability to approximate intricate patterns and fluctuations in water demand, making them highly valuable in dynamic and unpredictable urban environments. In one study, an RBF model was developed using 17 different water demand prediction factors, such as population growth, weather patterns, and industrial usage. This comprehensive input set allowed the model to capture the diverse influences on water demand accurately. Over a period of three consecutive years, the model demonstrated outstanding performance, achieving prediction accuracy with relative errors of less than 3.5%, underscoring its reliability and precision. This level of accuracy is critical for urban water management, where even small errors in demand forecasting can lead to overuse of resources or shortfalls in supply. By leveraging RBF neural networks, water systems can optimize operations, reduce inefficiencies, and ensure that supply meets demand more precisely. This approach directly aligns with the document's objective of promoting smarter, more adaptive, and sustainable water management solutions that utilize advanced AI techniques to enhance operational efficiency and resource conservation (Chang & Liu, 2009).

The combination of neural networks with heuristic optimization algorithms, such as genetic algorithms (GA) and particle swarm optimization (PSO), represents a significant advancement in water demand forecasting, especially in urban environments. Traditional backpropagation (BP) neural networks, while effective, often struggle with local minima and require extensive fine-tuning to achieve optimal performance. By integrating heuristic optimization algorithms like GA and PSO, these combined models enhance the global search capability of BP networks, allowing them to better navigate complex, non-linear data and avoid getting stuck in suboptimal solutions. This results in significantly improved prediction accuracy and reduced errors compared to standard BP models. In the context of urban water demand, where forecasting precision is crucial for efficient resource allocation and management, these hybrid approaches have demonstrated superior performance. They are particularly valuable in handling the dynamic and multifaceted nature of water consumption patterns in cities, where factors like population density, climate variability, and infrastructure usage must be accounted for. By lowering prediction errors and enhancing accuracy, these combined approaches contribute to more efficient and sustainable water management practices, directly supporting the document's goal of leveraging advanced AI techniques to optimize water supply systems and meet the growing demands of urban populations (Xing et al., 2017).

3.1.2. Water Supply Optimization Algorithms

Water supply optimization is essential for achieving both operational efficiency and long-term system sustainability, particularly as cities grow and water demands become more complex. One of the most effective methods for optimizing water supply systems is the use of particle swarm optimization (PSO) algorithms. PSO, a nature-inspired optimization technique, excels in finding optimal solutions to complex design and operational problems by simulating the behavior of swarms, such as flocks of birds or schools of fish, searching for the best path in a given environment. In water supply systems, PSO has been applied to optimize factors like pump scheduling, pipe network configurations, and resource allocation, ensuring that water is distributed efficiently while minimizing energy consumption and operational costs. Recent studies have demonstrated the practical success of PSO, notably in large-scale applications such as the Hanoi water distribution network and the New York City water supply tunnel system. These applications have shown significant improvements in operational efficiency and cost reductions, highlighting the potential of PSO to address the challenges of managing large and complex water infrastructure. This approach directly supports the document's goal of promoting advanced, AI-driven solutions to enhance the sustainability and efficiency of urban water management systems, ensuring that water supply networks can meet growing demands while optimizing resource use (Montalvo et al., 2008).

Pump scheduling optimization is an innovative approach that significantly enhances the energy efficiency of water supply systems by using advanced techniques like genetic algorithms. This method optimizes the operation of pumps by predicting water demand and determining the best times and conditions for pump usage, ensuring that the system meets demand while minimizing energy consumption. By simulating natural evolutionary processes, genetic algorithms explore various pump operation schedules and select the optimal configuration that reduces operational costs and lowers carbon dioxide emissions. In one study, the application of this method led to improvements in energy efficiency of up to 25%, underscoring the substantial impact optimization algorithms can have on reducing both financial and environmental costs. This increase in efficiency is particularly important in large-scale water distribution networks, where energy consumption for pumping often represents a significant portion of operating expenses. The successful implementation of pump scheduling optimization aligns directly with the document's goal of promoting AI-driven solutions that enhance the sustainability of water supply systems. By reducing energy use and greenhouse gas emissions, this approach contributes to a more sustainable and cost-effective management of water resources in urban environments (Luna et al., 2019).

In addition to particle swarm optimization (PSO) and genetic algorithms, nonlinear optimization models have proven to be highly effective in managing water supply systems, particularly due to their ability to handle the complexities inherent in such systems. Water distribution networks often face challenges such as varying water demand, friction losses in pipes, and dynamic changes in flow rates, which traditional linear models may struggle to address. Nonlinear optimization models are better equipped to manage these factors by accurately capturing the intricate relationships between system components, such as pumps, reservoirs, and distribution networks. For instance, a nonlinear optimization model applied to a water supply system consisting of a pumping station, an elevated reservoir, and an extensive distribution network demonstrated marked improvements in both operational efficiency and cost reduction. By optimizing how the system handles pressure, flow, and energy usage, these models allow for more efficient resource allocation and operational control. This approach aligns perfectly with the document's objective of promoting more sustainable and efficient water management solutions. By leveraging nonlinear optimization, water managers can better navigate the complexities of modern water systems, ensuring that supply meets demand with minimal resource waste and lower operational costs (Kher et al., 1979).

Optimization of water supply systems can be further enhanced by the use of hybrid algorithms that combine multiple optimization techniques, leveraging the strengths of each to create more powerful solutions. A prime example of this is the combination of genetic algorithms with the hill climber method, a hybrid approach that has been shown to significantly outperform traditional optimization techniques when applied to complex water distribution systems. Genetic algorithms excel at exploring a broad range of potential solutions by simulating evolutionary processes, while the hill climber method fine-tunes these solutions by quickly converging to local optima. This combination not only allows for the discovery of optimal solutions more efficiently, but also improves the responsiveness and robustness of the system. In water distribution networks, where variables like pressure, flow, and demand fluctuate frequently, this hybrid approach enables faster, more reliable adjustments to optimize system performance. By reducing the time needed to identify the best operational configurations, hybrid algorithms contribute to both energy savings and enhanced system reliability. This directly supports the document's goal of promoting advanced, AI-driven solutions that improve the efficiency, sustainability, and adaptability of water supply systems, ensuring that these systems can meet the demands of growing urban populations while minimizing operational costs and resource waste (Zyl et al., 2004).

Finally, the application of generalized network optimization techniques can significantly improve the operation of large-scale water supply systems. These algorithms allow optimizing water distribution in systems with multiple supply sources and planning objectives, such as reducing operating costs and meeting demand. In one study, the implementation of a generalized network algorithm in the Algarve multi-municipal water supply system demonstrated its ability to optimize system operation, considering both economic and non-economic objectives (Sun et al., 1995).

These optimization techniques demonstrate the potential of advanced algorithms to improve the efficiency and sustainability of water supply systems, addressing complex challenges and promoting more effective water resource management (See Figure 5).

3.2. Experimental Design

3.2.1. Historical and Synthetic Data Collection

The collection of both historical and synthetic data is a critical foundation for building effective AI models in water demand forecasting. The first phase of this process involves gathering historical water consumption data, which provides detailed insights into how water usage varies across different time periods—daily, weekly, and monthly. This data is typically sourced from the records of water supply companies, offering a comprehensive view of past consumption patterns. By analyzing these historical records, AI models can identify recurring trends, seasonal fluctuations, and any anomalies that may impact future water demand. Establishing this baseline is essential for the development of accurate predictive models, as it allows the AI system to "learn" from past behavior and make informed predictions about future consumption. Additionally, this historical data serves as a benchmark for validating the model's predictions, ensuring that it can accurately forecast water demand across various timeframes and conditions. By incorporating such a robust data collection process, water management systems can optimize their operations, reducing inefficiencies and ensuring that resources are distributed effectively. This directly aligns with the document's goal of promoting smarter, data-driven, and sustainable water management practices that can adapt to the evolving demands of urban populations (Jain & Ormsbee, 2002).

The second phase of data collection, involving the generation of synthetic data, plays a crucial role in the development of AI models for water demand forecasting, particularly when historical data is insufficient or incomplete. Synthetic data is generated through simulations based on mathematical or statistical models, creating additional datasets that reflect various possible water consumption scenarios under different conditions. This approach is especially valuable in situations where historical records may lack the depth or detail required to train robust AI models. By introducing synthetic data, water managers can simulate a wide range of potential scenarios, including extreme weather events, population growth, or infrastructure changes, which may not be well-represented in historical data alone. These simulated datasets allow AI models to train on more diverse inputs, improving their ability to generalize and predict accurately across a broader spectrum of conditions. Moreover, synthetic data enhances the validation process, ensuring that the AI models can handle real-world complexities and variabilities in water demand. This comprehensive approach to data collection supports the document's objective of promoting advanced, AI-driven solutions for sustainable and efficient water management. By combining historical and synthetic data, AI models become more adaptive and resilient, capable of delivering accurate forecasts that optimize resource use and support long-term water supply planning (Romano & Kapelan, 2014).

A crucial aspect of the data collection process for building accurate AI models in water demand forecasting is the integration of diverse data sources, such as smart meters, weather data, and demographic information. Smart meters are particularly valuable because they provide highly granular, real-time data on water consumption at the individual user level, capturing detailed patterns of usage that can vary by household, time of day, or specific activities. This real-time data greatly enhances the precision of predictive models, allowing them to adjust forecasts dynamically based on actual consumption trends. When this smart meter data is integrated with external factors, such as weather conditions (e.g., temperature, rainfall) and demographic data (e.g., population density, socio-economic trends), the AI models become even more robust. These additional inputs help capture a wider range of influences on water demand, including seasonal variations, behavioral changes, and long-term growth trends. The combination of these data sources strengthens the models' ability to accurately predict water demand in diverse and complex urban environments. This comprehensive data integration directly supports the document's objective of promoting smarter, more adaptive, and sustainable water management practices by leveraging AI-driven solutions that can respond to the multifaceted demands of modern water systems (Pesantez et al., 2020).

Preprocessing the collected data is a crucial step in the experimental design process for developing accurate AI models in water demand forecasting. This stage involves several key tasks, including data cleaning, normalization, and transformation, all of which are essential to ensure that the data is of high quality and ready for training AI models. Data cleaning involves identifying and removing errors or outliers that could distort the predictive accuracy of the model. Additionally, techniques like missing data interpolation help fill in gaps where data may be incomplete, ensuring that the models can still function effectively with a full dataset. Wavelet denoising is another important preprocessing technique that removes noise from the data, allowing the model to focus on meaningful patterns rather than irrelevant fluctuations. Time series segmentation is used to divide long sequences of water consumption data into smaller, more manageable segments, which makes it easier for the AI models to learn and make accurate predictions. By thoroughly preprocessing the data, it becomes more reliable and better suited for training AI models, leading to improved predictive accuracy and robustness. This comprehensive approach to data preparation aligns with the document's goal of promoting the use of advanced AI-driven solutions for efficient and sustainable water management, as it ensures that models are built on clean, high-quality data that reflects real-world complexities (Duerr et al., 2018).

Finally, data validation and verification are essential steps in the data collection process, ensuring that the data used to train and validate AI models for water demand forecasting is accurate, reliable, and representative of actual water consumption patterns. Validation involves comparing the collected data against external or independent data sources, such as government databases, industry reports, or other verified datasets. This comparison helps to confirm the accuracy and consistency of the data, ensuring that it aligns with real-world conditions. Verification, on the other hand, focuses on using statistical and visualization techniques to identify and correct any potential errors or inconsistencies within the dataset itself. Techniques such as outlier detection, trend analysis, and visual data inspection help uncover anomalies that may otherwise skew the performance of the AI models. By rigorously validating and verifying the data, water managers can be confident that the models they develop will be based on high-quality inputs, leading to more reliable and accurate predictions. This process is crucial for building robust AI models that effectively optimize water management systems, directly supporting the document's goal of promoting sustainable, data-driven solutions for managing water resources in increasingly complex urban environments (Bakker et al., 2013).

These approaches and techniques in collecting historical and synthetic data are essential for building accurate and robust AI models, which in turn enable better management and optimization of water supply.

3.2.2. Data Preparation

Data preparation is a critical step in experimental design for building AI models in water demand forecasting. This process involves collecting, cleaning, transforming, and normalizing data before it is used for training and validating models. The first step is data cleaning, which includes removing duplicate data, correcting errors, and managing outliers. Outlier data can distort model performance if not handled properly, so techniques such as outlier detection and interpolation are used to smooth the data. One study showed that data cleaning can significantly improve the accuracy of water demand forecasting models (Bakker et al., 2013).

The next step in data preparation is data transformation, which includes converting data into formats suitable for analysis. This may involve transforming time series, aggregating daily data to monthly, or segmenting data into specific periods for detailed analysis. Data transformation can also include creating additional features from the original data, such as daily mean temperature or accumulated precipitation, that may be relevant to the predictive model. Including these additional features has been shown to improve model accuracy in water demand forecasting studies (Romano & Kapelan, 2014).

Data normalization is another crucial step in preparing data for AI models. Normalization involves scaling the data to a standard range, such as between 0 and 1, to ensure that all features contribute equally to the model and to prevent features with larger values from dominating the model's learning. Common normalization techniques include Min-Max normalization and Z-score standardization. Data normalization has been shown to improve the stability and speed of convergence of neural network models in water demand forecasting (Tiwari et al., 2016).

Splitting data into training, validation, and test sets is standard practice in data preparation. This split allows for the independent evaluation of model performance and ensures that the model is not overfitted to the training data. A common approach is to split the data into 70% for training, 15% for validation, and 15% for testing. This practice has been used in several studies to ensure the robustness and generalizability of AI models in water demand forecasting (Ghiassi et al., 2008).

Finally, the creation of synthetic datasets is a technique used to increase the volume of data available for model training. Synthetic data is generated using simulation techniques and can help overcome the limitations of insufficient historical data. This data allows AI models to be trained and validated on scenarios not found in the historical data, improving the model's ability to generalize to new data. A recent study showed that the inclusion of synthetic data can significantly improve the performance of water demand forecasting models (Xu et al., 2018).

These steps in data preparation are critical to building robust and accurate AI models for water demand forecasting, ensuring that the data used is of high quality and adequately represents consumption patterns.

3.2.3. Model Training and Validation

Model training and validation are critical steps in building effective AI models for water demand forecasting, ensuring that the model can accurately predict future water consumption based on historical data. The training process involves fine-tuning the internal parameters of the model using a training dataset, with the goal of minimizing prediction errors and improving the model's ability to generalize to new, unseen data. This fine-tuning is achieved through iterative adjustments, where the model learns from its mistakes and continuously refines its predictions. One of the most common techniques used in neural network models is the backpropagation algorithm. This algorithm adjusts the weights of the connections between neurons in the network by comparing the predicted outputs to the actual values, calculating the error, and then propagating this error backward through the network to modify the weights. This iterative process allows the model to learn and become more accurate over time, reducing errors in its predictions. By improving the model's ability to generalize and predict accurately across a wide range of scenarios, this process directly supports the document's objective of promoting smarter, AI-driven solutions for optimizing water resource management. Training and validating the models effectively ensures that the AI can deliver reliable, accurate forecasts, enabling more efficient and sustainable water distribution in both urban and rural environments (Cubero, 1991).

Once the model is trained, it is crucial to validate it to ensure that it is not overfitted to the training data and that it can generalize to unseen data. Cross-validation is a commonly used technique that splits the data into several subsets and trains the model multiple times, each time using a different subset as the test set and the rest as the training set. This helps to assess the robustness and generalization ability of the model. A study showed that k-fold cross-validation is effective in assessing the accuracy and stability of water demand forecasting models (Tiwari et al., 2016).

During the training process, it is important to monitor model performance using appropriate evaluation metrics such as mean square error (MSE), mean absolute error (MAE), and mean absolute percentage error (MAPE). These metrics provide a quantitative view of model accuracy and allow for comparing different models or training configurations. In a study on water demand forecasting, these metrics were used to assess the accuracy of several machine learning models, showing that the combination of neural networks and wavelet decomposition techniques significantly improved forecast accuracy (Campisi-Pinto et al., 2012).

In addition to evaluation metrics, it is essential to use regularization techniques during training to avoid overfitting. Regularization adds a penalty term to the model optimization objective, which helps to keep model parameters at reasonable values and improves its generalization ability. Common regularization techniques include L1 and L2, which penalize large weights in the model. One study highlighted that L2 regularization significantly improved the robustness of neural network models in water demand forecasting (Romano & Kapelan, 2014).

Finally, evaluating models on an independent test data set is crucial to ensure that the model performs well on unseen data. This test set should be representative of the data that the model will encounter in production and should not have been used during training or validation. Evaluation on an independent test set provides a realistic estimate of model performance and helps to identify any overfitting or underfitting issues. A recent study used this technique to evaluate a deep learning model for water demand forecasting, demonstrating that evaluation on an independent test set is essential to ensure model robustness and accuracy (Salloom et al., 2021).

These steps in model training and validation are critical to ensuring that AI models built for water demand forecasting are accurate, robust, and able to generalize to new data, which is essential for efficient water supply management (See Figure 6).

Figure 6. Experimental design for AI models in water demand forecasting.

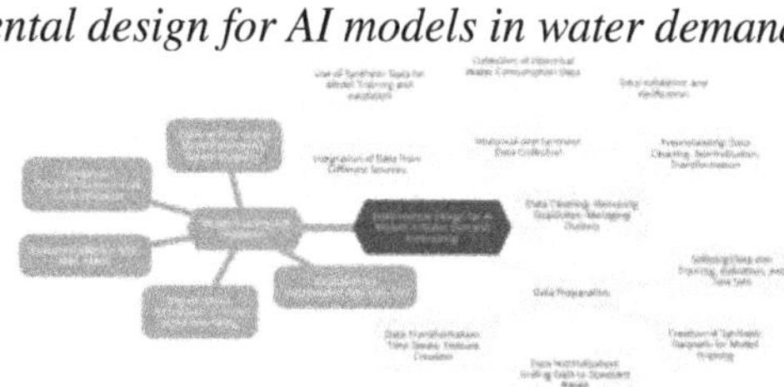

4. INTEGRATIVE EXAMPLE

The following code provided implements a water demand prediction system using various artificial intelligence methods. First, it loads and prepares a relevant dataset by scaling the selected features. Then, it constructs two models: an Artificial Neural Network (ANN) and a hybrid CNN-LSTM model that combines convolutional layers to extract spatial features and LSTM layers to capture temporal dependencies. To optimize the ANN model's accuracy, optimization techniques such as Particle Swarm Optimization (PSO) and Genetic Algorithms (GA) are applied. Finally, both models are evaluated using metrics like Mean Squared Error (MSE), Mean Absolute Error (MAE), and R^2, with results visualized through graphs and a summary table comparing the models performance (See Figure 7).

Figure 7. Description of the integrative example.

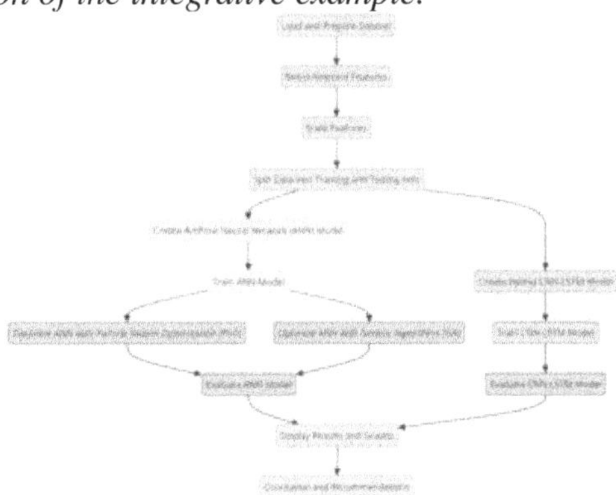

5. CONCLUSIONS AND RECOMMENDATIONS

5.1. Main Conclusions

The study successfully demonstrated that artificial intelligence (AI) models, particularly those incorporating advanced techniques such as Artificial Neural Networks (ANN), Convolutional Neural Networks (CNN), and Long Short-Term Memory (LSTM) networks, can significantly enhance the accuracy of water demand forecasting in urban areas. These models outperform traditional methods by better handling the non-linear and variable nature of urban water consumption patterns.

This improved accuracy is essential for efficient water supply management, as it allows for more precise planning and timely adjustments in operations.

Furthermore, the comparative analysis of various AI techniques, including ANN, Adaptive Neuro-Fuzzy Inference Systems, and hybrid models, revealed that hybrid models tend to deliver the most robust and accurate predictions. The integration of optimization algorithms, such as Particle Swarm Optimization (PSO) and Genetic Algorithms (GA), further enhances the performance of these models. This finding underscores the importance of using a combination of techniques to handle the complexity and variability of urban water demand effectively.

The research also highlighted the value of developing adaptive, real-time forecasting systems. Evolutionary Artificial Neural Networks (EANN) that continuously adjust to changing water demand patterns were particularly effective in real-time scenarios. Implementing such systems in water management practices enables quicker and more effective responses to fluctuations in demand, ultimately contributing to the resilience and sustainability of water supply systems in urban settings.

Finally, the study emphasizes the need for ongoing research to refine these AI models further and explore their application across different urban environments and water management contexts. Future work should focus on integrating additional data sources, such as real-time weather and demographic information, to enhance the models' adaptability and accuracy. By continuing to advance these AI-driven approaches, water resource managers can better meet the challenges of growing urban populations and increasingly variable environmental conditions.

5.2. Recommendations for Future Research

Future research should focus on enhancing the adaptability and precision of AI models for water demand forecasting. While current models such as ANN, CNN, and LSTM have shown significant promise, there is a need to explore more sophisticated hybrid models that can integrate additional variables such as real-time weather, social behaviors, and economic indicators. These models should be capable of learning from a broader range of data sources, which would allow them to better predict water demand under various scenarios, including extreme weather conditions and sudden population changes.

Another important direction for future research is the development of AI-driven decision support systems that can be seamlessly integrated into existing water management infrastructures. These systems should not only forecast water demand but also provide actionable insights for optimizing supply chains, reducing energy consumption, and preventing water shortages. Research should explore the application of AI in predictive maintenance, where machine learning algorithms could anticipate equipment failures and optimize maintenance schedules, thereby ensuring the continuous and efficient operation of water distribution systems.

Additionally, future studies should investigate the scalability and generalizability of AI models across different urban environments. While the models developed in this study are effective for the specific urban area under consideration, their applicability to other regions with different climatic, demographic, and infrastructural characteristics needs thorough evaluation. Research should also consider the ethical implications of AI in water management, particularly in terms of data privacy and the potential biases that could arise from automated decision-making processes.

Finally, there is a critical need for interdisciplinary research that brings together experts in AI, water resource management, environmental science, and urban planning. Collaborative efforts should aim to create comprehensive AI frameworks that address the multifaceted challenges of urban water management. This includes exploring the use of AI in emerging areas such as smart cities, where water management is closely integrated with other urban systems like energy and transportation. Such holistic approaches will be essential for developing sustainable and resilient urban water management solutions in the face of growing environmental and societal pressures.

REFERENCES

Aani, S., Bonny, T., Hasan, S., & Hilal, N. (2019). Can machine language and artificial intelligence revolutionize process automation for water treatment and desalination? *Desalination*, 458, 84–96. Advance online publication. DOI: 10.1016/j.desal.2019.02.005

Bakker, M., Vreeburg, J., Schagen, K., & Rietveld, L. (2013). A fully adaptive forecasting model for short-term drinking water demand. *Environmental Modelling & Software*, 48, 141–151. DOI: 10.1016/j.envsoft.2013.06.012

Bougadis, J., Adamowski, K., & Diduch, R. (2005). Short-term municipal water demand forecasting. *Hydrological Processes*, 19(1), 137–148. DOI: 10.1002/hyp.5763

Campisi-Pinto, S., Adamowski, J., & Oron, G. (2012). Forecasting Urban Water Demand Via Wavelet-Denoising and Neural Network Models. Case Study: City of Syracuse, Italy. *Water Resources Management*, 26, 3539–3558. DOI: 10.1007/s11269-012-0089-y

Chang, M., & Liu, J. (2009). Water Demand Prediction Model Based on Radial Basis Function Neural Network. *2009 First International Conference on Information Science and Engineering*, 5295-5298. https://doi.org/DOI: 10.1109/ICISE.2009.1343

Cubero, R. (1991). Neural Networks for Water Demand Time Series Forecasting. 453-460. https://doi.org/.DOI: 10.1007/BFb0035927

Duerr, I., Merrill, H., Wang, C., Bai, R., Boyer, M., Dukes, M., & Bliznyuk, N. (2018). Forecasting urban household water demand with statistical and machine learning methods using large space-time data: A Comparative study. *Environmental Modelling & Software*, 102, 29–38. DOI: 10.1016/j.envsoft.2018.01.002

Ghiassi, M., Zimbra, D., & Saidane, H. (2008). Urban Water Demand Forecasting with a Dynamic Artificial Neural Network Model. *Journal of Water Resources Planning and Management*, 134(2), 138–146. DOI: 10.1061/(ASCE)0733-9496(2008)134:2(138)

Gonçalves, R., Soares, J. J. M., & Lima, R. M. F. (2020). An IoT-Based Framework for Smart Water Supply Systems Management. *Future Internet*, 12(7), 114. DOI: 10.3390/fi12070114

Gurung, T., Stewart, R., Beal, C., & Sharma, A. (2015). Smart meter enabled water end-use demand data: Platform for the enhanced infrastructure planning of contemporary urban water supply networks. *Journal of Cleaner Production*, 87, 642–654. DOI: 10.1016/j.jclepro.2014.09.054

Herrera, M., Torgo, L., Izquierdo, J., & Pérez-García, R. (2010). Predictive models for forecasting hourly urban water demand. *Journal of Hydrology (Amsterdam)*, 387(1-2), 141–150. DOI: 10.1016/j.jhydrol.2010.04.005

Hu, P., Tong, J., Wang, J., Yang, Y., & Turci, L. (2019). A hybrid model based on CNN and Bi-LSTM for urban water demand prediction. *2019 IEEE Congress on Evolutionary Computation (CEC)*, 1088-1094. https://doi.org/DOI: 10.1109/CEC.2019.8790060

Jain, A., & Ormsbee, L. (2002). Short-term water demand forecast modeling techniques—CONVENTIONAL METHODS VERSUS AI. *Journal - American Water Works Association*, 94(7), 64–72. Advance online publication. DOI: 10.1002/j.1551-8833.2002.tb09507.x

Kher, L., Khanna, P., & Agarwal, S. (1979). Nonlinear Optimization of Water Supply Systems. *Journal of the Environmental Engineering Division*, 105(4), 781–784. DOI: 10.1061/JEEGAV.0000939

Krishnan, S., Nallakaruppan, M., Chengoden, R., Koppu, S., Iyapparaja, M., Sadhasivam, J., & Sethuraman, S. (2022). Smart Water Resource Management Using Artificial Intelligence—A Review. *Sustainability (Basel)*, 14(20), 13384. Advance online publication. DOI: 10.3390/su142013384

Li, L., Rong, S., Wang, R., & Yu, S. (2021). Recent advances in artificial intelligence and machine learning for nonlinear relationship analysis and process control in drinking water treatment: A review. *Chemical Engineering Journal*, 405, 126673. Advance online publication. DOI: 10.1016/j.cej.2020.126673

Lorente-Leyva, L., Pavón-Valencia, J., Montero-Santos, Y., Herrera-Granda, I., Herrera-Granda, E., & Peluffo-Ordóñez, D. (2019). Artificial Neural Networks for Urban Water Demand Forecasting: A Case Study. *Journal of Physics: Conference Series*, 1284(1), 012004. Advance online publication. DOI: 10.1088/1742-6596/1284/1/012004

Luan, Y., & Liu, J. (2017). A distributed model for urban water demand prediction. *Kexue Tongbao*, 62(24), 2770–2779. DOI: 10.1360/N972016-01154

Luna, T., Ribau, J., Figueiredo, D., & Alves, R. (2019). Improving energy efficiency in water supply systems with pump scheduling optimization. *Journal of Cleaner Production*, 213, 342–356. Advance online publication. DOI: 10.1016/j.jclepro.2018.12.190

Mahadeva, R., Patel, V., Ghosh, A., Dixit, S., Suthar, B., Gupta, V., Awaar, V., Bharadwaj, Y., Gupta, M., Haranatti, J., & Arora, R. (2023). Artificial Intelligence in Water Desalination: A Novel Approach for Global Sustainability. E3S Web of Conferences. https://doi.org/.DOI: 10.1051/e3sconf/202343001193

Mehmood, H., Liao, D., & Mahadeo, K. (2020). A Review of Artificial Intelligence Applications to Achieve Water-related Sustainable Development Goals. 2020 IEEE / ITU International Conference on Artificial Intelligence for Good (AI4G), 135-141. https://doi.org/DOI: 10.1109/AI4G50087.2020.9311018

Montalvo, I., Izquierdo, J., Pérez-García, R., & Tung, M. (2008). Particle Swarm Optimization applied to the design of water supply systems. *Computers & Mathematics with Applications (Oxford, England)*, 56(3), 769–776. DOI: 10.1016/j.camwa.2008.02.006

Msiza, I., Nelwamondo, F., & Marwala, T. (2008). Water Demand Prediction using Artificial Neural Networks and Support Vector Regression. *Journal of Computers*, 3(11), 1–8. DOI: 10.4304/jcp.3.11.1-8

Najah, A., El-shafie, A., Karim, O., Jaafar, O., & El-Shafie, A. (2011). An application of different artificial intelligences techniques for water quality prediction. *International Journal of Physical Sciences*, 6, 5298–5308. DOI: 10.5897/IJPS11.1180

Nordin, N., Mohd, N., Koting, S., Ismail, Z., Sherif, M., & El-Shafie, A. (2021). Groundwater quality forecasting modelling using artificial intelligence: A review. *Groundwater for Sustainable Development*, 14, 100643. Advance online publication. DOI: 10.1016/j.gsd.2021.100643

Oyebode, O., & Ighravwe, D. (2019). Urban Water Demand Forecasting: A Comparative Evaluation of Conventional and Soft Computing Techniques. *Resources*, 8(3), 156. Advance online publication. DOI: 10.3390/resources8030156

Perea, R., Poyato, E., Montesinos, P., & Díaz, J. (2015). Irrigation Demand Forecasting Using Artificial Neuro-Genetic Networks. *Water Resources Management*, 29(15), 5551–5567. DOI: 10.1007/s11269-015-1134-4

Pérez-Padillo, J., García Morillo, J., Ramirez-Faz, J., Torres Roldán, M., & Montesinos, P. (2020). Design and Implementation of a Pressure Monitoring System Based on IoT for Water Supply Networks. *Sensors (Basel)*, 20(15), 4247. DOI: 10.3390/s20154247 PMID: 32751617

Pesantez, J., Berglund, E., & Kaza, N. (2020). Smart meters data for modeling and forecasting water demand at the user-level. *Environmental Modelling & Software*, 125, 104633. DOI: 10.1016/j.envsoft.2020.104633

Ponte, B., Fuente, D., Parreño, J., & Pino, R. (2016). Intelligent Decision Support System for Real-Time Water Demand Management. *Int. J. Comput. Intell. Syst.*, 9(1), 168–183. DOI: 10.1080/18756891.2016.1146533

Rajakumar, A., Anthony, A., & Kumar, V. (2021). Application of deep learning methods for urban water demand forecast modelling. https://doi.org/DOI: 10.5194/egusphere-egu21-9572

Rojek, I. (2014). Models for Better Environmental Intelligent Management within Water Supply Systems. *Water Resources Management*, 28(12), 3875–3890. DOI: 10.1007/s11269-014-0654-7

Romano, M., & Kapelan, Z. (2014). Adaptive water demand forecasting for near real-time management of smart water distribution systems. *Environmental Modelling & Software*, 60, 265–276. DOI: 10.1016/j.envsoft.2014.06.016

Salloom, T., Kaynak, O., & He, W. (2021). A novel deep neural network architecture for real-time water demand forecasting. *Journal of Hydrology (Amsterdam)*, 126353, 126353. Advance online publication. DOI: 10.1016/j.jhydrol.2021.126353

Sun, Y., Yeh, W., Hsu, N., & Louie, P. (1995). Generalized Network Algorithm for Water-Supply-System Optimization. *Journal of Water Resources Planning and Management*, 121, 392–398. DOI: 10.1061/(ASCE)0733-9496(1995)121:5(392)

Tiwari, M., Adamowski, J., & Adamowski, K. (2016). Water demand forecasting using extreme learning machines. *Journal of Water and Land Development*, 28(1), 37–52. DOI: 10.1515/jwld-2016-0004

Xiang, X., Li, Q., Khan, S., & Khalaf, O. (2021). Urban water resource management for sustainable environment planning using artificial intelligence techniques. *Environmental Impact Assessment Review*, 86, 106515. DOI: 10.1016/j.eiar.2020.106515

Xing, Y., You, B., & Yang, M. (2017). City Water Demand Forecasting Based on Improved BP Neural Network. *Journal of Residuals Science & Technology*, 14(S1), S111–S117. Advance online publication. DOI: 10.12783/issn.1544-8053/14/S1/15

Xu, Y., Zhang, J., Long, Z., & Chen, Y. (2018). A Novel Dual-Scale Deep Belief Network Method for Daily Urban Water Demand Forecasting. *Energies*, 11(5), 1068. DOI: 10.3390/en11051068

Yaseen, Z., El-Shafie, A., Jaafar, O., Afan, H., & Sayl, K. (2015). Artificial intelligence based models for stream-flow forecasting: 2000-2015. *Journal of Hydrology (Amsterdam)*, 530, 829–844. DOI: 10.1016/j.jhydrol.2015.10.038

Zyl, J., Savić, D., & Walters, G. (2004). Operational Optimization of Water Distribution Systems using a Hybrid Genetic Algorithm. *Journal of Water Resources Planning and Management*, 130(2), 160–170. DOI: 10.1061/(ASCE)0733-9496(2004)130:2(160)

Chapter 8
Optimising Water Use Through Smart Models and Artificial Intelligence

Jaime Aguilar Ortiz
Universidad Politécnica de Pachuca, Mexico

Víctor Manuel Zamudio-García
Universidad Politécnica Metropolitana de Hidalgo, Mexico

ABSTRACT

This study examines the use of advanced AI techniques to optimize water management in three key areas: water quality prediction, leak detection, and water distribution. By applying ensemble learning models like Random Forest, Gradient Boosting, AdaBoost, and Bagging, the research addresses the complexities of managing water resources, such as non-linear data patterns and the need for high predictive accuracy. Integrating AI models with real-time data and IoT technologies enhances the adaptability of water management, enabling real-time monitoring and decision-making for efficient and sustainable resource use. These AI-driven approaches improve operational efficiency by optimizing water distribution and minimizing losses from leaks while ensuring accurate water quality predictions. This contributes to better decision-making, crucial for public health and environmental sustainability. The study highlights the transformative potential of AI in water management, advocating for its broader adoption to meet the challenges posed by urbanization, population growth, and climate change.

DOI: 10.4018/979-8-3693-8074-1.ch008

1. INTRODUCTION

1.1. Context and Relevance of Water Management

Water management is crucial due to increasing urbanization and rising water demand in cities. Proper management of water resources in urban contexts is essential to ensure global sustainability, as it directly affects quality of life, economic development and environmental health. More than half of the world's population lives in urban areas, making it imperative to ensure adequate supply and quality of water to support social well-being and sustainable development.

Water management is also facing a significant paradigm shift, moving from purely technical solutions to integrated approaches that consider social, cultural and governance factors. The importance of cultural adaptation and social learning has become crucial in water resources management, highlighting the need for a more holistic approach that integrates technical aspects with human behaviors and beliefs to achieve sustainable and adaptive water resources management (Pahl-Wostl et al., 2008).

In addition, water demand management is being redefined to include not only reducing the amount of water used, but also improving efficiency and managing the timing of use to mitigate demand peaks. This integrated approach considers water use efficiency, equity, environmental protection, and public participation, making water demand management an essential governance strategy to address water scarcity and promote sustainable development at various stages of economic development (Brooks, 2006).

1.2. Objectives of the Study

The main objective of this study is to improve water management by applying artificial intelligence (AI) models in three key areas: water quality prediction, leak detection and location, and water distribution optimization. For water quality prediction, ensemble models such as decision trees and Random Forest will be used, which have proven to be effective in handling non-linear data and improving the accuracy of predictions (Shah et al., 2021).

In water leak detection and location, the study will apply ensemble methods such as Gradient Boosting and AdaBoost. These methods have been successfully used in various studies to improve the accuracy and speed of leak identification, which is critical to reducing water losses and the costs associated with repairs (Hu et al., 2014).

For the optimization of water distribution, models based on Bagging and Boosting will be used, which are capable of handling large volumes of data and diverse variables, thus optimizing the use of water in complex distribution networks. The application of these models will allow a more efficient and sustainable management of water resources, ensuring that water is distributed optimally and reducing the environmental impact (Yetilmezsoy et al., 2011).

2. LITERATURE REVIEW

2.1. Water Quality Prediction

Water quality prediction has emerged as a crucial tool for effective water resources management and environmental protection. The use of artificial intelligence (AI) algorithms enables accurate modeling and prediction of water quality parameters such as water quality index (WQI) and water quality classification (WQC). Artificial neural network models such as NARNET and deep learning algorithm LSTM have shown superior performance in predicting WQI values (Aldhyani et al., 2020).

Adaptive Neuro-Fuzzy Inference System (ANFIS) model has been developed to predict WQI, showing superiority compared to other models. Furthermore, the use of feed-forward neural networks (FFNN) and K-nearest neighbors (K-NN) has been shown to be effective for water quality classification. ANFIS and FFNN models have demonstrated remarkable accuracy in predicting and classifying water quality, highlighting their robustness and efficiency (Al-Adhaileh & Alsaade, 2021).

The use of machine learning techniques, such as the wavelet-decomposition-enhanced adaptive neuro-fuzzy inference system (WDT-ANFIS), has significantly improved the accuracy of water quality predictions. This approach addresses complex nonlinear relationships between input and output data, outperforming traditional models in accuracy (Ahmed et al., 2019).

The multilayer perceptron neural network (MLP-ANN) approach and ensemble neural network (E-ANN) models have been investigated to predict water quality parameters such as dissolved oxygen (DO), biochemical oxygen demand (BOD), and chemical oxygen demand (COD). The ability of these models to handle complex and non-linear data makes them particularly suitable for water quality prediction applications (Najah et al., 2011).

Gene expression programming (GEP) and artificial neural networks (ANN) have proven to be reliable techniques for modeling total dissolved solids (TDS) and specific conductivity (EC) in rivers. These models have shown high correlation with real data, highlighting their ability to accurately predict water quality parameters (Shah et al., 2021).

The use of probabilistic neural networks (PNN) and multilayer perceptron neural networks (MLP-ANN) has been explored for the estimation of water quality indicators such as dissolved oxygen (DO) and five-day biochemical oxygen demand

(BOD5). These models have shown superior performance in terms of estimation accuracy, being valuable tools for water quality management (Allawi et al., 2022).

The integration of Internet of Things (IoT) devices and AI techniques has enabled the real-time capture and analysis of water quality data. These technologies have proven effective in modeling and predicting water quality parameters quickly and accurately, facilitating decision making and water resource management (Mustafa et al., 2021).

Figure 1. AI applications in water demand forecasting.

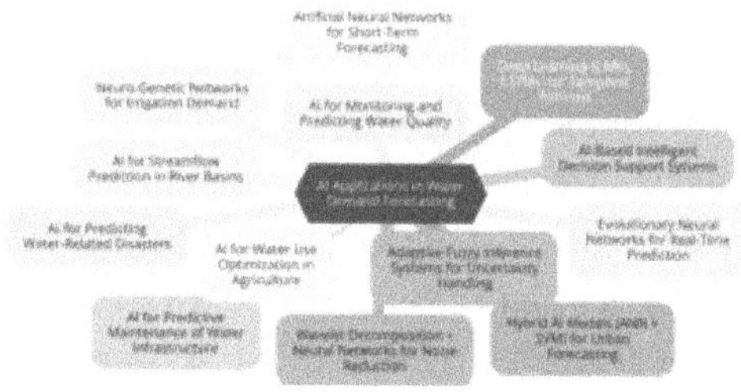

The use of recurrent neural networks (RNN) and long short-term memory (LSTM) architecture has shown remarkable performance in predicting water quality parameters in wastewater treatment plants. These models are capable of handling sequential data, providing accurate and robust predictions (Wongburi & Park, 2023).

Radial basis neural network (RBFNN) and backpropagation neural network models have been applied to predict the water quality index (WQI) in tropical environments. These models have demonstrated high accuracy and efficiency, offering an effective alternative to traditional manual WQI calculation methods (Hameed et al., 2017).

In summary, the application of artificial intelligence techniques in water quality prediction has been shown to be extremely effective in improving the accuracy and efficiency of predictions. These technological advances enable better management of water resources and help ensure water quality, contributing significantly to environmental protection and public health (Mehmood et al., 2020) (See Figure 1).

2.2. Leak Detection and Localization

Leak detection and localization in water distribution networks is a critical challenge for water resource managers, due to the economic losses and environmental risks that leaks can cause. Traditional signal processing methods and statistical models are being replaced by artificial intelligence (AI)-based models, capable of recognizing complex leak patterns and handling large data sets. A recent approach systematically reviews the application of AI in water leak management, highlighting the techniques adopted and the equipment used to collect data (Abdelmageed et al., 2022).

The combination of Internet of Things (IoT) monitoring devices and AI technology is one of the most promising directions for leak detection in industrial-scale infrastructures and smart cities. An IoT-enabled water transmission system and a series of simulated leak experiments allowed the evaluation of several anomaly detection methods such as isolation forest (iForest), support vector classification (SVC), and recurrent neural network with long short-term memory (RNN-LSTM) models (Ebisi et al., 2023).

A hybrid method combining AI algorithms and hydraulic relations has been proposed for leak detection and localization in large water distribution networks. This method uses inexpensive flow sensors and hydraulic formulas such as Hazen-Williams and Darcy-Weisbach to generate influential features in leak detection, thereby achieving faster and more accurate detection (Fereidooni et al., 2020).

The use of machine learning algorithms to detect leaks in water pipelines has been examined using pressure and flow sensors. The data obtained was used to develop an AI algorithm that detects leaks based on the acquired data, evaluating several machine learning methods to determine the most effective one (Badar et al., 2023).

Artificial neural networks (ANN) and genetic algorithms (GA) have been successfully applied to optimize the positioning of hydrophonic sensors, improving coverage and leak detection in drinking water distribution networks. These methods have proven to be effective and fast in optimal sensor placement, significantly reducing costs and improving detection efficiency (Lechgar et al., 2016).

A cloud-based system for smart water management has been proposed to detect and locate leaks in pipelines. This system uses water flow rates in the pipelines to determine leaks and applies machine learning techniques to predict leak locations, achieving high accuracy in detection and localization (Shravani et al., 2019).

The use of wireless sensor networks for leak detection has also been explored, highlighting technical challenges and areas for future research. The practical application of these techniques in large-scale water distribution networks remains an area of significant concern, although research has shown promising progress (Adedeji et al., 2017).

Vibration sensors in water pipe networks have been used to perform comparative studies of leak detection techniques. The application of machine learning techniques has enabled the design of leak classification models, showing high accuracy in detecting and classifying leaks in water supply pipelines (Kang et al., 2023).

Figure 2. Leak detection and location in water distribution networks.

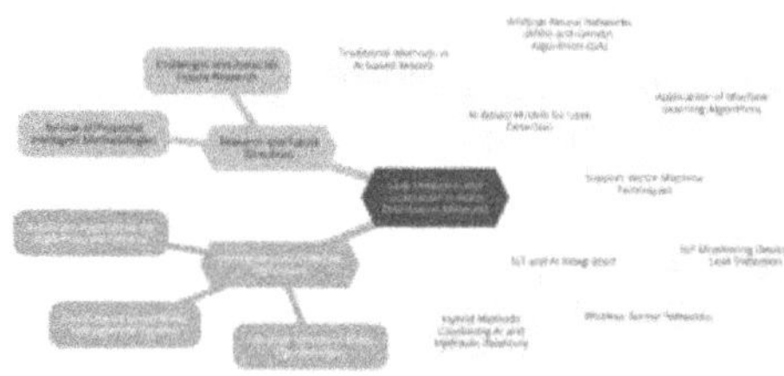

A fully automatic approach for leak detection in high-pressure and fluid-filled cable systems has combined various AI and data processing techniques. This system has shown high detection capability for various leak rates, including small leaks, demonstrating the effectiveness of AI in accurately detecting leaks (Tylman et al., 2010).

Finally, a review of the proposed intelligent methodologies for leak detection in water distribution networks highlights support vector machine techniques, neural networks and convolutional neural networks as promising approaches. These methodologies have shown significant improvements in leak detection and localization accuracy (Chan et al., 2018) (See Figure 2).

2.3. Water Distribution Optimization

Water distribution optimization is essential to improve the efficiency and sustainability of water resource management systems. Artificial intelligence (AI) has emerged as a powerful tool in this domain, enabling the development of models that optimize water distribution across complex distribution networks. These models not only improve operational efficiency but also reduce costs and minimize water waste (Krishnan et al., 2022).

A common approach in water distribution optimization is the use of artificial immune systems (AIS), which employ algorithms inspired by the human immune system to solve complex problems. These systems have proven to be highly efficient in minimizing costs and improving the quality of solutions in water distribution network design problems (Lin & Chu, 2008).

Machine learning models also play a crucial role in water distribution optimization. These models can analyse large volumes of data in real time, providing accurate predictions and decision support. The application of deep learning techniques has enabled the development of smarter and more adaptive water management systems, which optimise the use of water resources and improve sustainability (Drogkoula et al., 2023).

The implementation of heuristic optimisation algorithms, such as harmonic search and genetic algorithms, has shown significant improvements in the efficiency of water distribution networks. These algorithms can effectively manage pressure and flow constraints in networks, ensuring optimal water distribution and reducing leakage losses (Eryiğit, 2015).

The optimisation of water distribution systems has also benefited from advances in complex network analysis (CNA). This approach allows large distribution networks to be analysed and optimised with a lower computational burden, facilitating the identification of optimal solutions in terms of costs and performance (Sitzenfrei et al., 2020).

Figure 3. Water distribution optimization using AI.

Differential optimization algorithms, such as fuzzy C-means adaptive differential evolution (FCADE), have proven effective in solving water distribution network design problems. These algorithms combine clustering and optimization techniques to improve the accuracy and efficiency of solutions, allowing for better management of water resources (Bilal et al., 2021).

The integration of AI in the automation and optimization of processes in water treatment plants has made it possible to improve the operation and management of these systems. AI-based models are capable of performing water quality diagnostics, making autonomous decisions, and optimizing operational processes, contributing to more efficient and sustainable water management (Li et al., 2021).

The use of systems based on artificial neural networks (ANN) in water quality management has been shown to be an effective tool for the optimization of water treatment processes. These systems can model and predict critical water quality parameters, improving efficiency and reducing operating costs (O'Reilly et al., 2018).

The application of artificial intelligence techniques in water resource management also includes irrigation optimization and the prediction of water levels in rivers and reservoirs. These models allow for better planning and management of water resources, contributing to sustainability and efficiency in water use (Yaseen et al., 2018).

In summary, artificial intelligence offers powerful tools for the optimization of water distribution, improving efficiency, reducing costs and contributing to the sustainability of water resource management systems. These technological advances allow for smarter and more adaptive water management, facing the challenges of scarcity and increasing demand for this vital resource (Mahardhika & Putriani, 2023) (See Figure 3).

3. ENSEMBLE MODELS TO BE APPLIED

3.1. Decision Tree Ensemble and Random Forest for Water Quality Prediction

Decision tree ensemble and Random Forest model are powerful techniques used in water quality prediction. Decision trees are models that divide data into subsets based on specific features to make predictions. However, single decision trees can be prone to overfitting and drift due to small changes in the training data. To overcome these limitations, decision tree ensemble is used, combining multiple trees to improve the accuracy and stability of the model (Talekar, 2020).

Random Forest is an ensemble technique that combines multiple decision trees trained on different subsets of the dataset. This approach reduces overfitting and improves the overall accuracy of the model. Random Forest has been successfully applied in the prediction of several water quality parameters such as dissolved oxygen, biochemical oxygen demand, and turbidity, showing consistent and accurate results (Scornet et al., 2014).

In recent studies, hybrid models combining Random Forest with empirical mode decomposition (EMD) techniques have proven to be particularly effective. These models can decompose water quality time series into simpler components before applying Random Forest, significantly improving the accuracy of predictions (Lu & Ma, 2020).

Furthermore, the use of Random Forest allows for the identification of key parameters that influence water quality. By analyzing the importance of variables within the model, researchers can determine which are the most critical factors that need to be monitored to maintain and improve water quality (Chen et al., 2019).

The stability and robustness of Random Forest make it suitable for use in various climatic and geographical conditions. This method has been applied in studies ranging from tropical regions to more temperate climates, demonstrating its ability to handle significant variations in input data (Galelli & Castelletti, 2013).

Combining Random Forest with other ensemble techniques, such as Gradient Boosting and AdaBoost, has been shown to further improve the accuracy of water quality predictions. These approaches combine the strengths of multiple algorithms to produce a more robust and accurate model (Shaik & Srinivasan, 2018).

Using Random Forest also makes it easier to interpret results, as it allows users to visualize and understand how different variables contribute to model predictions. This is especially useful in water management applications, where transparency and interpretability are crucial for decision making (Zhao et al., 2019).

Studies have shown that Random Forest is capable of handling large volumes of data and providing real-time predictions. This is particularly useful in water quality monitoring systems, where rapid and accurate data analysis is required to ensure a timely response to water quality issues (Iyer et al., 2023).

Figure 4. Ensemble models for water quality prediction.

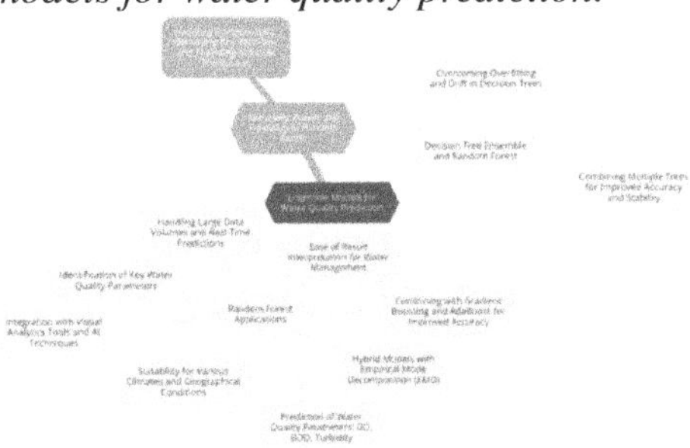

The integration of Random Forest with visual analytics tools and other advanced artificial intelligence techniques has expanded its applications and improved its effectiveness in water quality prediction. These tools allow water researchers and managers to explore results in more detail and make informed decisions based on the data (Dedja et al., 2022).

In summary, the ensemble of decision trees and Random Forest provides a powerful and flexible approach to water quality prediction. Its ability to handle complex data, identify key variables, and provide accurate predictions makes it an invaluable tool in water resources management (Sokolov, 2022) (See Figure 4).

3.2. Ensemble Methods Such as Gradient Boosting and Adaboost For Water Leak Detection

Water leak detection in distribution systems is a critical challenge due to the need to minimize water losses and avoid significant damage. Ensemble methods such as Gradient Boosting and AdaBoost have proven extremely effective in this area, combining multiple weak models to create a strong and accurate model. Gradient Boosting, for example, builds models in a sequence of steps, adjusting for errors from previous models, while AdaBoost assigns weights to misclassified observations to improve model performance (Seo et al., 2023).

Gradient Boosting is particularly effective for leak detection due to its ability to handle complex and non-linear data. In a recent study, a Gradient Boosting model was used to identify and locate leaks in a water distribution network, showing high accuracy in predicting the exact location of leaks. The approach was based on collecting pressure data at network nodes using hydraulic simulations, which enabled fast and accurate leak detection (Marvin et al., 2023).

AdaBoost has also been widely used in water leak detection, due to its ability to improve the performance of base models by combining multiple weak classifiers. A comparative study showed that AdaBoost outperformed other leak detection models in terms of accuracy and false positive rate, highlighting its effectiveness in identifying real leaks under on-site test conditions (Shen & Cheng, 2022).

The integration of Gradient Boosting with Bayesian optimization techniques has allowed further improvements in the accuracy and efficiency of leak detection models. This approach optimizes Gradient Boosting model parameters using Bayesian optimization methods, resulting in improved leak detection accuracy and a significant reduction in false positives (Paul et al., 2022).

In the realm of real-time prediction and detection, ensemble methods such as Gradient Boosting and AdaBoost have proven to be particularly useful. These methods enable real-time analysis of large volumes of data, which is crucial for rapid and accurate leak detection in water distribution networks. The ability of these models to adapt and learn from new data continuously improves their performance (Bahad & Saxena, 2019).

The combination of Gradient Boosting and AdaBoost with IoT sensor technologies has revolutionized water leak detection. These sensors collect real-time data that is processed by AI models to identify and locate leaks with high accuracy. The integration of these technologies has enabled more efficient management of water resources and faster response to leak issues (Ebisi et al., 2023).

A recent study highlighted the effectiveness of Gradient Boosting and AdaBoost models in leak detection under different climatic and geographical conditions. These models have proven to be robust and adaptable, effectively handling variations in input data and providing accurate predictions in diverse situations (Khammari et al., 2005).

In addition, ensemble methods such as Gradient Boosting and AdaBoost have been successfully applied in leak identification in critical infrastructures such as water treatment plants and urban distribution networks. These models have enabled more accurate detection and faster localization of leaks, improving operational efficiency and reducing costs associated with repairs (Ravichandran et al., 2021) (See Figure 5).

Figure 5. Ensemble methods for water leak detection.

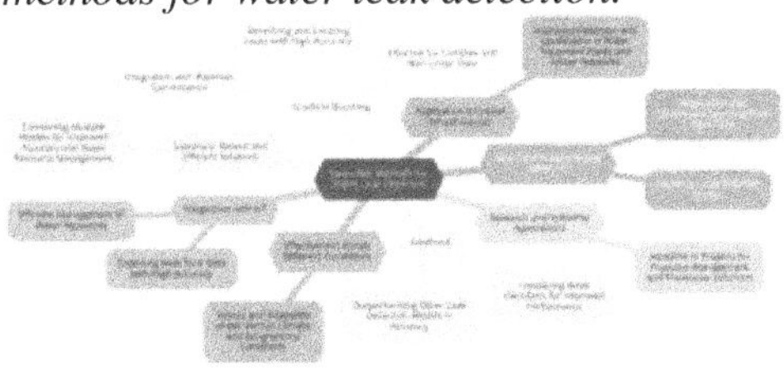

The ability of Gradient Boosting and AdaBoost to handle large data sets and provide detailed analysis has led to their adoption in research projects and industrial applications. These methods enable more proactive management of water distribution networks, facilitating early leak identification and implementation of preventive solutions (Wang et al., 2021).

In summary, ensemble methods such as Gradient Boosting and AdaBoost offer robust and efficient solutions for water leak detection. Their ability to combine multiple models and continuously improve their accuracy makes them essential tools for water resource management and optimization of water distribution systems (Kanamori et al., 2007).

3.3. Bagging and Boosting-Based Model Ensemble for Water Distribution Optimization

Ensemble methods such as Bagging and Boosting are powerful tools used to improve the accuracy and robustness of prediction models. In water distribution optimization, these methods allow combining the predictions of several weak models to form a strong and accurate model. Bagging (Bootstrap Aggregating) reduces model variance by averaging the predictions of multiple models trained on subsets of data, while Boosting improves performance by correcting the errors of previous models in an iterative sequence (Erdal & Karakurt, 2013).

The application of Bagging in water distribution optimization has been shown to be effective in handling data with high variability and noise. By creating multiple versions of a model using different subsets of data, Bagging reduces the sensitivity of the model to the training data and improves overall accuracy. In a study on water demand prediction, the use of Bagging with decision trees showed a significant improvement in accuracy compared to individual models (Sumana & Santhanam, 2015).

Boosting, on the other hand, has been widely used to improve the accuracy of prediction models in water resource management. AdaBoost and Gradient Boosting are two of the most popular techniques. AdaBoost adjusts the weights of misclassified observations, while Gradient Boosting optimizes a loss function at each iteration to improve the model. These methods have proven effective in predicting water distribution, improving the ability of models to handle complex and non-linear data (Malek et al., 2022) (Figure 6).

Figure 6. Bagging and boosting based model ensemble for water distribution optimization.

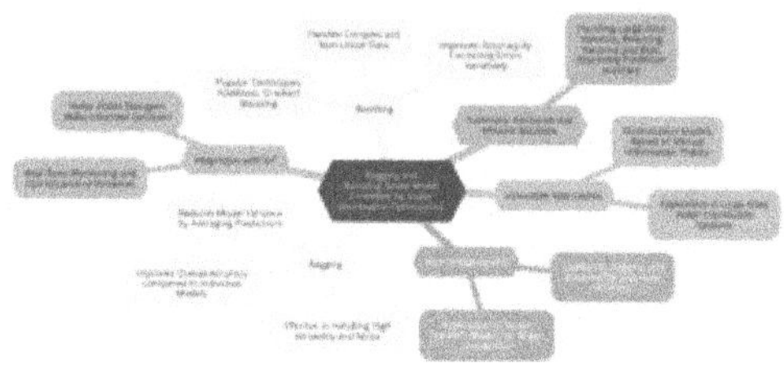

In practice, combining Bagging and Boosting can offer additional benefits. A recent study showed that integrating both methods in a hybrid approach significantly improved the accuracy of water quality predictions. This approach allowed to take advantage of the strengths of both methods, reducing both the variance and bias of the model (Dou et al., 2019).

Water distribution optimization has also benefited from the application of ensemble techniques in neural network models. By combining neural networks with Bagging and Boosting, researchers have managed to improve the accuracy and robustness of predictions of critical parameters such as water pressure and flow in distribution networks (Shahhosseini et al., 2019).

An innovative approach is the use of Bagging and Boosting in optimization models based on mutual information theory. This approach has proven effective in improving the accuracy of water demand predictions and identifying key patterns in water distribution data. Mutual information theory helps to select the most relevant features and build more accurate and efficient models (Hasan & Twala, 2015).

The application of Bagging and Boosting has also been shown to be beneficial in the optimization of large-scale water distribution systems. In a study on the management of water distribution networks in cities, the combination of these techniques allowed for better management of water resources, reducing water losses and improving operational efficiency (Simidjievski et al., 2015) (See Figure 6).

Advances in the integration of ensemble models with IoT sensor technologies have enabled real-time monitoring and optimization of water distribution networks. These systems collect real-time data that is processed by Bagging and Boosting models, providing accurate predictions and helping water managers make informed decisions (Ravichandran et al., 2021).

In the optimization of water distribution, the use of ensemble techniques has allowed for better management of uncertainty and greater resilience of distribution systems. Bagging and Boosting models help to handle data variability and provide more robust solutions, which is crucial for the long-term sustainability of water resources (Erdal et al., 2013).

In summary, Bagging and Boosting-based models offer advanced and efficient solutions for water distribution optimization. Their ability to handle large data volumes, reduce variance and bias, and improve prediction accuracy makes them essential tools for water resources management and water distribution system optimization (Mosavi et al., 2020).

4. EXPERIMENTAL DESIGN

4.1. Collection and Preparation of Real and Synthetic Data

Data collection is a crucial first step in developing AI models for water management. Data can be obtained from a variety of sources, including water quality sensors, databases from governments and environmental organizations, and scientific studies. It is important to ensure that the collected data are representative of real-world conditions to ensure the accuracy of the models (Li et al., 2021).

Data preparation includes several steps such as data cleaning, transformation, and normalization. These processes are essential to handle incomplete, noisy, and inconsistent data. Techniques such as imputation of missing values, removal of duplicates, and data normalization ensure that the data is suitable for use in machine learning models (Whang et al., 2021).

In addition to real data, generating synthetic data can be useful to expand the available dataset. Synthetic data is created using statistical models or simulation techniques that mimic the characteristics of real data. This is especially useful in situations where real data is scarce or difficult to obtain (Ahmed et al., 2019).

Integrating data from multiple sources is another important aspect of data preparation. Combining data from sensors, historical databases, and scientific studies allows for the creation of a more complete and robust data set. This integration may require advanced data management techniques to ensure consistency and compatibility between different sources (Lowe et al., 2022).

Data validation is a critical step to ensure the quality and reliability of data used in AI models. This includes verifying data accuracy, detecting anomalies, and correcting errors. Data validation techniques can include statistical methods and machine learning algorithms to detect and correct problems in the data (William et al., 2023).

Using data mining techniques can help uncover hidden patterns and relationships in the collected data. These techniques include clustering, regression, and classification analysis, which can provide valuable insights for creating predictive models. Data mining is especially useful for identifying key variables that affect water quality and other important parameters (Blaya et al., 2001).

Data preparation also includes creating training and testing data sets to validate AI models. This involves splitting the data into subsets that are used to train the model and evaluate its performance. The cross-validation technique is commonly used to ensure that the model generalizes well to new data not seen during training (Suchetana et al., 2023).

Creating synthetic data can involve simulating specific scenarios to evaluate the model's performance under extreme or unusual conditions. This can include generating data under different weather conditions, pollution levels, and water use patterns, providing a more complete view of how the model may behave in various situations (Chen et al., 2022).

The use of technologies such as the Internet of Things (IoT) facilitates the continuous collection of real-time data. IoT devices can continuously monitor water quality and other important parameters, providing a constant stream of data that can be used to dynamically improve and update predictive models (Krishnan et al., 2022).

Finally, collaboration across different disciplines and entities is crucial for the collection and preparation of high-quality data. This includes cooperation between scientists, engineers, water managers, and government agencies to share data and knowledge, ensuring a rich and diverse database that can significantly improve the accuracy and utility of AI models (Sun & Scanlon, 2019).

4.2. Model Training and Validation

Training artificial intelligence (AI) models begins with the selection of a representative dataset. This data is typically split into two subsets: one for training and one for validation. During the training phase, the model learns to identify patterns and relationships within the data, adjusting its parameters to minimize prediction error (Shah et al., 2021).

Cross-validation is a commonly used technique to evaluate the performance of AI models. This method splits the training data into several subsets, training and validating the model multiple times with different partitions of the data. This helps ensure that the model generalizes well to new data not seen during training (Salvaris, Dean and Tok, 2018).

AI models can also be evaluated using specific performance metrics, such as coefficient of determination (R^2), root mean square error (RMSE), and mean absolute error (MAE). These metrics provide a quantitative measure of model accuracy and help identify areas where the model may need improvement (Dehghani & Poudeh, 2021).

The use of advanced techniques such as deep learning and recurrent neural networks (RNN) has proven especially useful in predicting water quality parameters. Long-short-term memory (LSTM) neural networks are capable of handling sequential data and capturing long-term dependencies, making them ideal for time series analysis in water management (Wongburi & Park, 2023).

Hyperparameter tuning is a critical step in training AI models. Hyperparameters control aspects of the model that are not directly learned from the data, such as the learning rate and the number of layers in a neural network. Techniques such as grid search and Bayesian optimization are used to find the optimal combination of hyperparameters that maximizes model performance (Ahmed et al., 2019).

Model validation is not limited to cross-validation. It also includes external validation, where the model is tested on data that was not used during training or in cross-validation. This provides an additional measure of the model's ability to generalize to new data sets (William et al., 2023).

Model robustness and stability are also assessed using stress validation techniques, where the model is tested under extreme conditions or anomalous data. This helps to identify the model's resilience and its ability to handle unforeseen situations (Yetilmezsoy et al., 2011).

The use of synthetic data generated from statistical models or simulations can also be useful in the validation process. These data can complement real data, providing additional scenarios to test the model and evaluate its performance under a variety of conditions (Casini et al., 2019).

The interpretation of model results is crucial for its practical application. Visualization tools and model interpretability techniques, such as feature importance plots, can help users understand how the model makes decisions and which variables are most influential in predictions (Fleming et al., 2021).

Finally, continuous feedback and updating of the model is essential to maintain its relevance and accuracy over time. As new data is collected, the model must be periodically retrained and validated to ensure that it continues to provide accurate and useful predictions for water management (Lowe et al., 2022).

4.3. Implementation Using Python and Recent Libraries

Implementing artificial intelligence (AI) models for water management using Python is a common practice due to the flexibility and broad ecosystem of tools that this language offers. Python, with its advanced libraries such as NumPy, pandas, and scikit-learn, facilitates data manipulation and machine learning model building. These libraries allow developers to preprocess data, build models, train them, and evaluate them efficiently (Ahmed et al., 2019).

One of the most widely used libraries in Python for building AI models is scikit-learn, which provides a variety of supervised and unsupervised learning algorithms. This library is especially useful for classification and regression tasks in water quality prediction and leak detection. Furthermore, scikit-learn facilitates the process of cross-validation and hyperparameter tuning, which is essential for optimizing model performance (Vijayaragavan et al., 2022).

Another fundamental library is TensorFlow, which is widely used to build and train deep neural network models. TensorFlow enables the development of complex models such as convolutional neural networks (CNN) and recurrent neural networks (RNN), which are particularly useful for time series analysis and prediction of water quality parameters (Aldhyani et al., 2020).

Keras, a high-level API for neural networks that runs on top of TensorFlow, makes it easy to build and train deep learning models with a simple and clear syntax. Keras is ideal for developers who want to quickly prototype AI models without worrying about low-level implementation details (Tomlinson et al., 2020).

The PyTorch library is another powerful tool for deploying deep learning models. PyTorch is known for its flexibility and ease of use, allowing researchers to experiment with different neural network architectures and optimization techniques. PyTorch also makes it easy to deploy models on GPUs, which significantly speeds up the training process (Zancan et al., 2023).

For the optimization and simulation of water distribution networks, the Pywr library can be used, which allows the modeling of complex water networks and the simulation of water distribution operations under different scenarios. Pywr uses linear programming to optimize the allocation of water resources, facilitating informed decision making in water management (Tomlinson et al., 2020).

Using parallelization techniques with the Dask library can significantly improve the efficiency of processing large volumes of data and running AI models. Dask allows for distributing computational tasks across multiple CPU cores or even a cluster of computers, which is crucial for handling large data sets in water management applications (Zhang et al., 2013).

Figure 7. Experimental design in AI models for water management.

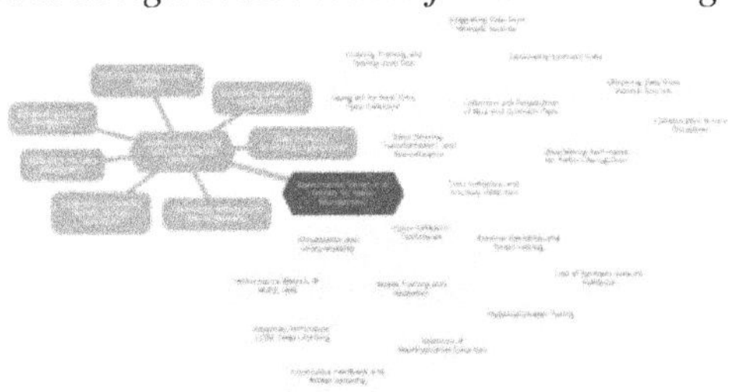

The matplotlib library is commonly used for data and model output visualization. Clear and effective visualizations are essential for interpreting and communicating AI model results to decision makers in water management (Ahmed et al., 2019).

Finally, integrating database management systems such as SQLite with Python allows for storing and retrieving large volumes of data efficiently. This is critical for handling the historical and real-time data required to train and validate AI models in water management (Shah et al., 2021).

In summary, the implementation of AI models for water management using Python and its recent libraries provides a robust and flexible platform to address the challenges of water quality prediction, leak detection, and water distribution optimization. These tools enable water researchers and practitioners to develop innovative and efficient solutions for sustainable water resources management (Lowe et al., 2022) (See Figure 7).

In this section, we present three examples developed in Python. In each case, an explanation of what is being done is provided.

5. INTEGRATIVE EXAMPLES

Example 1. The provided code performs a comprehensive analysis of water quality data by loading the dataset, handling missing values, and normalizing the data to prepare it for machine learning. After preprocessing, the code splits the data into training and testing sets, which are then used to train two different machine learning models: a Random Forest classifier and an Ensemble Decision Trees model using Bagging. The Random Forest model combines multiple decision trees to enhance predictive accuracy, while the Bagging classifier further boosts decision trees by training multiple models on different subsets of the data. Following the training phase, the code evaluates both models by calculating accuracy scores, generating detailed classification reports, and visualizing the results using confusion matrices. These matrices provide insight into the models' prediction capabilities. Additionally, the code compares the accuracy of both models through a bar chart, making it easier to interpret which model performs better in terms of predictive power. Finally, side-by-side confusion matrices are plotted to allow for a more granular comparison of the models' performance, facilitating a deeper understanding of where each model excels or falls short (See Figures 8 and 9).

Figure 8. Description of example 1.

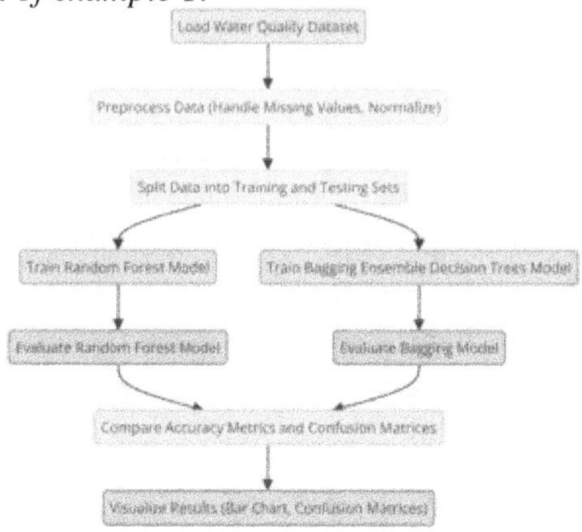

Figure 9. In the QR code you have access to the script.

Example 2. The provided code is designed to train and evaluate multiple machine learning models—specifically, Random Forest, Gradient Boosting, AdaBoost, and Bagging—on a dataset related to water leak detection. Initially, the dataset is loaded and undergoes preprocessing, which includes handling missing values and normalizing the features to ensure consistency and improve model performance. The dataset is then split into training and testing sets. Four different classifiers are trained using the training data: Random Forest, which builds an ensemble of decision trees; Gradient Boosting, which sequentially builds trees to correct the errors of the previous ones; AdaBoost, which focuses on misclassified instances by assigning them higher weights; and Bagging, which trains multiple instances of the same model on different subsets of the data to reduce variance. After training, the models are evaluated on the test set by calculating accuracy scores and generating classification reports, which detail the precision, recall, and F1-score for each class. Confusion matrices are also visualized to provide a detailed view of the models' predictive performance. The results from all models are compared, both in tabular form and with a bar plot, highlighting the accuracy of each model. Additionally, side-by-side confusion matrices are displayed to allow for a comprehensive comparison of how each model handles the classification task, offering insights into their strengths and weaknesses in detecting water leaks (See Figures 10 and 11).

Example 3. The provided code is designed to train and evaluate several regression models—specifically, Random Forest, Gradient Boosting, AdaBoost, and Bagging—using a dataset related to water distribution. The process begins by loading the dataset and performing preprocessing steps, which include handling missing values through imputation and normalizing the features to ensure consistent scale across all data points. Once the data is prepared, it is split into training and testing sets. The four regression models are then trained on the training data: Random Forest Regressor, which builds multiple decision trees to improve prediction accuracy; Gradient Boosting Regressor, which builds trees sequentially to correct previous errors; AdaBoost Regressor, which adjusts model weights to focus on difficult-to-predict cases; and Bagging Regressor, which reduces variance by training multiple instances of the same model on different subsets of the data. After training, each model is evaluated on the test data using two key metrics: R^2 Score, which indicates how well the model explains the variance in the target variable, and Mean Squared Error (MSE), which measures the average squared difference between the predicted and actual values. The results are visualized through scatter plots that compare predicted versus true values, providing a clear picture of model performance. Finally, the code compares the performance of all models by compiling a results table and creating bar plots that highlight the R^2 Scores and MSEs, making it easy to identify which model performs best in predicting water distribution metrics (See Figures 12 and 13).

Figure 10. Description of example 2.

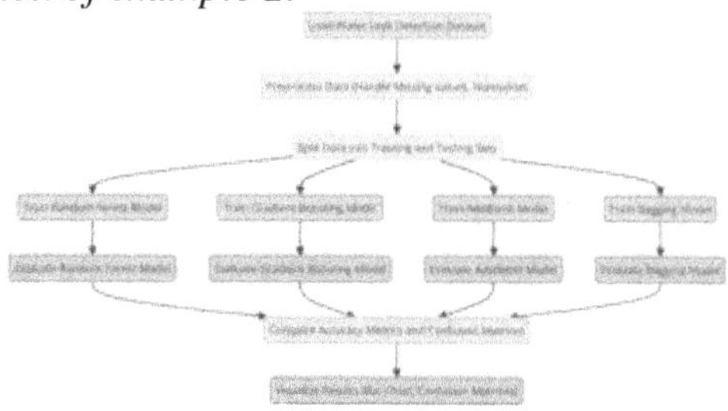

Figure 11. In the QR code you have access to the script.

Figure 12. Description of example 3.

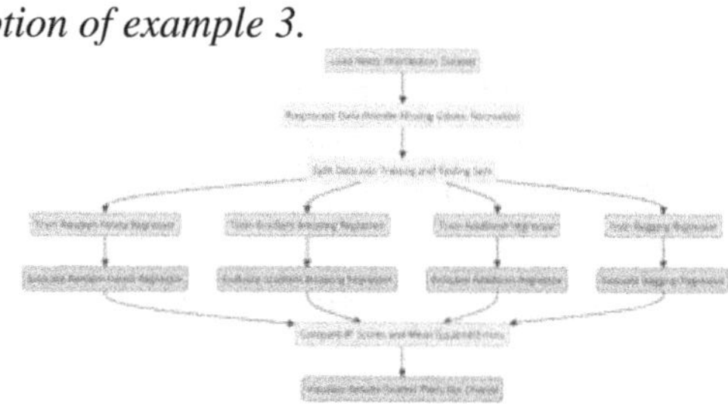

Figure 13. In the QR code you have access to the script.

6. CONCLUSIONS AND RECOMMENDATIONS

6.1. Main Conclusions

This study has demonstrated the significant potential of applying artificial intelligence (AI) techniques in the optimization of water management processes, specifically in the areas of water quality prediction, leak detection and localization, and water distribution optimization. The integration of ensemble models, such as Random Forest, Gradient Boosting, AdaBoost, and Bagging, has been shown to enhance predictive accuracy and operational efficiency across these domains. By leveraging these advanced AI methods, it is possible to address the complexities inherent in water management, such as handling non-linear data patterns, improving the speed and accuracy of leak detection, and optimizing the distribution of water resources in complex networks. The results indicate that AI can be a transformative tool in achieving more sustainable and resilient water management practices.

Furthermore, the study highlights the importance of data preprocessing and model validation in achieving reliable outcomes. The careful handling of missing data, normalization of features, and the use of cross-validation techniques were crucial in ensuring the robustness of the AI models applied. The research also underscores the value of integrating real-time data and synthetic data to enhance model performance, especially in scenarios where real data may be scarce or incomplete. The use of AI not only improved the accuracy of predictions but also allowed for the identification of key factors influencing water quality and distribution, providing valuable insights for water resource managers.

In conclusion, the findings of this study advocate for the broader adoption of AI technologies in water management systems. The demonstrated success of ensemble models in improving the accuracy and efficiency of water quality predictions, leak detection, and distribution optimization underscores the potential of AI to revolutionize this sector. By continuing to refine these models and incorporating them into practical applications, water management can be made more efficient, reducing waste, and ensuring that resources are utilized in a sustainable manner. These advancements are critical in meeting the challenges posed by increasing water demand and climate change, ultimately contributing to the resilience and sustainability of water resource management systems.

6.2. Recommendations for Future Research

Future research in the field of water management using artificial intelligence (AI) should focus on enhancing the scalability and adaptability of AI models across different geographical regions and climatic conditions. While this study has demonstrated the effectiveness of ensemble models like Random Forest, Gradient Boosting, and AdaBoost in specific contexts, there is a need to test and adapt these models to diverse environments. This will involve collecting and integrating large-scale datasets from various regions to improve the generalizability of the models. Additionally, researchers should explore the potential of transfer learning and domain adaptation techniques to facilitate the application of AI models developed in one region to be effectively utilized in another, with minimal retraining required.

Another critical area for future research is the integration of AI models with Internet of Things (IoT) technologies for real-time monitoring and management of water resources. The study has shown the benefits of using AI in combination with IoT for leak detection and water quality monitoring. However, further research is needed to develop more sophisticated AI-driven systems that can autonomously manage water resources in real time, responding to changes in water quality, demand, and distribution dynamically. This includes the development of predictive maintenance systems for water infrastructure, capable of anticipating and preventing failures before they occur, thereby minimizing water loss and optimizing resource allocation.

Lastly, future research should also address the ethical and social implications of deploying AI in water management. While AI offers significant benefits, it is essential to consider issues such as data privacy, the potential for algorithmic bias, and the impact of automation on employment within the water management sector. Researchers should work towards developing transparent and explainable AI models that allow stakeholders to understand and trust the decisions made by these systems. Additionally, interdisciplinary collaboration between AI experts, water resource managers, policymakers, and social scientists will be crucial in ensuring that AI-driven water management solutions are not only technically sound but also socially equitable and environmentally sustainable. This holistic approach will help ensure that the advancements in AI contribute positively to the global challenge of sustainable water management.

REFERENCES

Abdelmageed, S., Tariq, S., Boadu, V., & Zayed, T. (2022). Criteria-based critical review of artificial intelligence applications in water-leak management. *Environmental Reviews*, 30(2), 280–297. Advance online publication. DOI: 10.1139/er-2021-0046

Adedeji, K., Hamam, Y., Abe, B., & Abu-Mahfouz, A. (2017). Towards Achieving a Reliable Leakage Detection and Localization Algorithm for Application in Water Piping Networks: An Overview. *IEEE Access : Practical Innovations, Open Solutions*, 5, 20272–20285. DOI: 10.1109/ACCESS.2017.2752802

Ahmed, A., Othman, F., Afan, H., Ibrahim, R., Fai, C., Hossain, S., Ehteram, M., & Elshafie, A. (2019). Machine learning methods for better water quality prediction. *Journal of Hydrology (Amsterdam)*, 578, 124084. Advance online publication. DOI: 10.1016/j.jhydrol.2019.124084

Al-Adhaileh, M., & Alsaade, F. (2021). Modelling and Prediction of Water Quality by Using Artificial Intelligence. *Sustainability (Basel)*, 13(8), 4259. DOI: 10.3390/su13084259

Aldhyani, T., Al-Yaari, M., Alkahtani, H., & Maashi, M. (2020). Water Quality Prediction Using Artificial Intelligence Algorithms. *Applied Bionics and Biomechanics*, 2020, 1–12. Advance online publication. DOI: 10.1155/2020/6659314 PMID: 33456498

Allawi, M., Salih, S., Kassim, M., Ramal, M., Mohammed, A., & Yaseen, Z. (2022). Application of Computational Model Based Probabilistic Neural Network for Surface Water Quality Prediction. *Mathematics*, 10(21), 3960. Advance online publication. DOI: 10.3390/math10213960

Badar, S., Labghough, S., Al-Abdulghani, A., Mohammed, E., Bouhali, O., & Qaraqe, K. (2023). Machine Learning Assisted Approach for Water Leaks Detection. *2023 International Conference on Information Networking (ICOIN)*, 433-437. https://doi.org/DOI: 10.1109/ICOIN56518.2023.10048954

Bahad, P., & Saxena, P. (2019). Study of AdaBoost and Gradient Boosting Algorithms for Predictive Analytics., 235-244. https://doi.org/. DOI: 10.1109/ACCESS.2021.3052032

Bilal, P., Pant, M., & Snasel, V. (2021). Design Optimization of Water Distribution Networks through a Novel Differential Evolution. *IEEE Access : Practical Innovations, Open Solutions*, 9, 16133–16151. DOI: 10.1109/ACCESS.2021.3052032

Blaya, J., Gómez-Skarmeta, A., Valdés-Vela, M., & Padilla, A. (2001). Data Mining Applied to Irrigation Water Management., 547-554. https://doi.org/. DOI: 10.1109/ACCESS.2021.3052032

Brooks, D. B. (2006). An Operational Definition of Water Demand Management. *International Journal of Water Resources Development*, 22(4), 521–528. DOI: 10.1080/07900620600779699

Casini, L., Delnevo, G., Roccetti, M., Zagni, N., & Cappiello, G. (2019). Deep Water: Predicting Water Meter Failures Through a Human-Machine Intelligence Collaboration., 688-694. https://doi.org/. DOI: 10.1109/ACCESS.2021.3052032

Chan, T., Chin, C., & Zhong, X. (2018). Review of Current Technologies and Proposed Intelligent Methodologies for Water Distributed Network Leakage Detection. *IEEE Access : Practical Innovations, Open Solutions*, 6, 78846–78867. DOI: 10.1109/ACCESS.2018.2885444

Chen, J., Wu, Q., & Ahmadi, C. (2022). AIoT@Water - An Intelligent Water Resources Management System. 2022 5th International Conference on Information and Communications Technology (ICOIACT), 199-204. https://doi.org/DOI: 10.1109/ICOIACT55506.2022.9971964

Chen, K., Chen, H., Zhou, C., Huang, Y., Qi, X., Shen, R., Liu, F., Zuo, M., Zou, X., Wang, J., Zhang, Y., Chen, D., Chen, X., Deng, Y., & Ren, H. (2019). Comparative analysis of surface water quality prediction performance and identification of key water parameters using different machine learning models based on big data. *Water Research*, 171, 115454. DOI: 10.1016/j.watres.2019.115454 PMID: 31918388

Dedja, K., Nakano, F., Pliakos, K., & Vens, C. (2022). Explaining random forest prediction through diverse rulesets. ArXiv, abs/2203.15511. https://doi.org//arXiv.2203.15511.DOI: 10.48550

Dehghani, R., & Poudeh, H. (2021). Application of novel hybrid artificial intelligence algorithms to groundwater simulation. *International Journal of Environmental Science and Technology*. Advance online publication. DOI: 10.1007/s13762-021-03596-5

Dou, J., Yunus, A., Bui, D., Merghadi, A., Sahana, M., Zhu, Z., Chen, C., Han, Z., & Pham, B. (2019). Improved landslide assessment using support vector machine with bagging, boosting, and stacking ensemble machine learning framework in a mountainous watershed, Japan. *Landslides*, 17(3), 641–658. DOI: 10.1007/s10346-019-01286-5

Drogkoula, M., Kokkinos, K., & Samaras, N. (2023). A Comprehensive Survey of Machine Learning Methodologies with Emphasis in Water Resources Management. *Applied Sciences (Basel, Switzerland)*, 13(22), 12147. Advance online publication. DOI: 10.3390/app132212147

Ebisi, F., Nikolakakos, I., Karunamurthi, J., Alnuaimi, A., Buraimi, E., & Alblooshi, S. (2023). Machine Learning Schemes for Leak Detection in IoT-enabled Water Transmission System. 2023 International Conference on IT Innovation and Knowledge Discovery (ITIKD), 1-7. https://doi.org/DOI: 10.1109/ITIKD56332.2023.10100175

Erdal, H., & Karakurt, O. (2013). Advancing monthly streamflow prediction accuracy of CART models using ensemble learning paradigms. *Journal of Hydrology (Amsterdam)*, 477, 119–128. DOI: 10.1016/j.jhydrol.2012.11.015

Erdal, H., Karakurt, O., & Namli, E. (2013). High performance concrete compressive strength forecasting using ensemble models based on discrete wavelet transform. *Engineering Applications of Artificial Intelligence*, 26(4), 1246–1254. DOI: 10.1016/j.engappai.2012.10.014

Eryiğit, M. (2014). Cost optimization of water distribution networks by using artificial immune systems. *Journal of Water Supply: Research & Technology - Aqua*, 64(1), 47–63. DOI: 10.2166/aqua.2014.031

Fereidooni, Z., Tahayori, H., & Bahadori-Jahromi, A. (2020). A hybrid model-based method for leak detection in large scale water distribution networks. *Journal of Ambient Intelligence and Humanized Computing*, 12(2), 1613–1629. DOI: 10.1007/s12652-020-02233-2

Fleming, S., Garen, D., Goodbody, A., McCarthy, C., & Landers, L. (2021). Assessing the new Natural Resources Conservation Service water supply forecast model for the American West: A challenging test of explainable, automated, ensemble artificial intelligence. *Journal of Hydrology (Amsterdam)*, 602, 126782. DOI: 10.1016/j.jhydrol.2021.126782

Galelli, S., & Castelletti, A. (2013). Assessing the predictive capability of randomized tree-based ensembles in streamflow modelling. *Hydrology and Earth System Sciences*, 17(7), 2669–2684. DOI: 10.5194/hess-17-2669-2013

Hameed, M., Sharqi, S., Yaseen, Z., Afan, H., Hussain, A., & El-Shafie, A. (2017). Application of artificial intelligence (AI) techniques in water quality index prediction: A case study in tropical region, Malaysia. *Neural Computing & Applications*, 28(S1), 893–905. DOI: 10.1007/s00521-016-2404-7

Hasan, A., & Twala, B. (2015). Improving single classifiers prediction accuracy for underground water pump station in a gold mine using ensemble techniques. IEEE EUROCON 2015 - International Conference on Computer as a Tool (EUROCON), 1-7. https://doi.org/DOI: 10.1109/EUROCON.2015.7313694

Hu, Y., Zhang, Y., Jian, D., & Li, J. (2023). Prediction of Coagulant Dosage in Waterworks Based on Ensemble Learning. 2023 12th International Conference of Information and Communication Technology (ICTech), 184-189. https://doi.org/ DOI: 10.1109/ICTech58362.2023.00046

Iyer, S., Kaushik, S., & Nandal, P. (2023). *Water Quality Prediction Using Machine Learning*. MR International Journal of Engineering and Technology., DOI: 10.58864/mrijet.2023.10.1.8

Kanamori, T., Takenouchi, T., Eguchi, S., & Murata, N. (2007). Robust Loss Functions for Boosting. *Neural Computation*, 19(8), 2183–2244. DOI: 10.1162/ neco.2007.19.8.2183 PMID: 17571942

Kang, M., Kim, K., Yu, J., Bae, H., & Jang, I. (2023). Comparison of Leak Detection Techniques Using Water Pipeline Vibration Sensor Data. 2023 IEEE International Conference on Consumer Electronics-Asia (ICCE-Asia), 1-4. https://doi.org/DOI: 10.1109/ICCE-Asia59966.2023.10326371

Khammari, A., Nashashibi, F., Abramson, Y., & Laurgeau, C. (2005). Vehicle detection combining gradient analysis and AdaBoost classification. Proceedings. 2005 IEEE Intelligent Transportation Systems, 2005., 66-71. https://doi.org/.DOI: 10.1109/ITSC.2005.1520202

Krishnan, S., Nallakaruppan, M., Chengoden, R., Koppu, S., Iyapparaja, M., Sadhasivam, J., & Sethuraman, S. (2022). Smart Water Resource Management Using Artificial Intelligence—A Review. *Sustainability (Basel)*, 14(20), 13384. Advance online publication. DOI: 10.3390/su142013384

Lechgar, H., Mallouk, A., Malaainine, M., Nahhal, T., & Rhinane, H. (2016). Artificial Intelligence (AI) Contribution to GIS in Optimal Positioning of Hydrophone Sensors Using Genetic Algorithm (Case Study: Water Network, Casablanca, Morocco). 11-19. https://doi.org/.DOI: 10.1007/978-3-319-30298-0_2

Li, L., Rong, S., Wang, R., & Yu, S. (2021). Recent advances in artificial intelligence and machine learning for nonlinear relationship analysis and process control in drinking water treatment: A review. *Chemical Engineering Journal*, 405, 126673. Advance online publication. DOI: 10.1016/j.cej.2020.126673

Lin, M., & Chu, C. (2008). Applying Artificial Immune System to Minimize Construction Cost of Water Distribution Networks. *2008 Fourth International Conference on Natural Computation*, 6, 628-632. https://doi.org/DOI: 10.1109/ICNC.2008.376

Lowe, M., Qin, R., & Mao, X. (2022). A Review on Machine Learning, Artificial Intelligence, and Smart Technology in Water Treatment and Monitoring. *Water (Basel)*, 14(9), 1384. Advance online publication. DOI: 10.3390/w14091384

Lu, H., & Ma, X. (2020). Hybrid decision tree-based machine learning models for short-term water quality prediction. *Chemosphere*, 249, 126169. DOI: 10.1016/j.chemosphere.2020.126169 PMID: 32078849

Mahardhika, S., & Putriani, O. (2023). Deployment and use of Artificial Intelligence (AI) in water resources and water management. *IOP Conference Series. Earth and Environmental Science*, 1195(1), 012056. Advance online publication. DOI: 10.1088/1755-1315/1195/1/012056

Malek, N., Yaacob, W., Wah, Y., Nasir, S., Shaadan, N., & Indratno, S. (2022). Comparison of ensemble hybrid sampling with bagging and boosting machine learning approach for imbalanced data. *Indonesian Journal of Electrical Engineering and Computer Science*, 29(1), 598. Advance online publication. DOI: 10.11591/ijeecs.v29.i1.pp598-608

Marvin, G., Grbčić, L., Družeta, S., & Kranjčević, L. (2023). Water distribution network leak localization with histogram-based gradient boosting histogram-based gradient boosting water network leak localization. *Journal of Hydroinformatics*, 25(3), 663–684. Advance online publication. DOI: 10.2166/hydro.2023.102

Mehmood, H., Liao, D., & Mahadeo, K. (2020). A Review of Artificial Intelligence Applications to Achieve Water-related Sustainable Development Goals. 2020 IEEE / ITU International Conference on Artificial Intelligence for Good (AI4G), 135-141. https://doi.org/DOI: 10.1109/AI4G50087.2020.9311018

Mosavi, A., Hosseini, F., Choubin, B., Goodarzi, M., Dineva, A., & Sardooi, E. (2020). Ensemble Boosting and Bagging Based Machine Learning Models for Groundwater Potential Prediction. *Water Resources Management*, 35(1), 23–37. DOI: 10.1007/s11269-020-02704-3

Mustafa, H., Mustapha, A., Hayder, G., & Salisu, A. (2021). Applications of IoT and Artificial Intelligence in Water Quality Monitoring and Prediction: A Review. 2021 6th International Conference on Inventive Computation Technologies (ICICT), 968-975. https://doi.org/DOI: 10.1109/ICICT50816.2021.9358675

Najah, A., El-shafie, A., Karim, O., Jaafar, O., & El-Shafie, A. (2011). An application of different artificial intelligences techniques for water quality prediction. *International Journal of Physical Sciences*, 6, 5298–5308. DOI: 10.5897/IJPS11.1180

O'Reilly, G., Bezuidenhout, C. C., & Bezuidenhout, J. J. (2018). Artificial neural networks: Applications in the drinking water sector. *Water Science and Technology: Water Supply*, 18(6), 1869–1887. DOI: 10.2166/ws.2018.016

Pahl-Wostl, C., Tàbara, D., Bouwen, R., Craps, M., Dewulf, A., Mostert, E., Ridder, D., & Taillieu, T. (2008). The importance of social learning and culture for sustainable water management. *Ecological Economics*, 64(3), 484–495. DOI: 10.1016/j.ecolecon.2007.08.007

Paul, D., Goswami, A., Chetri, R., Roy, R., & Sen, P. (2022). Bayesian Optimization-Based Gradient Boosting Method of Fault Detection in Oil-Immersed Transformer and Reactors. *IEEE Transactions on Industry Applications*, 58(2), 1910–1919. DOI: 10.1109/TIA.2021.3134140

Ravichandran, T., Gavahi, K., Ponnambalam, K., Burtea, V., & Mousavi, S. (2021). Ensemble-based machine learning approach for improved leak detection in water mains. *Journal of Hydroinformatics*, 23(2), 307–323. Advance online publication. DOI: 10.2166/hydro.2021.093

Salvaris, M., Dean, D., & Tok, W. (2018). Training AI Models., 211-241. https://doi.org/.DOI: 10.1007/978-1-4842-3679-6_9

Scornet, E., Biau, G., & Vert, J. (2014). Consistency of Random Forests. *Annals of Statistics*, 43(4), 1716–1741. DOI: 10.1214/15-AOS1321

Seo, Y., Choi, K., Lim, Y., Lee, B., & Choi, Y. (2023). Application of Machine Learning Models for Water Pipeline Leakage Detection. Crisis and Emergency Management: Theory and Praxis. https://doi.org/.DOI: 10.14251/crisisonomy.2023.19.4.45

Shah, M. I., Alaloul, W. S., Alqahtani, A., Aldrees, A., Musarat, M. A., & Javed, M. F. (2021). Predictive Modeling Approach for Surface Water Quality: Development and Comparison of Machine Learning Models. *Sustainability (Basel)*, 13(14), 7515. DOI: 10.3390/su13147515

Shahhosseini, M., Hu, G., & Pham, H. (2019). Optimizing Ensemble Weights and Hyperparameters of Machine Learning Models for Regression Problems. ArXiv, abs/1908.05287. https://doi.org/.DOI: 10.1016/j.mlwa.2022.100251

Shaik, A., & Srinivasan, S. (2018). A Brief Survey on Random Forest Ensembles in Classification Model. *International Conference on Innovative Computing and Communications*. https://doi.org/DOI: 10.1007/978-981-13-2354-6_27

Shen, Y., & Cheng, W. (2022). A Tree-Based Machine Learning Method for Pipeline Leakage Detection. *Water (Basel)*, 14(18), 2833. Advance online publication. DOI: 10.3390/w14182833

Shravani, D., R, P., B, P., S, G., & G, S. (2019). Cloud-Based Water Leakage Detection and Localization. 2019 IEEE International Conference on Cloud Computing in Emerging Markets (CCEM), 86-91. https://doi.org/DOI: 10.1109/CCEM48484.2019.00018

Simidjievski, N., Todorovski, L., & Džeroski, S. (2015). Predicting long-term population dynamics with bagging and boosting of process-based models. *Expert Systems with Applications*, 42(22), 8484–8496. DOI: 10.1016/j.eswa.2015.07.004

Sitzenfrei, R., Wang, Q., Kapelan, Z., & Savić, D. (2020). Using Complex Network Analysis for Optimization of Water Distribution Networks. *Water Resources Research*, 56(8), e2020WR027929. Advance online publication. DOI: 10.1029/2020WR027929 PMID: 32999510

Sokolov, A. (2022). On expressive abilities of ensembles of decision trees. Proceedings of Academician O.B. Lupanov 14th International Scientific Seminar "Discrete Mathematics and Its Applications". https://doi.org/DOI: 10.20948/dms-2022-72

Suchetana, B., Srivastava, B., Gupta, H., & Saharia, M. (2023). Promoting Sustainable Water Usage and Management With Water Data, AI and Policy. Proceedings of the 6th Joint International Conference on Data Science & Management of Data (10th ACM IKDD CODS and 28th COMAD). https://doi.org/DOI: 10.1145/3570991.3571021

Sumana, B., & Santhanam, T. (2015). Optimizing the Prediction of Bagging and Boosting. *Indian Journal of Science and Technology*, 8(35), 1–13. DOI: 10.17485/ijst/2015/v8i35/78449

Sun, A., & Scanlon, B. (2019). How can Big Data and machine learning benefit environment and water management: A survey of methods, applications, and future directions. *Environmental Research Letters*, 14(7), 073001. Advance online publication. DOI: 10.1088/1748-9326/ab1b7d

Talekar, B. (2020). A Detailed Review on Decision Tree and Random Forest. *Bioscience Biotechnology Research Communications*, 13(14), 245–248. Advance online publication. DOI: 10.21786/bbrc/13.14/57

Tomlinson, J., Arnott, J., & Harou, J. (2020). A water resource simulator in Python. *Environmental Modelling & Software*, 126, 104635. DOI: 10.1016/j.envsoft.2020.104635

Tylman, W., Kolczynski, J., & Anders, G. (2010). Fully automatic AI-based leak detection system. *Energy*, 35(9), 3838–3848. DOI: 10.1016/j.energy.2010.05.038

Vijayaragavan, D., Praveen, N., Sudharsan, M., & Vijayan, P. (2022). Machine Learning Model for Water Quality Prediction using Python and AI framework. International Journal of Advanced Research in Science. *Tongxin Jishu*. Advance online publication. DOI: 10.48175/IJARSCT-3749

Wang, L., Wu, J., Zhang, W., Wang, L., & Cui, W. (2021). Efficient Seismic Stability Analysis of Embankment Slopes Subjected to Water Level Changes Using Gradient Boosting Algorithms. *Frontiers in Earth Science (Lausanne)*, 9, 807317. Advance online publication. DOI: 10.3389/feart.2021.807317

Whang, S., Roh, Y., Song, H., & Lee, J. (2021). Data collection and quality challenges in deep learning: A data-centric AI perspective. *The VLDB Journal*, •••, 1–23. DOI: 10.1007/s00778-022-00775-9

William, P., Oyebode, O., Ramu, G., Gupta, M., Alzubaidi, L., & Srivastava, A. (2023). Architecture for AI-Based Validation of Wastewater Management Using Open Data Exchange Technique. *2023 International Conference on Circuit Power and Computing Technologies (ICCPCT)*, 1502-1507. https://doi.org/DOI: 10.1109/ICCPCT58313.2023.10244857

Wongburi, P., & Park, J. (2023). *Prediction of Wastewater Treatment Plant Effluent Water Quality Using Recurrent Neural Network (RNN)*. Models. Water., DOI: 10.3390/w15193325

Yaseen, Z., Deo, R., Ebtehaj, I., & Bonakdari, H. (2018). Hybrid Data Intelligent Models and Applications for Water Level Prediction., 121-139. https://doi.org/.DOI: 10.4018/978-1-5225-4766-2.ch006

Yetilmezsoy, K., Ozkaya, B., & Cakmakci, M. (2011). ARTIFICIAL INTELLIGENCE-BASED PREDICTION MODELS FOR ENVIRONMENTAL ENGINEERING. *Neural Network World*, 21(3), 193–218. DOI: 10.14311/NNW.2011.21.012

Zancan, C., Passador, J., & Souza, C. (2023). *Brazilian Intermunicipal Water and Sewage Consortia Management: Harnessing the Power of Artificial Intelligence*. International Journal of Business & Management Studies., DOI: 10.56734/ijbms.v4n8a5

Zhang, X., Beeson, P., Link, R., Manowitz, D., Izaurralde, R., Sadeghi, A., Thomson, A., Sahajpal, R., Srinivasan, R., & Arnold, J. (2013). Efficient multi-objective calibration of a computationally intensive hydrologic model with parallel computing software in Python. *Environmental Modelling & Software*, 46, 208–218. DOI: 10.1016/j.envsoft.2013.03.013

Zhao, X., Wu, Y. L., & Cui, W. (2019). iForest: Interpreting Random Forests via Visual Analytics. *IEEE Transactions on Visualization and Computer Graphics*, 25(1), 407–416. DOI: 10.1109/TVCG.2018.2864475 PMID: 30188822

Chapter 9
Precision Agriculture and AI-Driven Resource Optimization for Sustainable Land and Resource Management

Mrutyunjay Padhiary
https://orcid.org/0000-0002-2236-568X
Assam University, Silchar, India

Azmirul Hoque
https://orcid.org/0009-0002-7220-9818
Assam University, Silchar, India

Gajendra Prasad
https://orcid.org/0000-0001-5521-4960
Assam University, Silchar, India

Kundan Kumar
https://orcid.org/0009-0001-6831-0676
Assam University, Silchar, India

Bhabashankar Sahu
https://orcid.org/0009-0008-2253-9361
Parala Maharaja Engineering College, India

ABSTRACT

Amidst the escalating global challenges of climate change, limited resources, and

DOI: 10.4018/979-8-3693-8074-1.ch009

population growth, the adoption of sustainable land and resource management has become imperative to ensure food security and environmental conservation. Precision agriculture enhances process efficiency, reduces environmental impact, and improves agricultural productivity through the integration of artificial intelligence technologies, including machine learning, deep learning, and computer vision. Key findings indicate a reduction of 10–20% in input costs and an increase of 15–25% in crop yields through efficient resource utilisation. Furthermore, precision irrigation systems can achieve water savings of up to 50%, while targeted pesticide treatments reduce chemical usage by 30–40%. This chapter examines the economic and environmental benefits, highlighting a 20% reduction in CO_2 *emissions. Recent advancements underscore the potential of AI to foster sustainable agriculture, promoting environmental conservation and economic viability.*

1. INTRODUCTION

1.1 Background

In order to satisfy present needs without jeopardizing the capacity of future generations to fulfill their own needs, it is imperative to sustainably manage land and natural resources. This entails the integration of economic, environmental, and social objectives to guarantee productivity, ecological sustainability, and community well-being. Precision agriculture is an agricultural approach that uses sophisticated technologies such as the Global Positioning System (GPS), sensors, drones, and data analytics to strategically monitor and regulate agricultural inputs, including water, fertilizers, and pesticides (Farooqui et al., 2024). Artificial intelligence (AI)-driven resource optimization utilizes machine learning algorithms and data from various sources to make well-informed decisions regarding the allocation of resources (Jawad & Balázs, 2024). This strategy mitigates resource overexploitation, safeguards ecosystems, minimizes pollution, and averts soil deterioration. Through the provision of real-time data and predictive analytics, these technologies empower farmers to adjust to dynamic climatic conditions and make proactive and strategic decisions (Hoque et al., 2025). These methods facilitate the achievement of long-term sustainability by guaranteeing the effective utilization of land and resources, minimizing the environmental impact, and preserving the optimal health of agricultural systems for future generations (Figure 1).

Figure 1. Overview of sustainable land and resource management practices

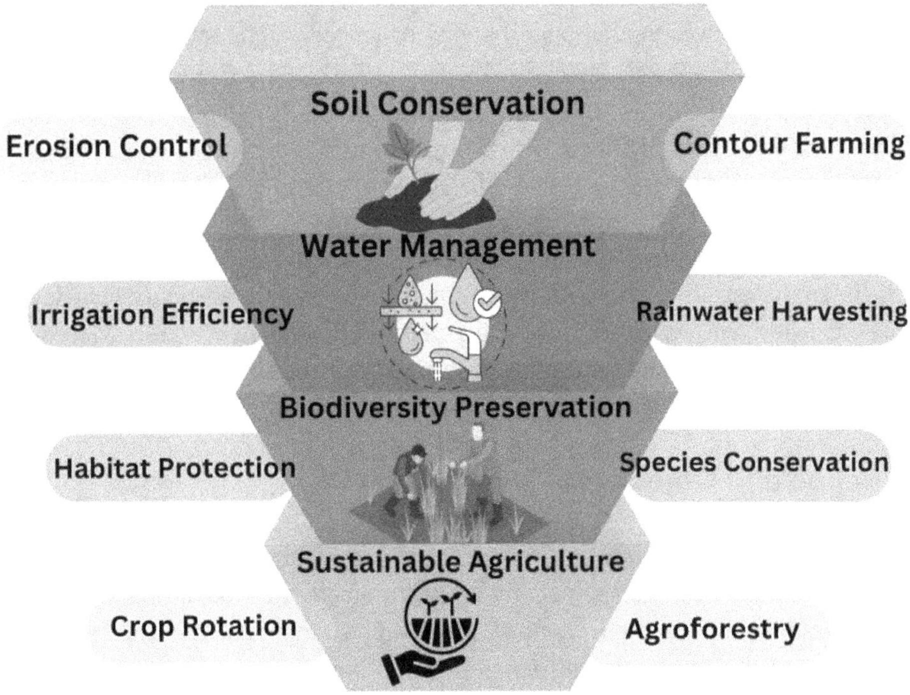

1.2 Historical Evolution

The domestication of plants and animals initiated the establishment of sedentary agricultural communities. Initially, agricultural methods emphasized crop rotation and basic irrigation to enhance crop production (Edwards, 2020). The advent of machinery such as the plow, reaper, and tractor during the Industrial Revolution resulted in monumental transformations in agriculture. The advent and extensive application of chemical fertilizers and pesticides during the 19th century not only considerably enhanced agricultural productivity but also sparked apprehensions over soil deterioration and environmental contamination (Dhananjayan et al., 2020). The Green Revolution brought about the introduction of high-yielding cultivars of staple crops, such as wheat and rice, as well as novel irrigation methods and chemical inputs. The prevalence of monoculture, or the cultivation of a single crop over extensive areas, contributed to the increased vulnerability of crops to pests and diseases, hence necessitating further chemical inputs (Begna, 2020). This century has witnessed the incorporation of AI and big data analytics into the field of agriculture. AI algorithms analyze vast amounts of data from sensors, weather forecasts, satellite imagery, and more, providing farmers with real-time insights and predictive analytics. Artificial intelligence-powered automation, such as drones for crop surveillance and robots for planting and harvesting, has significantly enhanced efficiency and accuracy in agriculture (Padhiary, Saha, et al., 2024).

1.3 Current Challenges

The integration of precision agriculture technologies, such as sensors, drones, and satellite imagery, poses challenges due to overwhelming data and the need for analysis. High costs limit accessibility for small-scale farmers, particularly in developing countries, while a lack of technical expertise hinders effective use. Moreover, resource demands raise sustainability concerns (Balkrishna et al., 2023; Saiz-Rubio & Rovira-Más, 2020; Sharma et al., 2023).

1.4 Scope and Objectives

This chapter examines how AI-driven precision agriculture can enhance sustainable land and resource management by optimizing activities like crop monitoring, irrigation, soil health, and pest control. It highlights technical advances, practical examples, and the economic and environmental benefits of AI, emphasizing its potential to transform agriculture, ensure food security, and promote sustainability (Jararweh et al., 2023). The study offers guidance for researchers, policymakers, and practitioners in adopting and expanding AI solutions in agriculture.

2. PRINCIPLES OF SUSTAINABLE AGRICULTURE

2.1 Environmental, Economic, and Social Dimensions

Sustainable agriculture aims to efficiently use natural resources like water, soil, and energy to minimize environmental harm. Technologies such as Variable Rate Technology (VRT) reduce runoff and pollution, while AI algorithms optimize crop management and promote biodiversity (Hassan et al., 2023). It also focuses on reducing greenhouse gas emissions and boosting soil carbon sequestration. Economically, it enhances profitability by minimizing waste and optimizing resource use. AI-driven analytics help farmers identify optimal planting and harvesting times, improving productivity and mitigating risks from climate variability and market dynamics (Padhiary et al., 2023).

From a social perspective, sustainable agriculture advocates for the equitable availability of resources, technology, and opportunities for all farmers, irrespective of their scale or geographical position. Agronomic platforms powered by artificial intelligence enable the exchange of knowledge and collaboration among farmers, therefore fostering more inclusive farming methods (Foster et al., 2023).

Enhancing the livelihoods of rural communities is achieved through the creation of employment opportunities, improvement of incomes, and stimulus of local economies. Implementing new technology can generate employment opportunities in rural regions, ranging from data analysis to machinery operation, therefore fostering economic growth. Precision agriculture and artificial intelligence enhance crop production systems, guaranteeing a consistent and nourishing food supply and making a significant contribution to global food security (Shaikh et al., 2022).

2.2 Conservation Practices

Soil conservation techniques include no-till farming, contour farming, reduced tillage, and terracing. No-till farming scatters crop residues and sows seeds without plowing, while contour farming aligns crops with natural contours, and reduced tillage minimizes plowing depth or frequency. AI analyzes soil data to optimize these practices. Precision irrigation, using technologies like drip or sprinkler systems, minimizes water waste, with AI refining schedules based on weather forecasts (Ali et al., 2024). Rainwater harvesting reduces reliance on external sources, with AI and GIS identifying optimal collection sites (Hashim & Sayl, 2021). Inorganic nitrogen management combines organic and inorganic fertilizers, with AI optimizing nutrient mixes and application rates for sustainability.

2.3 Soil Health and Water Conservation

Optimal soil health is vital for agriculture, supporting crop growth, nutrient cycles, and beneficial microbes. AI analyzes soil data to detect issues like nutrient deficiencies or erosion risks, offering targeted solutions. VRT enables precise application of soil amendments, reducing excess use and runoff (Morgan & Kadyampakeni, 2020). AI also optimizes tillage timing and depth to minimize disturbance and improve soil preparation. Effective water management ensures crops receive adequate water, with AI adjusting flow rates using real-time data, irrigation systems, and rainfall collection, reducing reliance on external sources.

3. PRECISION AGRICULTURE: AN OVERVIEW

3.1 Definition and Key Components

Precision agriculture applies technology and data analytics to monitor and adapt to changes in crop, soil, and environmental conditions, enhancing productivity, reducing inefficiencies, and promoting sustainability (Mondal & Palit, 2022). Key tools include satellites, drones, hyperspectral cameras, GIS, GPS, and VRT. Remote sensing provides data on crop health, soil quality, and moisture, while GIS creates detailed field maps (Figure 2). GPS ensures precise input application, and VRT applies inputs as needed. AI and big data identify patterns and predict outcomes, with Decision Support Systems (DSS) offering actionable insights for farm management (Ahmad et al., 2023). Autonomous vehicles further advance efficient, scalable practices (Padhiary et al., 2024).

Figure 2. Technological components of precision agriculture

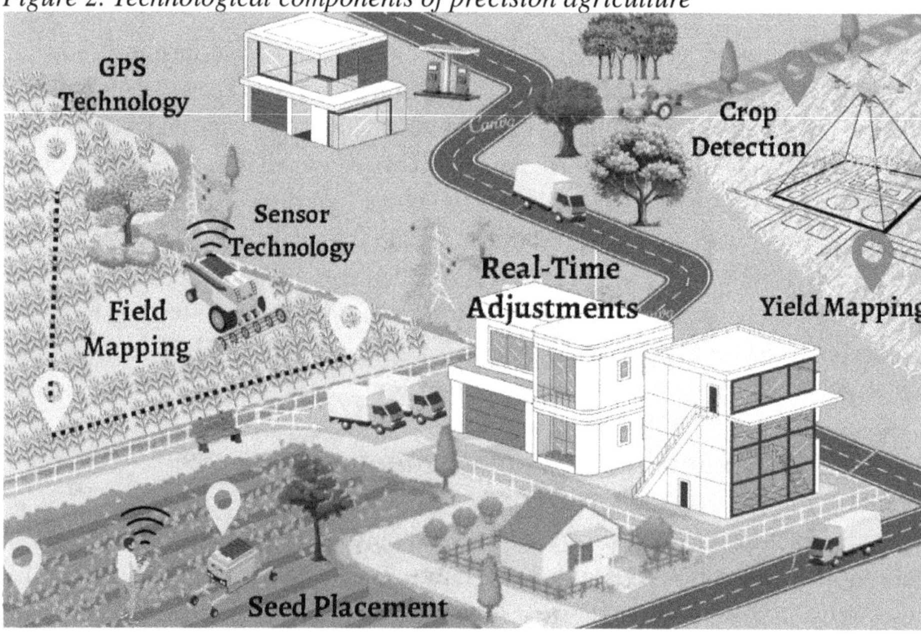

Table 1 highlights the key differences between traditional and precision agriculture practices, showcasing how precision agriculture enhances resource efficiency, reduces environmental impact, and improves overall productivity. The concise comparison illustrates the benefits of adopting precision agriculture, particularly in water use, fertilizer application, and cost efficiency.

Table 1. Comparison of traditional vs. precision agriculture practices

Practice/Aspect	Traditional Agriculture	Precision Agriculture	References
Resource use (water)	Uniform irrigation; overuse and runoff common.	Smart irrigation; up to 50% water savings.	(Sidhu et al., 2021)
Fertilizer application	Uniform application; risk of over- or under-fertilization.	Variable Rate Technology; 15-30% reduction in fertilizer use.	(Manasa et al., 2023)
Pesticide usage	Fixed schedules; increased pesticide use and resistance.	AI-driven targeted application; 30-40% reduction in use.	(Talaviya et al., 2020)
Crop yield	Variable yields; dependent on uniform practices.	15-25% yield increase through optimized practices.	(Lu et al., 2020)
Cost efficiency	Higher long-term costs due to inefficient resource use.	10-20% input cost reduction through precision management.	(Ahmad et al., 2023)
Environmental impact	Higher impact due to overuse of resources and chemicals.	20% reduction in CO2 emissions; improved conservation.	(Wang & Azam, 2024)
Labor efficiency	Labor-intensive; higher costs.	Automated operations; up to 30% reduction in labor costs.	(Ronzoni et al., 2021)
Field variability management	Limited management of variability; inconsistent performance.	Site-specific management; uniform crop performance.	(Maestrini & Basso, 2018)
Energy consumption	High energy use due to inefficient machinery.	15-25% reduction in energy use with optimized machinery.	(Bhunia et al., 2021)
Decision-making	Relies on experience; potential delays.	AI tools enable real-time, data-driven decisions.	(Kliestik et al., 2023)

3.2 Technological Advancements

Remote sensing and UAVs provide critical data on crop conditions, soil moisture, and land use. Satellite imagery tracks vegetation changes, while UAVs assess crop health, pests, and irrigation efficiency (Barbedo, 2019). GIS maps visualize spatial variability, aiding decisions on planting, fertilization, and irrigation. GPS ensures precise machinery guidance, reducing overlaps and gaps. Soil sensors monitor moisture for optimized irrigation. IoT devices collect weather data, enhancing decision-making (Rajak et al., 2023). AI algorithms analyze past and real-time data to predict outcomes.

3.3 Benefits and Limitations

Precision agricultural technology such as VRT and GPS-guided machinery enables farmers to precisely administer seeds, fertilizers, and pesticides, therefore minimizing waste and maximizing resource utilization. Artificial intelligence models assist farmers in predicting pest disease outbreaks and determining the best times to plant crops, resulting in improved agricultural management and increased yield (Shaikh et al., 2022). Nevertheless, small and medium-sized farms may face significant upfront expenses, and the massive volume of data produced necessitates very efficient data handling strategies. Furthermore, the vulnerability of agricultural systems to cyberattacks escalates, therefore exacerbating the possibility of operational disruptions, compromise of data integrity, or financial losses.

4. ROLE OF AI IN PRECISION AGRICULTURE

4.1 AI Technologies in Agriculture

AI systems leverage large datasets to forecast agricultural outcomes, optimizing planting dates, crop selection, and yield predictions. Drone and sensor images enable early detection of pests and diseases, while machine learning analyzes remote sensing data to monitor crop health and identify stressors like nutrient deficiencies or water scarcity (Figure 3). Deep learning models, such as CNNs, excel in image analysis, identifying crops, detecting weeds, and assessing conditions. These models support autonomous harvesting and create detailed maps using multispectral images (Khan et al., 2022). Real-time computer vision systems enhance precision spraying and plant phenotyping, while Raspberry Pi systems enable cost-effective plant identification (Padhiary et al., 2023). Additionally, 3D printing has streamlined the customization of all-terrain vehicles for precision farming, reducing costs and prototyping time (Padhiary & Roy, 2024).

Figure 3. AI-driven crop management system

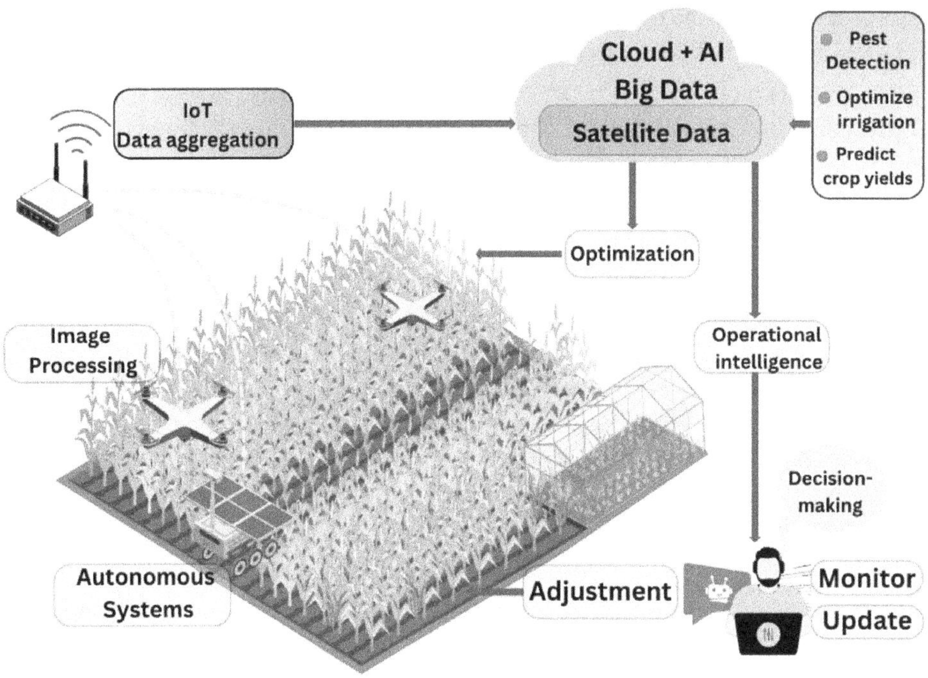

4.2 Predictive Analytics for Crop Management

Predictive analytics utilizes data from heterogeneous sources such as satellite imaging, drones, IoT sensors, and historical weather patterns to estimate soil conditions, crop health, and rainfall. Intelligent algorithms, namely machine learning models, examine this data to detect patterns and correlations, forecasting the most favorable planting dates, watering schedules, and fertilization requirements. The models provide farmers with practical information, such as forecasting insect infestations or disease outbreaks, thereby enabling the implementation of preventive measures. Additionally, they assist in the prediction of crop production, the optimization of irrigation schedules, the determination of appropriate fertilizer quantities and types, and the prediction of pest and disease emergencies. This contributes to the reduction of chemical usage and mitigation of ecological damage.

4.3 AI-Driven Pest and Disease Detection

Agronomic pest and disease detection systems powered by artificial intelligence utilize a range of data sources, such as satellites, drones, Internet of Things sensors, and historical records. These systems utilize crop photos to detect visual indications of pests and diseases, distinguish between several categories, and forecast possible epidemics by using environmental variables and past data (Zhang et al., 2019). Based on the nature and degree of the infestation, these systems offer farmers actionable insights that enable prompt interventions and the suggestion of specialized treatments or integrated pest management plans.

Table 2 summarizes the various AI technologies used in agriculture, along with their specific applications and benefits, such as improved yield prediction, pest detection, and farm management. It provides a clear overview of how AI is integrated into agricultural practices, contributing to more informed and efficient farming operations.

Table 2. AI technologies and their applications in agriculture

AI Technology	Application Area	Benefits	Examples	References
Machine Learning	Crop Yield Prediction	Accurate yield forecasts.	Models predict yields with up to 90% accuracy.	(Paudel et al., 2021)
Deep Learning	Pest and Disease Detection	Early, accurate identification.	CNNs detect pests with 95% accuracy.	(Türkoğlu & Hanbay, 2019)
Computer Vision	Crop Monitoring	Real-time crop health analysis.	Drones assess crop health, detecting issues early.	(Abbas et al., 2023; Padhiary & Kumar, 2024)
Natural Language Processing	Agricultural Data Analysis	Extract insights from vast data.	AI analyzes research and trends for actionable insights.	(Misra et al., 2020)
Robotics and Automation	Autonomous Farm Machinery	Reduces labor costs and increases efficiency.	Autonomous tractors increase operational efficiency by 20-30%.	(Lowenberg-DeBoer et al., 2021)
Reinforcement Learning	Precision Irrigation	Optimizes irrigation for water savings.	AI models adapt to conditions, reducing water use.	(Lakhiar et al., 2024)
Predictive Analytics	Weather Forecasting	Improves weather prediction accuracy.	Models predict extreme weather with 85% accuracy.	(Fathi et al., 2022)
Sensor Fusion	Integrated Farm Management	Comprehensive real-time farm management.	Integrates data from multiple sensors for holistic management.	(Barrile et al., 2022)

5. DATA COLLECTION AND MANAGEMENT IN PRECISION AGRICULTURE

5.1 Sources of Agricultural Data

Soil sensors measure moisture, temperature, pH, and nutrients, providing real-time data for precise irrigation and fertilization. Climate sensors track weather parameters like temperature and humidity, predicting crop growth and pest risks. Satellites offer high-resolution imagery to monitor crop health and land use, using indices like NDVI to assess vegetation (Adhikary et al., 2022). Advanced drones with high-resolution cameras collect detailed aerial data, analyzed through software to create maps, ortho mosaics, and 3D models for decision-making in precision agriculture.

5.2 Data Integration and Analysis

Precision agriculture relies on data from sensors, satellites, and drones, integrated through data harmonization and GIS systems for comprehensive analysis and visualization (Sharma et al., 2018). Machine learning and AI analyze this data to predict crop yields, pest infestations, and diseases, identifying anomalies like nutrient deficiencies. Advanced visualization methods aid in interpreting complex datasets, such as variations in soil moisture and crop health (Kharel et al., 2020).

5.3 Challenges in Data Management and Privacy Concerns

Synthesizing data from sensors, satellites, and drones is challenging due to varied formats and protocols, requiring sophisticated algorithms for consistency. Advanced precision agriculture technologies generate large data volumes, demanding effective management solutions (Shafi et al., 2019). Farmers retain ownership and control over their data, with access restricted to authorized users. Ensuring data security involves protecting against cyber threats through robust encryption and mitigating vulnerabilities in collection and storage systems.

6. PRECISION IRRIGATION AND WATER MANAGEMENT

6.1 Smart Irrigation Systems

An intelligent irrigation system is a technological solution that generates real-time information on soil conditions, weather forecasts, and crop health using data obtained from sensors and aerial imagery. Sophisticated algorithms examine this data to ascertain exact water requirements for crops, taking into account variables such as soil composition, plant development phase, climatic conditions, and past water consumption (Pereira et al., 2020). The system optimizes the watering schedule and volume automatically, enabling farmers to remotely monitor and regulate it. This enables rapid modifications and interactive control without the requirement of being physically present in the field. The system iteratively acquires knowledge and enhances its performance based on the gathered data, continuously optimizing irrigation patterns. The application of machine learning algorithms enables the prediction of future water requirements and subsequent adjustment of irrigation schedules.

6.2 AI-Based Water Usage Predictions

Artificial intelligence and machine learning algorithms are employed in AI-based water usage modeling to precisely anticipate the water needs of crops. This technology refines precision irrigation by facilitating more knowledgeable decision-making regarding the timing, location, and quantity of water application. Artificial intelligence systems include data from a multitude of sources, such as soil moisture sensors, meteorological stations, satellite imaging, drone data, historical weather data, and agricultural growth models (Subeesh & Mehta, 2021a). These algorithms use real-time data analysis to detect patterns and trends that impact water requirements, including soil moisture levels, meteorological conditions, plant growth phases, and evaporation flows. The development of IoT-based solenoid-controlled pressure regulation systems has improved the precision of sprayers, reducing waste and ensuring uniform application of agrochemicals (Saha et al., 2023). Data-driven machine learning algorithms, such as regression models, neural networks, and decision trees, are trained using past data to forecast future water needs. Advanced artificial intelligence systems have the capability to predict water requirements several days, weeks, or even months ahead of time, consistently revising their forecasts using up-to-date data. Efficient irrigation schedules can be optimized by identifying the optimal periods for watering crops using forecasts of future weather conditions, soil moisture levels, and plant water requirements. Artificial intelligence systems harness feedback loops to acquire knowledge from results and consistently improve their models. The ability to identify irregularities in data enables prompt modifications to irrigation strategies, therefore mitigating crop stress and minimizing water loss.

6.3 Impact on Water Conservation and Crop Yield

Advanced irrigation and water management systems, which integrate artificial intelligence and machine learning, greatly enhance water saving and agricultural productivity. By optimizing water use and ensuring crops receive the appropriate amount of water at the correct time; these technologies conserve water resources and enhance agricultural production. Precision irrigation optimizes water efficient use and mitigates evaporation losses by utilizing data obtained from soil moisture sensors, weather forecasts, and crop growth phases. conditions, making real-time adjustments to irrigation schedules. This facilitates water Artificial intelligence-driven intelligent irrigation systems continuously monitor soil and weather conservation during periods of excess rainfall or when soil moisture levels are adequate. Precision irrigation mitigates excessive demand on groundwater resources, therefore preserving aquifer levels and guaranteeing sustained water availability for agricultural use (Sabir et al., 2024). The utilization of AI and machine learning models enables the optimization of irrigation schedules by considering the distinct water requirements at various growth stages, therefore enhancing growth and productivity at crucial periods like blooming and fruit sets. Integration of real-time meteorological data into precision irrigation systems enables the anticipation and adjustment for severe weather occurrences, therefore safeguarding crops against potential harm and ensuring production stability.

7. SOIL HEALTH AND FERTILITY MANAGEMENT

7.1 Soil Monitoring Techniques

Soil monitoring is crucial for sustainable agriculture to guarantee ideal plant development and reduce environmental consequences. As part of strategies, soil samples are taken and analyzed in a lab. Electronic instruments like tensiometers, capacitance probes, and neutron probes are also used, along with satellite photos, drone photos, or aerial photos, and GIS data is added to help managers make better decisions (Guan, 2023). Sensors measure the electrical conductivity of soil, which depends on various elements like soil texture, moisture content, salinity, and organic matter. Organic matter serves as a crucial determinant of soil health, influencing the availability of nutrients, retention of water, and physical structure of the soil. Real-time detection of specific nutrients by these sensors enables precise fertilization, enhancing nutrient use efficiency and minimizing environmental effects. Electrochemical methods for in-situ soil fertilizer measurement offer a precise, real-time approach to monitoring and optimizing soil nutrient levels, enhancing crop productivity (Padhiary, Kyndiah, et al., 2024). The monitoring of soil is of utmost importance in order to guarantee the best possible plant development and minimum environmental consequences.

7.2 AI-Driven Soil Analysis and Recommendations

In order to enhance soil health and fertility management, AI-driven soil analysis and suggestions employ sophisticated algorithms and machine learning techniques. These systems utilize data obtained from soil sensors, remote sensing technologies, and historical soil data to quantitatively examine soil characteristics, moisture content, nutrient composition, and environmental circumstances. This data is consolidated into a single system, where artificial intelligence computers analyze the data, frequently employing machine learning models to detect patterns and correlations. Artificial intelligence models have the capability to forecast soil health patterns and anticipate future nutrient requirements, therefore predicting possible problems like nutrient insufficiencies, soil erosion, or fluctuations in moisture availability. Artificial intelligence systems have the capability to consistently monitor soil conditions and make real-time adjustments to recommendations, therefore ensuring the maintenance of ideal soil management practices (Shaikh et al., 2022). Artificial intelligence systems frequently incorporate dashboards and visualization capabilities, which empower farmers and land managers to make well-informed decisions.

7.3 Sustainable Fertilizer Application

Ensuring the sustainable application of fertilizers is essential for preserving soil health and fertility while reducing environmental consequences. Systematic soil testing facilitates the assessment of nutrient concentrations, pH levels, and organic matter content, permitting well-informed choices on fertilizer application. Opting for fertilizers that align with the specific requirements of the soil and crops, such as slow-release or controlled-release alternatives, can enhance effectiveness. Utilizing both organic and inorganic fertilizers, in conjunction with soil amendments, maximizes the availability of nutrients and minimizes the environmental footprint (Palansooriya et al., 2023). The implementation of cover crops and crop rotation can effectively enhance natural nutrient cycling, therefore mitigating the need for exogenous fertilizers. The implementation of measures such as buffer strips or controlled drainage systems serves to mitigate nutrient runoff and leaching, therefore safeguarding water quality and negating environmental consequences.

8. CROP MONITORING AND YIELD PREDICTION

8.1 Remote Sensing Technologies

The application of remote sensing technologies for crop monitoring and yield prediction is growing owing to their capacity to offer real-time data across extensive regions (Gao & Zhang, 2021). These technologies encompass the acquisition of data from objects or geographic regions without direct physical interaction, such as satellites, drones, or aircraft. Earth-orbiting satellites offer extensive coverage for vast agricultural regions. Drones or unmanned aerial vehicles achieve close proximity to the ground, therefore providing high-resolution imagery and facilitating meticulous data collection over more limited geographical regions (Padhiary, Kumar, et al., 2024). Manned airplanes outfitted with remote sensing equipment have the capability to survey extensive regions and deliver data of exceptional resolution. These technologies acquire images using visible and near-infrared light to evaluate the health of crops, identify plant species, and detect problems such as nutrient deficits or pest contamination. Furthermore, they quantify fluctuations in surface temperature, therefore monitoring the levels of agricultural water stress and irrigation efficiency. The penetration of cloud cover by radar waves enables the acquisition of data pertaining to crop structure and soil moisture, particularly in regions characterized by recurrent cloud cover or difficult weather conditions.

8.2 AI Algorithms for Crop Growth Prediction

Artificial intelligence algorithms are crucial in the prediction of crop growth for the purposes of monitoring and forecasting yield. The operational mechanisms and agricultural uses of these algorithms are as follows: Data and images obtained from satellites, drones, and sensors offer valuable insights into the health of crops, their growth stages, and the prevailing environmental conditions. Data pertaining to soil characteristics, moisture content, and nutrient concentrations. Meteorological parameters such as temperature, precipitation, humidity, and other associated variables. Historical agricultural production and development trends.

Preprocessed data is refined and converted into a structure appropriate for the purpose of analysis. These tasks encompass managing missing values, standardizing data, and extracting features. Artificial intelligence algorithms are trained by utilizing past data to acquire knowledge about the correlations between input factors (such as soil moisture and temperature) and outputs (such as crop production) (Akkem et al., 2023). Training is the process of fine-tuning model parameters in order to minimize prediction mistakes. Validating and testing the trained model on brand-new or unheard-of data allows for an evaluation of its performance and accuracy. This stage guarantees that the model can effectively generalize to various environmental situations. After undergoing training and validation, the model has the capability to forecast future crop growth and production by utilizing existing data inputs. Specifically, it can predict the impact of present weather conditions on the growth of crops.

8.3 Optimizing Harvest Timing

Optimal harvest timing is essential for maximizing both crop productivity and quality. Artificial intelligence and data-driven methodologies assist in determining the optimal timing for harvesting. Dissemination of real-time data on crop health, development phases, and field conditions by satellites and drones allows for the evaluation of ripeness and maturity (Inoue, 2020). Accurate monitoring of soil moisture and nutrient levels facilitates comprehension of crop preparedness. Historical and contemporary meteorological data forecast future weather conditions, while observations made on the ground offer further information. Artificial intelligence algorithms employ both historical and current data to forecast the maturity of crops and estimate the best time for harvesting. These models take into account variables like growth patterns, weather conditions, and soil health. Machine learning algorithms, namely convolutional neural networks, evaluate agricultural photos obtained from remote sensing technology to determine the level of ripeness of crops and identify indications of maturity (Garillos-Manliguez & Chiang, 2021). Indices such as NDVI are used to assess the health and developmental phases of plants. Synthesizing diverse data sources and artificial intelligence models yields practical suggestions on the most favorable harvest timing. Simulating several scenarios in response to evolving conditions is crucial for achieving optimal harvest timing.

9. PEST AND DISEASE MANAGEMENT

9.1 Early Detection Through AI and IoT

The rapid identification of pests and diseases is essential for efficient crop management, enabling prompt response and reducing any harm. The synergistic integration of AI and IoT technology enhances this process. Industrial Internet of Things (IoT) sensors track environmental parameters including temperature, humidity, soil moisture, and leaf wetness, which can impact the growth of pests and diseases. Specialized cameras and imaging equipment acquire high-resolution photos, which are then examined to identify the first indications of infestation or disease (Pakeerathan, 2023). Meteorological stations equipped with IoT technology offer up-to-date information on weather conditions, enabling the prediction of pest and disease outbreaks by analyzing past trends. IoT devices provide instantaneous data transmission to AI systems, therefore facilitating prompt analysis and early identification of possible issues. Advanced artificial intelligence algorithms have the capability to immediately notify farmers or agronomists using smartphones or other devices, offering precise treatment suggestions. By accurately identifying the precise location and scope of infestations or diseases, this focused strategy reduces the usage of chemicals, decreases expenses, and mitigates environmental consequences (Subeesh & Mehta, 2021b).

9.2 Integrated Pest Management (IPM)

In agriculture, Integrated Pest Management (IPM) is a sustainable strategy for managing pests and diseases by combining different control technologies and practices to minimize damage and decrease the usage of chemical pesticides. The fundamental objective is to sustain pest populations at levels that do not result in substantial economic harm while safeguarding the environment and fostering advantageous ecosystems. IPM is founded on fundamental concepts such as crop rotation, careful selection of resistant varieties, and diligent management of healthy soil. Systematic surveillance and precise detection of pests and illnesses are essential for efficient control. The IPM system uses economic thresholds to determine the appropriate timing for taking action, and assuring that the financial burden of harm is above the expenses associated with control measures. IPM combines cultural, biological, mechanical, and chemical methods to efficiently and sustainably control pests (Angon et al., 2023). Following the implementation of control measures, IPM places great importance on assessing their efficacy and making any required modifications.

9.3 Reducing Chemical Usage

Minimizing the use of synthetic pesticides and fertilizers is essential for sustainable agriculture by reducing chemical intervention in pest and disease control. This strategy safeguards ecosystems by implementing measures to avoid pollution of soil, water, and air, as well as non-target species such as beneficial insects and birds. Implementing pesticide reduction strategies also decreases the amount of hazardous chemicals that farmworkers, consumers, and local communities are exposed to, therefore enhancing their health results. Reduced consumption can help prevent the development of chemically resistant pest populations caused by excessive reliance on chemical pesticides. Reducing chemical inputs can also lead to a long-term reduction in farmers' expenses by minimizing spending on pesticides and fertilizers.

10. CASE STUDIES OF AI IN PRECISION AGRICULTURE

10.1 Success Stories from Different Regions

Examples of successful implementation of artificial intelligence in precision agriculture from various locations demonstrate the capacity of sophisticated technologies to enhance farming methods, boost crop production, lower expenses, and foster sustainability. The presented case studies exemplify the successful integration of AI into agricultural systems globally. These efforts offer significant insights and lessons for farmers and areas seeking to embrace comparable methodologies (Linaza et al., 2021).

A John Deere subsidiary called Blue River Technology developed the AI-driven "See & Spray" system. This system uses computer vision and machine learning techniques to accurately detect and effectively control weeds. This technique utilizes high-resolution cameras affixed to tractors to scan crops and dynamically differentiate between crops and weeds in real-time. Instead of dispersing chemicals across the whole field, the device selectively sprays herbicides specifically at weeds (Singh & Jain, 2022). By improving precision, herbicide usage is reduced by up to 90%, resulting in substantial economic and environmental effect reductions. Artificial intelligence systems are trained to identify many categories of weeds and crops, enabling exceptional precision in targeting. Over time, the system enhances its performance by assimilating knowledge from additional data and systematically identifying particular plant species (Su, 2020).

Small-scale farmers in India encounter substantial obstacles arising from pests, diseases, and volatile climatic circumstances. AI-controlled systems such as Plantix, designed by the German agricultural technology (Agri-tech) firm PEAT, offer farmers immediate and accurate identification of pests and diseases using a smartphone application. When farmers share images of impacted crops, the application uses artificial intelligence to detect the problem and recommend suitable remedies (Maitra & Damle, 2024). The application is available to small-scale farmers equipped with cell phones, offering them professional guidance without any charge. The democratization of access to agricultural knowledge serves to diminish reliance on expensive pesticides. Artificial intelligence models are built using data that is relevant to each region, therefore guaranteeing that recommendations are customized to the distinct local conditions and crop kinds. Thus, the relevancy and efficacy of the advice given are heightened.

Cotton producers in Australia encounter difficulties pertaining to the control of pests and control of water consumption. The Cotton Research and Development Corporation (CRDC) has formed partnerships with technology firms to create AI-based systems for monitoring pests such as the cotton bollworm and enhancing water efficiency. Artificial intelligence systems conduct analysis on data obtained from in-field sensors and satellite photography in order to promptly identify pest infestations. This enables farmers to implement focused pest management strategies, therefore minimizing the usage of chemicals and safeguarding the population of beneficial insects. By minimizing pesticide application and enhancing water efficiency, the AI-powered technology enables farmers to save expenses and improve sustainability, in accordance with Australia's rigorous environmental standards. Semi-autonomous vehicle sprayers equipped with IoT systems represent a new frontier in precision agriculture, offering both efficiency and scalability for diverse farming operations (Padhiary, Tikute, et al., 2024).

10.2 Lessons Learned and Best Practices

Utilizing artificial intelligence in precision agriculture has demonstrated considerable potential in upgrading farming techniques, increasing crop productivity, minimizing resource use, and advancing sustainability. Unmanned agricultural vehicles have demonstrated significant potential in enhancing the efficiency of hill farming by providing tailored solutions for difficult terrain (Padhiary, Sethi, et al., 2024). The many success stories from different locations showcase significant insights gained and exemplary methods that can direct future endeavors to integrate AI technologies in agriculture. Outlined below are several fundamental insights and optimal strategies obtained from these case studies.

It is imperative to customize AI solutions to align with the particular environmental, climatic, and soil requirements of a given area. For example, the strategies that are effective in the vineyards of Bordeaux may not be directly relevant to maize farms in Kenya because of variations in temperature, soil composition, and pest infestations. An AI model's efficacy is greatly dependent on the quality and volume of the accessible data. Precise data obtained from trustworthy sources such as sensors, satellite imaging, and weather observations is essential for constructing resilient models (Nishant et al., 2020). Disseminating knowledge to farmers on the advantages and functioning of AI technology is essential for their acceptance. Training programs should prioritize the development of digital literacy skills and equip farmers with the ability to comprehend and analyze recommendations given by artificial intelligence.

10.3 Scaling AI Solutions for Smallholder Farmers

The implementation of AI solutions is crucial for transforming worldwide agriculture, particularly in emerging areas where small-scale farmers constitute a substantial portion of the farming population. Small-scale farmers encounter distinct obstacles, including restricted technological resources, financial limitations, inadequate education, and susceptibility to climate change. In order to surmount these challenges, artificial intelligence solutions should be cost-effective, easily available on inexpensive smartphones and feature phones, and offer immediate information and suggestions (Dhillon & Moncur, 2023). Simple and intuitive interfaces, visual aids, voice commands, and detailed instructions are crucial for small-scale farmers who have limited knowledge of digital technology. Artificial intelligence models should undergo training using local data that accurately represents the unique characteristics of smallholder farms, including weather patterns, soil types, crop kinds, and insect pressures. It is imperative that AI solutions operate offline or utilize low-bandwidth networks to guarantee uninterrupted access to vital services for farmers. Ensuring continuous availability of connectivity allows for data synchronization, enabling farmers to continue reaping the advantages of AI technologies.

11. ECONOMIC AND ENVIRONMENTAL IMPACT OF PRECISION AGRICULTURE

11.1 Cost-Benefit Analysis

An evaluation of the economic and environmental effects of precision agriculture technologies using a cost-benefit analysis is crucial. Through a comparison of the costs of adoption with the possible benefits in terms of greater productivity, lower input consumption, and improved environmental outcomes, this research assists farmers, policymakers, and stakeholders in evaluating the value of investing in new technologies (Figure 4).

Figure 4. Economic and environmental impact of precision agriculture

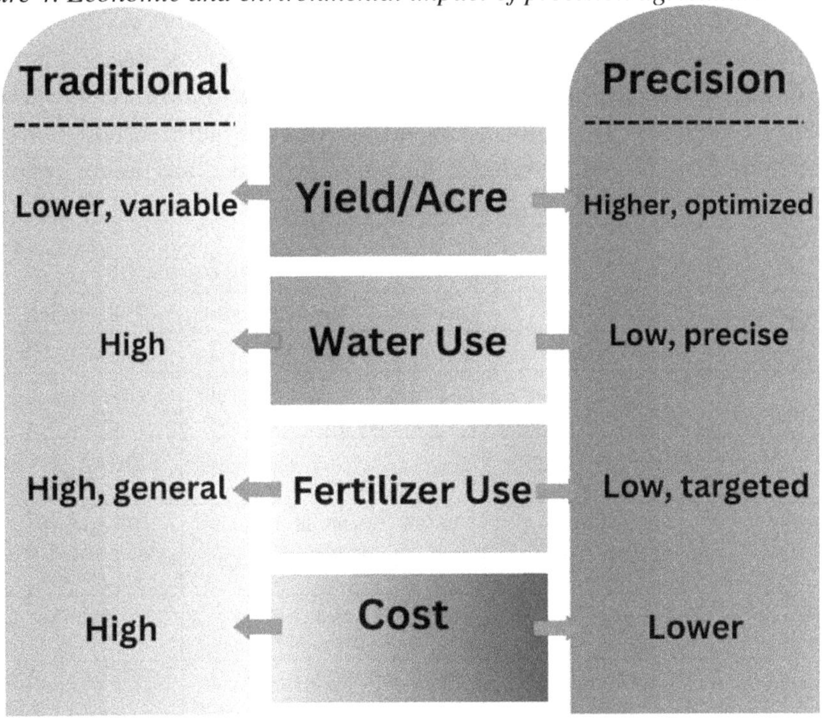

Precision agricultural technologies, such as VRT, enable accurate and targeted administration of fertilizers and pesticides according to the specific requirements of the soil and crops. This focused application minimizes the total input level needed, resulting in substantial cost reductions. For instance, a farmer employing VRT systems for fertilization might potentially decrease fertilizer application by 20–30%, resulting in significant long-term cost savings (Saleem et al., 2023). Precision agricultural technologies enhance crop management efficiency through the effective optimization of planting density, irrigation schedules, and nutrient application. These techniques have the potential to result in higher agricultural production and enhanced quality. Precision irrigation systems can achieve optimal water distribution to crops, therefore mitigating water stress and facilitating robust development. Advanced technologies such as remote sensing and drone-based monitoring enable farmers to promptly identify pest infestations, illnesses, and nutrient deficits. Preliminary identification enables prompt responses, hence minimizing agricultural losses and sustaining crop productivity. The upfront expenses associated with implementing precision agriculture technologies can be substantial, encompassing the acquisition of equipment, software, and training. For instance, a farmer may be required to make investments in tractors equipped with GPS technology, unmanned aerial vehicles, soil sensors, and sophisticated farm management software (Raj et al., 2021).

Table 3 presents the economic and environmental advantages of precision agriculture, including cost savings, yield improvements, and reductions in water, chemical usage, and greenhouse gas emissions. It underscores the long-term sustainability and profitability benefits that precision agriculture offers to farmers and the environment.

Table 3. Economic and environmental benefits of precision agriculture

Benefit Type	Description	Quantitative Metrics/Examples	References
Cost Savings	Reduced input costs.	10-20% reduction in input costs.	(Sishodia et al., 2020)
Yield improvement	Increased crop yields.	15-25% yield increase, especially in high-value crops.	(Balafoutis et al., 2017a)
Water conservation	Efficient water use.	Up to 50% reduction in water usage.	(Surendran et al., 2021)
Reduction in chemical usage	Reduced use of fertilizers and pesticides.	30-40% reduction in chemical inputs.	(Bongiovanni & Lowenberg-Deboer, 2004)
Labor cost reduction	Lower labor costs due to automation.	Up to 30% reduction in labor costs.	(Lowenberg-DeBoer et al., 2020)
Energy efficiency	Lower energy consumption.	15-25% reduction in energy use.	(Iddio et al., 2020)
Soil health improvement	Better soil management.	10-15% improvement in soil organic matter.	(Williams et al., 2020)
Carbon footprint reduction	Lower greenhouse gas emissions.	20% reduction in CO_2 emissions.	(Balafoutis et al., 2017b)
Improved crop quality	Better quality crops with higher market value.	10-15% increase in market value.	(Gyarmati & Mizik, 2020)
Economic resilience	Increased profitability and resilience.	20-30% increase in resilience to market and environmental changes.	(Loures et al., 2020)

11.2 Long-Term Sustainability Impacts

By enhancing the predictability of agricultural yields and decreasing the probability of crop failure, precision agriculture mitigates financial risks (Shaikh et al., 2022). Advanced technologies such as remote sensing and drone monitoring enable the early detection of pests, illnesses, and nutritional deficiencies (Padhiary, Kyndiah, et al., 2024), therefore preventing substantial losses. Furthermore, it decreases the requirement for chemical inputs, therefore significantly reducing their environmental consequences. Precision irrigation technologies, such as drip irrigation and soil moisture sensors, effectively preserve water resources, particularly in areas experiencing water shortages.

Precision agriculture advocates for sustainable land management techniques, mitigating land degradation through the optimization of crop rotations, preservation of soil cover, and reduction of soil disturbance (Landers et al., 2021). Thus, this guarantees the ongoing accessibility of cultivable land for subsequent generations.

11.3 Policy Implications and Support

Precision agriculture presents a multitude of economic and environmental advantages, but the attainment of these advantages necessitates government involvement (Barbosa Júnior et al., 2024). In order to expedite the use of these technologies, governments can offer financial incentives, such as subsidies or grants, to reduce the initial expenses associated with equipment and software. Furthermore, they have the capacity to improve agricultural extension services by providing farmers with the essential knowledge and skills. Government allocation of funds for research and development (R&D) can expedite the advancement of precision agriculture technology, therefore facilitating the creation of more cost-effective, user-friendly, and geographically tailored solutions (Lakhiar et al., 2024). Unique obstacles that smallholder and marginalized farmers may encounter include restricted availability of financial resources and technical proficiency. Governments have the ability to develop specialized support programs that offer focused financial aid, training, and access to cost-effective technological solutions specifically designed for these farmers.

12. FUTURE TRENDS IN PRECISION AGRICULTURE AND AI

12.1 Emerging Technologies

Ongoing advancements in technology are consistently transforming the fields of precision agriculture and AI. The aforementioned developments have the potential to significantly improve the effectiveness, output, and environmental friendliness of agricultural methods. Through the integration of novel technologies and approaches, the future of precision agriculture and artificial intelligence appears poised to address issues such as food security, resource management, and environmental sustainability.

Current research is focused on the development of sophisticated deep-learning models that can effectively analyze extensive datasets obtained from diverse sources, including satellite photography, drones, and IoT sensors. These models enable the identification of patterns and anomalies in crop health, soil conditions, and weather patterns with an unparalleled level of precision. Wireless sensor networks are comprised of interconnected sensors that are strategically placed throughout fields to observe and measure distinct environmental factors, including temperature, humidity, and soil moisture (Mowla et al., 2023) .These networks facilitate the capture and transfer of data in real-time to centralized systems for analysis. The integration of multispectral and hyperspectral sensors in drones enables the acquisition of high-resolution photographs of crops across a wide range of wavelengths. These images offer vital insights into crop health, nutrient insufficiencies, and pest infestations that are imperceptible to the unaided human vision. India's rapid adoption of farm automation technologies is driving significant advancements in precision agriculture, with a focus on increasing productivity and sustainability (Padhiary, 2024).

12.2 Integration with Climate Smart Agriculture

Climate-smart agriculture (CSA) is a strategic approach designed to enhance agricultural productivity and sustainability by simultaneously mitigating greenhouse gas emissions (Abhilash et al., 2021). Watershed management is the integration of precision agriculture technologies and artificial intelligence to optimize irrigation methods, minimize water wastage, and effectively manage water resources. AI-driven climate models and weather forecasting systems offer farmers precise forecasts, allowing them to strategically organize planting and harvesting, efficiently manage water resources, and stay prepared for severe weather phenomena. Implementing precision agriculture methods, such as variable rate fertilizer application and precision tillage, can greatly decrease greenhouse gas emissions by optimizing the use of resources and reducing soil disruption (Duff et al., 2021). Advanced artificial intelligence and precision agriculture technologies enhance pest and disease control through the provision of early warning systems and focused actions (Maraveas, 2022). Satellite imagery, unmanned aerial vehicles, and artificial intelligence algorithms can identify pest infestations and disease outbreaks at an early stage, enabling farmers to implement control measures with greater accuracy.

12.3 Global Trends and Predictions

Advanced precision agriculture and artificial intelligence are turning the agricultural sector by improving production, sustainability, and resilience. The market's needs, evolving farming practices, and technological advancements are driving these changes. AI and machine learning (ML) are facilitating sophisticated data analytics, empowering farmers to make informed decisions based on extensive information obtained from satellites, drones, and sensors (Fuentes-Peñailillo et al., 2024). IoT comprises intelligent sensors and wireless networks that enable the collection of real-time data. The integration of precision agriculture and artificial intelligence with climate-smart practices is an effort to strengthen resilience to climate change. This integration entails the optimization of resource utilization, the selection of climate-resistant crops, and the implementation of sustainable practices (Rivero et al., 2022). The adoption of precision agriculture and AI technology is anticipated to increase significantly, particularly in emerging regions. 3D printing is transforming smart farming and food processing by enabling customized equipment and efficient production workflows, thereby reducing costs and improving output quality (Padhiary, Barbhuiya, et al., 2024). With the reduction in technology prices and the enhancement of accessibility, farmers in various regions will reap the advantages of these advancements. Anticipated advancements could include more advanced models for agricultural prediction, pest control, and climate adjustment. The utilization of collaborative platforms and data-sharing networks will enhance the interchange of knowledge and best practices among farmers, researchers, and technology providers (Janssen et al., 2017).

13. CONCLUSION

AI-driven precision agriculture greatly increases crop yields by 15–25% and improves resource usage efficiency through the analysis of data from sensors and satellite photography. This approach minimizes the ecological footprint of agriculture by reducing chemical usage by 30–40% and attaining water savings of up to 50%, therefore fostering improved preservation of natural resources. The utilization of AI-driven decision-making provides farmers with immediate access to data, therefore facilitating prompt interventions and enhanced planning. Although the initial expenses may be prohibitively high, the long-term economic advantages encompass a 10-20% decrease in input costs, higher yields, and improved resource management. Precision agricultural technologies are highly scalable and versatile, making them well-suited for a wide range of farming systems. Nevertheless, obstacles such as exorbitant initial expenses, specialized knowledge, and apprehensions about data protection impede universal acceptance. Advanced artificial intelligence and precision agriculture are crucial for the future of sustainable land and resource management. Emerging technologies such as autonomous machinery and remote sensing provide additional enhancements in this field. Given the challenges posed by climate change, limited resources, and a rising global population, the use of these technologies becomes progressively imperative for the agricultural industry.

REFERENCES

Abbas, A., Zhang, Z., Zheng, H., Alami, M. M., Alrefaei, A. F., Abbas, Q., Naqvi, S. A. H., Rao, M. J., Mosa, W. F., Abbas, Q., Hussain, A., Hassan, M. Z., & Zhou, L. (2023). Drones in plant disease assessment, efficient monitoring, and detection: A way forward to smart agriculture. *Agronomy (Basel)*, 13(6), 1524. DOI: 10.3390/agronomy13061524

Abhilash, R. A., Kumari, A., Singh, R. N., & Kumari, K. (2021). Climate-Smart Agriculture: An Integrated Approach for Attaining Agricultural Sustainability. In V. K. Hebsale Mallappa & M. Shirur (Eds.), *Climate Change and Resilient Food Systems* (pp. 141–189). Springer Singapore. DOI: 10.1007/978-981-33-4538-6_5

Adhikary, S., Biswas, B., Naskar, M. K., Mukherjee, B., Singh, A. P., & Atta, K. (2022). Remote sensing for agricultural applications. In *Arid Environment-Perspectives, Challenges and Management*. IntechOpen. https://www.intechopen.com/chapters/83417

Ahmad, A. A., Kasron, N., Noh, N. H. M., Azmon, H., & Chuang, T. C. (2023). *The acceptance of decision support system (DSS) in vegetable cultivation among farmers in Malaysia*. http://etmr.mardi.gov.my/Content/ETMR%20Vol.21(2023)/Aimi%20Athirah_Vol21.pdf

Akkem, Y., Biswas, S. K., & Varanasi, A. (2023). Smart farming using artificial intelligence: A review. *Engineering Applications of Artificial Intelligence*, 120, 105899. DOI: 10.1016/j.engappai.2023.105899

Ali, G., Mijwil, M. M., Buruga, B. A., Abotaleb, M., & Adamopoulos, I. (2024). A Survey on Artificial Intelligence in Cybersecurity for Smart Agriculture: State-of-the-Art, Cyber Threats, Artificial Intelligence Applications, and Ethical Concerns. *Mesopotamian Journal of Computer Science*, 2024, 71–121. DOI: 10.58496/MJCSC/2024/007

Angon, P. B., Mondal, S., Jahan, I., Datto, M., Antu, U. B., Ayshi, F. J., & Islam, M. (2023). S. (2023). Integrated Pest Management (IPM) in Agriculture and Its Role in Maintaining Ecological Balance and Biodiversity. *Advances in Agriculture*, 2023, 1–19. DOI: 10.1155/2023/5546373

Balafoutis, A., Beck, B., Fountas, S., Vangeyte, J., Wal, T., Soto, I., Gómez-Barbero, M., Barnes, A., & Eory, V. (2017a). Precision Agriculture Technologies Positively Contributing to GHG Emissions Mitigation, Farm Productivity and Economics. *Sustainability (Basel)*, 9(8), 1339. DOI: 10.3390/su9081339

Balafoutis, A., Beck, B., Fountas, S., Vangeyte, J., Wal, T., Soto, I., Gómez-Barbero, M., Barnes, A., & Eory, V. (2017b). Precision Agriculture Technologies Positively Contributing to GHG Emissions Mitigation, Farm Productivity and Economics. *Sustainability (Basel)*, 9(8), 1339. DOI: 10.3390/su9081339

Balkrishna, A., Pathak, R., Kumar, S., Arya, V., & Singh, S. K. (2023). A comprehensive analysis of the advances in Indian Digital Agricultural architecture. *Smart Agricultural Technology*, 5, 100318. DOI: 10.1016/j.atech.2023.100318

Barbedo, J. G. A. (2019). A review on the use of unmanned aerial vehicles and imaging sensors for monitoring and assessing plant stresses. *Drones (Basel)*, 3(2), 40. DOI: 10.3390/drones3020040

Barbosa Júnior, M. R., Moreira, B. R. D. A., Carreira, V. D. S., Brito Filho, A. L. D., Trentin, C., Souza, F. L. P. D., Tedesco, D., Setiyono, T., Flores, J. P., Ampatzidis, Y., Silva, R. P. D., & Shiratsuchi, L. S. (2024). Precision agriculture in the United States: A comprehensive meta-review inspiring further research, innovation, and adoption. *Computers and Electronics in Agriculture*, 221, 108993. DOI: 10.1016/j.compag.2024.108993

Barrile, V., Simonetti, S., Citroni, R., Fotia, A., & Bilotta, G. (2022). Experimenting agriculture 4.0 with sensors: A data fusion approach between remote sensing, UAVs and self-driving tractors. *Sensors (Basel)*, 22(20), 7910. DOI: 10.3390/s22207910 PMID: 36298261

Begna, T. (2020). Major challenging constraints to crop production farming system and possible breeding to overcome the constraints. [IJRSAS]. *International Journal of Research Studies in Agricultural Sciences*, 6(7), 27–46.

Bhunia, S., Karmakar, S., Bhattacharjee, S., Roy, K., Kanthal, S., Pramanick, M., Baishya, A., & Mandal, B. (2021). Optimization of energy consumption using data envelopment analysis (DEA) in rice-wheat-green gram cropping system under conservation tillage practices. *Energy*, 236, 121499. DOI: 10.1016/j.energy.2021.121499

Bongiovanni, R., & Lowenberg-Deboer, J. (2004). Precision Agriculture and Sustainability. *Precision Agriculture*, 5(4), 359–387. DOI: 10.1023/B:PRAG.0000040806.39604.aa

Dhananjayan, V., Jayanthi, P., Jayakumar, S., & Ravichandran, B. (2020). Agrochemicals Impact on Ecosystem and Bio-monitoring. In Kumar, S., Meena, R. S., & Jhariya, M. K. (Eds.), *Resources Use Efficiency in Agriculture* (pp. 349–388). Springer Singapore., DOI: 10.1007/978-981-15-6953-1_11

Dhillon, R., & Moncur, Q. (2023). Small-Scale Farming: A Review of Challenges and Potential Opportunities Offered by Technological Advancements. *Sustainability (Basel)*, 15(21), 15478. DOI: 10.3390/su152115478

Duff, H., Hegedus, P., Loewen, S., Bass, T., & Maxwell, B. (2021). Precision Agroecology. *Sustainability (Basel)*, 14(1), 106. DOI: 10.3390/su14010106

Edwards, P. C. (2020). The Beginnings of Agriculture. In Hollander, D., & Howe, T. (Eds.), *A Companion to Ancient Agriculture* (1st ed., pp. 119–148). Wiley., DOI: 10.1002/9781118970959.ch7

Farooqui, N. A. Haleem, Mohd., Khan, W., & Ishrat, M. (2024). Precision Agriculture and Predictive Analytics: Enhancing Agricultural Efficiency and Yield. In N. Singh, S. Birla, M. D. Ansari, & N. K. Shukla (Eds.), *Intelligent Techniques for Predictive Data Analytics* (1st ed., pp. 171–188). Wiley. DOI: 10.1002/9781394227990.ch9

Fathi, M., Haghi Kashani, M., Jameii, S. M., & Mahdipour, E. (2022). Big data analytics in weather forecasting: A systematic review. *Archives of Computational Methods in Engineering*, 29(2), 1247–1275. DOI: 10.1007/s11831-021-09616-4

Foster, L., Szilagyi, K., Wairegi, A., Oguamanam, C., & de Beer, J. (2023). Smart farming and artificial intelligence in East Africa: Addressing indigeneity, plants, and gender. *Smart Agricultural Technology*, 3, 100132. DOI: 10.1016/j.atech.2022.100132

Fuentes-Peñailillo, F., Gutter, K., Vega, R., & Silva, G. C. (2024). Transformative Technologies in Digital Agriculture: Leveraging Internet of Things, Remote Sensing, and Artificial Intelligence for Smart Crop Management. *Journal of Sensor and Actuator Networks*, 13(4), 39. DOI: 10.3390/jsan13040039

Gao, F., & Zhang, X. (2021). Mapping Crop Phenology in Near Real-Time Using Satellite Remote Sensing: Challenges and Opportunities. *Yaogan Xuebao*, 2021/8379391, 2021/8379391. Advance online publication. DOI: 10.34133/2021/8379391

Garillos-Manliguez, C. A., & Chiang, J. Y. (2021). Multimodal deep learning and visible-light and hyperspectral imaging for fruit maturity estimation. *Sensors (Basel)*, 21(4), 1288. DOI: 10.3390/s21041288 PMID: 33670232

Guan, Y. (2023). *Development of UAS and Geophysical Techniques for Agriculture and Environmental Site Characterization* [PhD Thesis, Missouri University of Science and Technology]. https://search.proquest.com/openview/311861e9f4884b5d27fb61399e5ff507/1?pq-origsite=gscholar&cbl=18750&diss=y

Gyarmati, G., & Mizik, T. (2020). The present and future of the precision agriculture. *2020 IEEE 15th International Conference of System of Systems Engineering (SoSE)*, 593–596. DOI: 10.1109/SoSE50414.2020.9130481

Hashim, H. Q., & Sayl, K. N. (2021). Detection of suitable sites for rainwater harvesting planning in an arid region using geographic information system. *Applied Geomatics*, 13(2), 235–248. DOI: 10.1007/s12518-020-00342-3

Hassan, M., Kowalska, A., & Ashraf, H. (2023). Advances in deep learning algorithms for agricultural monitoring and management. *Applied Research in Artificial Intelligence and Cloud Computing*, 6(1), 68–88.

Hoque, A., Padhiary, M., Prasad, G., & Kumar, K. (2025). Real-Time Data Processing in Agricultural Robotics. In Bora, D. J., & Bania, R. K. (Eds.), *Computer Vision Techniques for Agricultural Advancements* (pp. 431–468). IGI Global Scientific Publishing.

Iddio, E., Wang, L., Thomas, Y., McMorrow, G., & Denzer, A. (2020). Energy efficient operation and modeling for greenhouses: A literature review. *Renewable & Sustainable Energy Reviews*, 117, 109480. DOI: 10.1016/j.rser.2019.109480

Inoue, Y. (2020). Satellite- and drone-based remote sensing of crops and soils for smart farming – a review. *Soil Science and Plant Nutrition*, 66(6), 798–810. DOI: 10.1080/00380768.2020.1738899

Janssen, S. J. C., Porter, C. H., Moore, A. D., Athanasiadis, I. N., Foster, I., Jones, J. W., & Antle, J. M. (2017). Towards a new generation of agricultural system data, models and knowledge products: Information and communication technology. *Agricultural Systems*, 155, 200–212. DOI: 10.1016/j.agsy.2016.09.017 PMID: 28701813

Jararweh, Y., Fatima, S., Jarrah, M., & AlZu'bi, S. (2023). Smart and sustainable agriculture: Fundamentals, enabling technologies, and future directions. *Computers & Electrical Engineering*, 110, 108799. DOI: 10.1016/j.compeleceng.2023.108799

Jawad, Z. N., & Balázs, V. (2024). Machine learning-driven optimization of enterprise resource planning (ERP) systems: A comprehensive review. *Beni-Suef University Journal of Basic and Applied Sciences*, 13(1), 4. DOI: 10.1186/s43088-023-00460-y

Khan, A., Vibhute, A. D., Mali, S., & Patil, C. H. (2022). A systematic review on hyperspectral imaging technology with a machine and deep learning methodology for agricultural applications. *Ecological Informatics*, 69, 101678. DOI: 10.1016/j.ecoinf.2022.101678

Kharel, T. P., Ashworth, A. J., Owens, P. R., & Buser, M. (2020). Spatially and temporally disparate data in systems agriculture: Issues and prospective solutions. *Agronomy Journal*, 112(5), 4498–4510. DOI: 10.1002/agj2.20285

Kliestik, T., Nica, E., Durana, P., & Popescu, G. H. (2023). Artificial intelligence-based predictive maintenance, time-sensitive networking, and big data-driven algorithmic decision-making in the economics of Industrial Internet of Things. *Oeconomia Copernicana*, 14(4), 1097–1138. DOI: 10.24136/oc.2023.033

Lakhiar, I. A., Yan, H., Zhang, C., Wang, G., He, B., Hao, B., Han, Y., Wang, B., Bao, R., Syed, T. N., Chauhdary, J. N., & Rakibuzzaman, M. (2024). A Review of Precision Irrigation Water-Saving Technology under Changing Climate for Enhancing Water Use Efficiency, Crop Yield, and Environmental Footprints. *Agriculture*, 14(7), 1141. DOI: 10.3390/agriculture14071141

Landers, J. N., De Freitas, P. L., De Oliveira, M. C., Da Silva Neto, S. P., Ralisch, R., & Kueneman, E. A. (2021). Next Steps for Conservation Agriculture. *Agronomy (Basel)*, 11(12), 2496. DOI: 10.3390/agronomy11122496

Linaza, M. T., Posada, J., Bund, J., Eisert, P., Quartulli, M., Döllner, J., Pagani, A. G., Olaizola, I., Barriguinha, A., Moysiadis, T., & Lucat, L. (2021). Data-Driven Artificial Intelligence Applications for Sustainable Precision Agriculture. *Agronomy (Basel)*, 11(6), 1227. DOI: 10.3390/agronomy11061227

Loures, L., Chamizo, A., Ferreira, P., Loures, A., Castanho, R., & Panagopoulos, T. (2020). Assessing the Effectiveness of Precision Agriculture Management Systems in Mediterranean Small Farms. *Sustainability (Basel)*, 12(9), 3765. DOI: 10.3390/su12093765

Lowenberg-DeBoer, J., Franklin, K., Behrendt, K., & Godwin, R. (2021). Economics of autonomous equipment for arable farms. *Precision Agriculture*, 22(6), 1992–2006. DOI: 10.1007/s11119-021-09822-x PMID: 34075304

Lowenberg-DeBoer, J., Huang, I. Y., Grigoriadis, V., & Blackmore, S. (2020). Economics of robots and automation in field crop production. *Precision Agriculture*, 21(2), 278–299. DOI: 10.1007/s11119-019-09667-5

Lu, P., Jiang, B., & Weiner, J. (2020). Crop spatial uniformity, yield and weed suppression. *Advances in Agronomy*, 161, 117–178. DOI: 10.1016/bs.agron.2019.12.003

Maestrini, B., & Basso, B. (2018). Drivers of within-field spatial and temporal variability of crop yield across the US Midwest. *Scientific Reports*, 8(1), 14833. DOI: 10.1038/s41598-018-32779-3 PMID: 30287846

Maitra, A., & Damle, M. (2024). Revolutionizing Plant Health Management with Technological Digital Transformation to Enhance Disease Control & Fortifying Plant Resilience. *2024 3rd International Conference for Innovation in Technology (INOCON)*, 1–8. DOI: 10.1109/INOCON60754.2024.10511728

Manasa, S., Anusha, K. N., Sharma, S., Thakur, A., & Sood, A. (2023). An Overview of Precision Farming. *International Journal of Environment and Climate Change*, 13(12), 441–456. DOI: 10.9734/ijecc/2023/v13i123701

Maraveas, C. (2022). Incorporating Artificial Intelligence Technology in Smart Greenhouses: Current State of the Art. *Applied Sciences (Basel, Switzerland)*, 13(1), 14. DOI: 10.3390/app13010014

Misra, N. N., Dixit, Y., Al-Mallahi, A., Bhullar, M. S., Upadhyay, R., & Martynenko, A. (2020). IoT, big data, and artificial intelligence in agriculture and food industry. *IEEE Internet of Things Journal*, 9(9), 6305–6324. DOI: 10.1109/JIOT.2020.2998584

Mondal, S., & Palit, D. (2022). Challenges in natural resource management for ecological sustainability. In *Natural Resources Conservation and Advances for Sustainability* (pp. 29–59). Elsevier., https://www.sciencedirect.com/science/article/pii/B9780128229767000041 DOI: 10.1016/B978-0-12-822976-7.00004-1

Morgan, K. T., & Kadyampakeni, D. M. (2020). Nutrition of Florida citrus trees. *University of Florida Institute of Food and Agricultural Sciences: Gainesville, FL, USA, 27.* https://journals.flvc.org/edis/article/download/117225/115342/170141

Mowla, M., Mowla, N., Shah, A. F. M. S., Rabie, K. M., & Shongwe, T. (2023). Internet of Things and Wireless Sensor Networks for Smart Agriculture Applications: A Survey. *IEEE Access : Practical Innovations, Open Solutions*, 11, 145813–145852. DOI: 10.1109/ACCESS.2023.3346299

Nishant, R., Kennedy, M., & Corbett, J. (2020). Artificial intelligence for sustainability: Challenges, opportunities, and a research agenda. *International Journal of Information Management*, 53, 102104. DOI: 10.1016/j.ijinfomgt.2020.102104

Padhiary, M. (2024). Status of Farm Automation, Advances, Trends, and Scope in India. [IJSR]. *International Journal of Science and Research (Raipur, India)*, 13(7), 737–745. DOI: 10.21275/SR24713184513

Padhiary, M., Barbhuiya, J. A., Roy, D., & Roy, P. (2024). 3D Printing Applications in Smart Farming and Food Processing. *Smart Agricultural Technology*, 100553, 100553. Advance online publication. DOI: 10.1016/j.atech.2024.100553

Padhiary, M., & Kumar, R. (2024). Enhancing Agriculture Through AI Vision and Machine Learning: The Evolution of Smart Farming. In Thangam, D. (Ed.), *Advancements in Intelligent Process Automation* (pp. 295–324). IGI Global Scientific Publishing.

Padhiary, M., Kumar, R., & Sethi, L. N. (2024). Navigating the Future of Agriculture: A Comprehensive Review of Automatic All-Terrain Vehicles in Precision Farming. *Journal of The Institution of Engineers (India): Series A, 105*(2). DOI: 10.1007/s40030-024-00816-2

Padhiary, M., Kyndiah, A. K., Kumar, R., & Saha, D. (2024). Exploration of electrode materials for in-situ soil fertilizer concentration measurement by electrochemical method. *International Journal of Advanced Biochemistry Research*, 8(4), 539–544. DOI: 10.33545/26174693.2024.v8.i4g.1011

Padhiary, M., Rani, N., Saha, D., Barbhuiya, J. A., & Sethi, L. N. (2023). Efficient Precision Agriculture with Python-based Raspberry Pi Image Processing for Real-Time Plant Target Identification. *International Journal of Research and Analytical Reviews*, 10(3), 539–545. http://doi.one/10.1729/Journal.35531

Padhiary, M., & Roy, P. (2024). Advancements in Precision Agriculture: Exploring the Role of 3D Printing in Designing All-Terrain Vehicles for Farming Applications. [IJSR]. *International Journal of Science and Research (Raipur, India)*, 13(5), 861–868. DOI: 10.21275/SR24511105508

Padhiary, M., Saha, D., Kumar, R., Sethi, L. N., & Kumar, A. (2024). Enhancing Precision Agriculture: A Comprehensive Review of Machine Learning and AI Vision Applications in All-Terrain Vehicle for Farm Automation. *Smart Agricultural Technology*, 8, 100483. DOI: 10.1016/j.atech.2024.100483

Padhiary, M., Sethi, L. N., & Kumar, A. (2024). Enhancing Hill Farming Efficiency Using Unmanned Agricultural Vehicles: A Comprehensive Review. *Transactions of the Indian National Academy of Engineering : an International Journal of Engineering and Technology*, 9(2), 253–268. DOI: 10.1007/s41403-024-00458-7

Padhiary, M., Tikute, S. V., Saha, D., Barbhuiya, J. A., & Sethi, L. N. (2024). Development of an IOT-Based Semi-Autonomous Vehicle Sprayer. *Agricultural Research*. Advance online publication. DOI: 10.1007/s40003-024-00760-4

Pakeerathan, K. (2023). Image Processing: A Smart Technology for Early Detection of Crop Pests and Diseases. In Pakeerathan, K. (Ed.), *Smart Agriculture for Developing Nations* (pp. 135–150). Springer Nature Singapore., DOI: 10.1007/978-981-19-8738-0_10

Palansooriya, K. N., Dissanayake, P. D., Igalavithana, A. D., Tang, R., Cai, Y., & Chang, S. X. (2023). Converting food waste into soil amendments for improving soil sustainability and crop productivity: A review. *The Science of the Total Environment*, 881, 163311. DOI: 10.1016/j.scitotenv.2023.163311 PMID: 37044338

Paudel, D., Boogaard, H., de Wit, A., Janssen, S., Osinga, S., Pylianidis, C., & Athanasiadis, I. N. (2021). Machine learning for large-scale crop yield forecasting. *Agricultural Systems*, 187, 103016. DOI: 10.1016/j.agsy.2020.103016

Pereira, L. S., Paredes, P., & Jovanovic, N. (2020). Soil water balance models for determining crop water and irrigation requirements and irrigation scheduling focusing on the FAO56 method and the dual Kc approach. *Agricultural Water Management*, 241, 106357. DOI: 10.1016/j.agwat.2020.106357

Raj, E. F. I., Appadurai, M., & Athiappan, K. (2021). Precision Farming in Modern Agriculture. In Choudhury, A., Biswas, A., Singh, T. P., & Ghosh, S. K. (Eds.), *Smart Agriculture Automation Using Advanced Technologies* (pp. 61–87). Springer Singapore., DOI: 10.1007/978-981-16-6124-2_4

Rajak, P., Ganguly, A., Adhikary, S., & Bhattacharya, S. (2023). Internet of Things and smart sensors in agriculture: Scopes and challenges. *Journal of Agriculture and Food Research*, 14, 100776. DOI: 10.1016/j.jafr.2023.100776

Rivero, R. M., Mittler, R., Blumwald, E., & Zandalinas, S. I. (2022). Developing climate-resilient crops: Improving plant tolerance to stress combination. *The Plant Journal*, 109(2), 373–389. DOI: 10.1111/tpj.15483 PMID: 34482588

Ronzoni, M., Accorsi, R., Botti, L., & Manzini, R. (2021). A support-design framework for Cooperative Robots systems in labor-intensive manufacturing processes. *Journal of Manufacturing Systems*, 61, 646–657. DOI: 10.1016/j.jmsy.2021.10.008

Sabir, R. M., Sarwar, A., Shoaib, M., Saleem, A., Alhousain, M. H., Wajid, S. A., Rasul, F., Adnan Shahid, M., Anjum, L., Safdar, M., Muhammad, N. E., Waqas, R. M., Zafar, U., & Raza, A. (2024). Managing Water Resources for Sustainable Agricultural Production. In Kanga, S., Singh, S. K., Shevkani, K., Pathak, V., & Sajan, B. (Eds.), *Transforming Agricultural Management for a Sustainable Future* (pp. 47–74). Springer Nature Switzerland., DOI: 10.1007/978-3-031-63430-7_3

Saha, D., Padhiary, M., Barbhuiya, J. A., Chakraborty, T., & Sethi, L. N. (2023). Development of an IOT based Solenoid Controlled Pressure Regulation System for Precision Sprayer. *International Journal for Research in Applied Science and Engineering Technology*, 11(7), 2210–2216. DOI: 10.22214/ijraset.2023.55103

Saiz-Rubio, V., & Rovira-Más, F. (2020). From smart farming towards agriculture 5.0: A review on crop data management. *Agronomy (Basel)*, 10(2), 207. DOI: 10.3390/agronomy10020207

Saleem, S. R., Zaman, Q. U., Schumann, A. W., & Abbas Naqvi, S. M. Z. (2023). Variable rate technologies. In *Precision Agriculture* (pp. 103–122). Elsevier., DOI: 10.1016/B978-0-443-18953-1.00010-6

Shafi, U., Mumtaz, R., García-Nieto, J., Hassan, S. A., Zaidi, S. A. R., & Iqbal, N. (2019). Precision agriculture techniques and practices: From considerations to applications. *Sensors (Basel)*, 19(17), 3796. DOI: 10.3390/s19173796 PMID: 31480709

Shaikh, T. A., Rasool, T., & Lone, F. R. (2022). Towards leveraging the role of machine learning and artificial intelligence in precision agriculture and smart farming. *Computers and Electronics in Agriculture*, 198, 107119. DOI: 10.1016/j.compag.2022.107119

Sharma, A., Sharma, A., Tselykh, A., Bozhenyuk, A., Choudhury, T., Alomar, M. A., & Sánchez-Chero, M. (2023). Artificial intelligence and internet of things oriented sustainable precision farming: Towards modern agriculture. *Open Life Sciences*, 18(1), 20220713. DOI: 10.1515/biol-2022-0713 PMID: 37854322

Sharma, R., Kamble, S. S., & Gunasekaran, A. (2018). Big GIS analytics framework for agriculture supply chains: A literature review identifying the current trends and future perspectives. *Computers and Electronics in Agriculture*, 155, 103–120. DOI: 10.1016/j.compag.2018.10.001

Sidhu, R. K., Kumar, R., Rana, P. S., & Jat, M. L. (2021). Automation in drip irrigation for enhancing water use efficiency in cereal systems of South Asia: Status and prospects. *Advances in Agronomy*, 167, 247–300. DOI: 10.1016/bs.agron.2021.01.002

Singh, S., & Jain, P. (2022). Applications of Artificial Intelligence for the Development of Sustainable Agriculture. In Kumar, P., Tomar, R. S., Bhat, J. A., Dobriyal, M., & Rani, M. (Eds.), *Agro-biodiversity and Agri-ecosystem Management* (pp. 303–322). Springer Nature Singapore., DOI: 10.1007/978-981-19-0928-3_16

Sishodia, R. P., Ray, R. L., & Singh, S. K. (2020). Applications of Remote Sensing in Precision Agriculture: A Review. *Remote Sensing (Basel)*, 12(19), 3136. DOI: 10.3390/rs12193136

Su, W.-H. (2020). Crop plant signaling for real-time plant identification in smart farm: A systematic review and new concept in artificial intelligence for automated weed control. *Artificial Intelligence in Agriculture*, 4, 262–271. DOI: 10.1016/j.aiia.2020.11.001

Subeesh, A., & Mehta, C. R. (2021a). Automation and digitization of agriculture using artificial intelligence and internet of things. *Artificial Intelligence in Agriculture*, 5, 278–291. DOI: 10.1016/j.aiia.2021.11.004

Subeesh, A., & Mehta, C. R. (2021b). Automation and digitization of agriculture using artificial intelligence and internet of things. *Artificial Intelligence in Agriculture*, 5, 278–291. DOI: 10.1016/j.aiia.2021.11.004

Surendran, U., Raja, P., Jayakumar, M., & Subramoniam, S. R. (2021). Use of efficient water saving techniques for production of rice in India under climate change scenario: A critical review. *Journal of Cleaner Production*, 309, 127272. DOI: 10.1016/j.jclepro.2021.127272

Talaviya, T., Shah, D., Patel, N., Yagnik, H., & Shah, M. (2020). Implementation of artificial intelligence in agriculture for optimisation of irrigation and application of pesticides and herbicides. *Artificial Intelligence in Agriculture*, 4, 58–73. DOI: 10.1016/j.aiia.2020.04.002

Türkoğlu, M., & Hanbay, D. (2019). Plant disease and pest detection using deep learning-based features. *Turkish Journal of Electrical Engineering and Computer Sciences*, 27(3), 1636–1651. DOI: 10.3906/elk-1809-181

Wang, J., & Azam, W. (2024). Natural resource scarcity, fossil fuel energy consumption, and total greenhouse gas emissions in top emitting countries. *Geoscience Frontiers*, 15(2), 101757. DOI: 10.1016/j.gsf.2023.101757

Williams, H., Colombi, T., & Keller, T. (2020). The influence of soil management on soil health: An on-farm study in southern Sweden. *Geoderma*, 360, 114010. DOI: 10.1016/j.geoderma.2019.114010

Zhang, J., Huang, Y., Pu, R., Gonzalez-Moreno, P., Yuan, L., Wu, K., & Huang, W. (2019). Monitoring plant diseases and pests through remote sensing technology: A review. *Computers and Electronics in Agriculture*, 165, 104943. DOI: 10.1016/j.compag.2019.104943

Chapter 10
Innovations and Efficiency in Wastewater Treatment Using Artificial Intelligence

Jaime Aguilar Ortiz
Universidad Politécnica de Pachuca, Mexico

Víctor Manuel Zamudio-García
Universidad Politécnica Metropolitana de Hidalgo, Mexico

ABSTRACT

This document explores the application of artificial intelligence (AI) in optimizing wastewater treatment processes, focusing on key areas such as chemical dosing, aeration control, and sedimentation. The integration of advanced AI techniques, including neural networks, genetic algorithms, and optimization methods like Particle Swarm Optimization (PSO) and Bayesian Optimization, has significantly enhanced the efficiency and reliability of these processes. Through the implementation of machine learning models and real-time control systems, AI has enabled more precise adjustments, leading to reduced operational costs, improved resource management, and better compliance with environmental regulations. The study also highlights the potential for AI-driven innovations to further enhance sustainability in wastewater management, presenting opportunities for future research to expand on these advancements and address emerging challenges in the sector.

DOI: 10.4018/979-8-3693-8074-1.ch010

1. INTRODUCTION

1.1. Context and Relevance of Wastewater Treatment

Wastewater treatment is essential for protecting human health and the environment, involving physical, chemical, and biological processes to remove contaminants. Treated water can be safely discharged or reused for purposes like agricultural irrigation and industrial applications, while sludge can generate biogas (Sonune & Ghate, 2004).

Growing urbanization and environmental challenges have driven advancements in wastewater treatment technologies over the past three decades. Modern plants combine methods to address chemical and biological contamination, ensuring treated water meets public health and environmental standards. These innovations enhance efficiency and promote long-term sustainability (Crini & Lichtfouse, 2018).

1.2. Current Challenges in Wastewater Management

Wastewater management faces challenges in implementing efficient, cost-effective systems to meet environmental regulations and performance expectations. In Japan, energy-intensive treatment processes highlight the need for sustainable, energy-saving solutions adaptable to socioeconomic changes. Sludge management through anaerobic digestion for biogas production offers potential but struggles with cost and scalability (Hosomi, 2016).

Urbanization and population growth strain existing wastewater infrastructure, especially in developing countries with limited financial and technical resources, leading to untreated wastewater discharge. Additionally, emerging pollutants like pharmaceuticals require innovative treatment methods, as traditional plants are ineffective in removing these compounds (Choudhary et al., 2019).

1.3. Chapter Objectives

The key objective is to examine practical applications and advanced process control methodologies in wastewater treatment plants using AI. This includes the implementation of recurrent neural networks (RNN) and other machine learning algorithms to optimize chemical dosing and other critical processes.

2. FUNDAMENTALS OF ARTIFICIAL INTELLIGENCE IN WASTEWATER TREATMENT

2.1. Importance of AI in Wastewater Management and Treatment

The importance of artificial intelligence (AI) in wastewater management and treatment is significant due to its ability to improve efficiency and accuracy in treatment processes. AI enables the analysis and optimization of water treatment systems through the use of predictive models and advanced algorithms, helping to manage variability and uncertainty in water quality. This is crucial to maintaining high treatment standards and minimizing negative environmental impacts (Zhao et al., 2020).

The integration of AI in wastewater treatment plants facilitates automation and real-time control of processes, resulting in more efficient and reliable operation. AI models such as artificial neural networks (ANN), fuzzy logic, and genetic algorithms are used to monitor critical parameters such as biological oxygen demand (BOD) and chemical oxygen demand (COD). These models allow for predicting and dynamically adjusting operating conditions, improving pollutant removal and optimizing resource use (Malviya & Jaspal, 2021).

Artificial intelligence (AI) enhances the efficiency and sustainability of wastewater treatment by optimizing key processes like energy and chemical usage. AI algorithms significantly reduce operating costs, particularly in energy-intensive operations such as aeration, while maintaining high treatment quality. Real-time monitoring and adaptability allow precise adjustments based on wastewater variability, ensuring compliance with environmental standards and improving economic sustainability (Hasan, 2021).

AI also supports treated water reuse by ensuring effluent quality meets agricultural, industrial, and municipal standards, improving contaminant removal accuracy and promoting sustainable water management (Martini & Roni, 2021). As AI advances, its role in achieving cost-effective, efficient, and sustainable wastewater treatment will continue to grow.

Artificial intelligence (AI) is transforming wastewater treatment by driving advancements in efficiency, innovation, and sustainability. Its ability to process vast datasets enables precise modeling, pattern recognition, and optimization of treatment processes, reducing costs, resource consumption, and inefficiencies. AI also powers smart monitoring systems and automated controls for real-time adjustments, ensuring optimal performance under varying conditions. These capabilities accelerate technological innovation, improve environmental compliance, and support sustainable water management solutions (Cairone et al., 2023).

AI enhances emergency response and risk management in wastewater treatment by enabling real-time monitoring and predictive analytics. It detects anomalies, such as contaminant spikes or equipment failures, and triggers immediate interventions, preventing environmental damage and public health risks. AI systems continuously analyze water quality, chemical levels, and operational conditions, learning from past data to improve predictive accuracy over time. This integration strengthens operational resilience and ensures compliance with environmental regulations, even during unforeseen challenges (Wang et al., 2023).

Finally, the use of AI in wastewater management fosters greater transparency and traceability in operations, which is crucial for regulation and compliance with environmental regulations. AI systems can provide detailed and accurate data on treatment performance, facilitating auditing and compliance with regulations. Furthermore, AI's ability to generate automatic reports and detailed analysis simplifies the administrative management of treatment plants (El Fels et al., 2023).

2.2. Brief History and Evolution of AI in the Wastewater Sector

The history of artificial intelligence (AI) in the wastewater sector begins in the 1950s with the development of the first AI systems and machine learning algorithms. Initially, these systems were rudimentary and limited to basic classification and prediction tasks. Over the following decades, growth in computational power and data availability allowed for the development of more sophisticated and accurate models. Beginning in the 1980s, the adoption of artificial neural networks and fuzzy logic algorithms in wastewater treatment enabled significant advances in the ability of treatment systems to handle variability and complexity in input data (Malviya & Jaspal, 2021).

In the 1990s, the introduction of deep learning techniques and genetic algorithms began to transform the ability of wastewater treatment plants to predict and optimize treatment processes. These advancements allowed for greater accuracy in predicting critical parameters such as biological oxygen demand (BOD) and chemical oxygen demand (COD), as well as the removal of specific nutrients and contaminants. Integrating these algorithms with real-time control systems allowed treatment plants to operate more efficiently and with lower operating costs (Kaul et al., 2020).

As the 21st century progressed, the use of AI in wastewater treatment expanded significantly. The evolution of microprocessor technology and the increase in data storage and processing capacity allowed for the development of more complex and powerful AI models. These models were able to handle large volumes of data in real time, improving the efficiency and effectiveness of treatment systems. The use of deep neural networks and machine learning algorithms became common practice in the sector, allowing for continuous improvement in the accuracy and efficiency of treatment systems (Jacobides, Brusoni, Candelon, 2021).

AI integration in wastewater treatment has enhanced operational flexibility and adaptability, enabling real-time responses to fluctuating water quality and critical factors. Technologies like fuzzy logic systems and optimization algorithms adjust parameters such as chemical dosing and aeration levels dynamically, ensuring consistent water treatment even during storms or industrial discharges. In emergencies, AI quickly detects influent quality deviations and adapts operations to prevent failures, maintaining treatment effectiveness. By continuously learning from data, AI improves resilience and ensures efficiency and compliance (Haupt, 2021).

AI integration in wastewater treatment is driven by the goals of sustainability and cost reduction. AI models analyze complex datasets to optimize processes like aeration and chemical dosing, minimizing energy waste and ensuring efficient resource use. These optimizations reduce operational costs, lower environmental footprints, and enable plants to comply with stringent environmental standards. By enhancing efficiency and sustainability, AI supports the broader goal of reducing the environmental impact of wastewater treatment (Dwivedi et al., 2019).

In summary, the evolution of AI in the wastewater sector has been a continuous process of innovation and improvement, driven by advances in technology and collaboration between experts. As technology continues to advance, AI is likely to play an increasingly important role in wastewater management and treatment, providing more effective, efficient and sustainable solutions to water management challenges (Lofrano & Brown, 2010) (See Figure 1).

Figure 1. Fundamentals of AI in wastewater treatment.

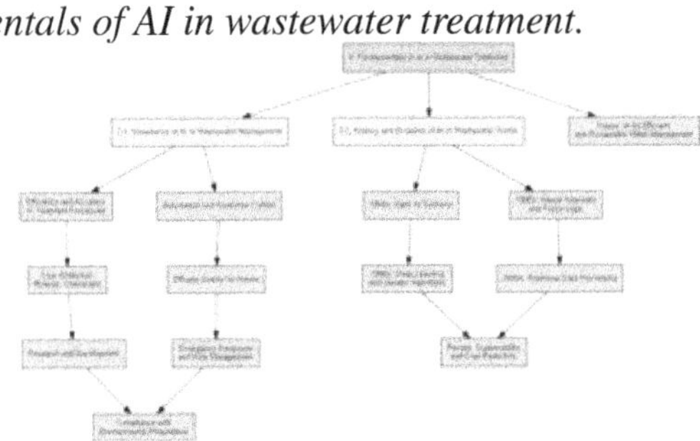

2.3. Concrete Application

Implementation of a wastewater load prediction system using machine learning. Example of how TensorFlow is used to create a predictive model based on historical data.

In the following algorithm, several artificial intelligence techniques are combined to enhance the prediction and optimization of wastewater treatment. First, the data is preprocessed using Machine Learning techniques to scale and split the dataset. Then, an Artificial Neural Network (ANN) model is implemented using TensorFlow to predict the wastewater load based on historical data. Additionally, Fuzzy Logic is integrated to make real-time decisions regarding chemical dosing based on water quality. Finally, the model's hyperparameters are optimized using Genetic Algorithms to find the most efficient and accurate configuration, ensuring more effective predictions and optimized operation of the wastewater treatment system (See Figures 2 and 3).

Figure 2. Concrete application flow.

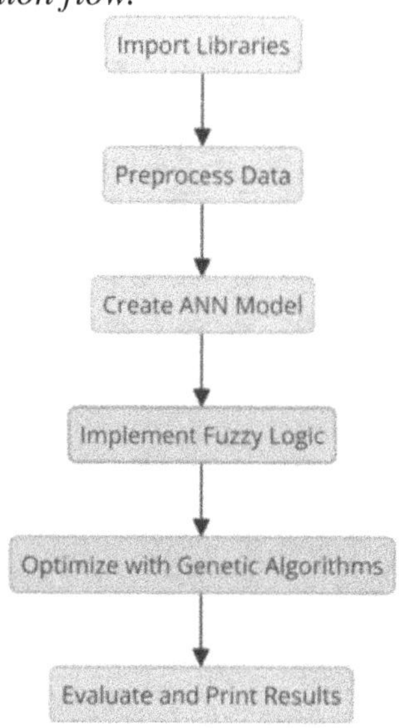

Figure 3. In the QR code you have access to the script.

3. PROCESS CONTROL IN WASTEWATER TREATMENT PLANTS USING AI

3.1 Importance of Process Control in Treatment Plants

Effective process control is fundamental to the successful operation of wastewater treatment plants, as it ensures that the plants run efficiently while meeting stringent environmental regulations. One of the key advantages of advanced process control is its ability to make real-time adjustments to operating conditions, which is essential for maintaining consistency in treatment performance. In a dynamic environment where influent flow and composition can vary due to factors such as weather changes or industrial activities, process control systems help optimize critical parameters like chemical dosing, aeration, and sedimentation. This adaptability allows plants to respond immediately to any deviations from optimal conditions, ensuring that treatment continues efficiently and the quality of the effluent remains within regulatory limits. Modern control systems, particularly those using fuzzy logic and neural networks, have been especially successful in enhancing the precision and reliability of these adjustments. Fuzzy logic systems are capable of handling uncertainty and non-linear relationships, making them well-suited for managing complex biological and chemical processes in wastewater treatment. Neural networks, on the other hand, excel at learning from historical data to predict optimal operating conditions, further improving process stability and resource management. These technologies not only reduce energy and chemical consumption, leading to lower operational costs, but also ensure that treatment plants operate within compliance, reducing the risk of environmental penalties. Process control in wastewater treatment plants is critical to ensuring operational efficiency and complying with environmental regulations. One of the main reasons is that proper process control allows for real-time adjustment and optimization of operating conditions, resulting in more efficient and consistent treatment. The implementation of advanced control systems, such as those based on fuzzy logic and neural networks, has been shown to significantly improve the stability and efficiency of treatment plants, allowing for better resource management and reduced operating costs (Hamilton et al., 2006).

Effective process control in wastewater treatment relies on quickly detecting and responding to influent variability caused by factors like rainfall and industrial discharges. Advanced AI techniques, including statistical models and neural networks, monitor real-time data to adjust aeration, chemical dosing, and other parameters dynamically. This ensures consistent treatment performance and compliance with regulations, even under unpredictable conditions, improving plant resilience and efficiency (West & Mangiameli, 2000).

The selection of an appropriate control architecture is a crucial factor in the effective operation of control systems within wastewater treatment plants. Different sections of a plant often experience varying conditions, such as fluctuating flow rates, chemical compositions, or microbial activity levels, which require tailored responses to maintain optimal treatment efficiency. A distributed control system (DeS) architecture offers significant advantages over centralized control systems by enabling localized decision-making and flexibility. Unlike centralized systems, which rely on a single point of control for all plant operations, distributed control allows each section of the plant to operate semi-independently, with control loops that can be fine-tuned according to the specific needs and conditions of that area. This level of autonomy is particularly beneficial in large or complex wastewater treatment facilities where operating conditions may differ significantly from one section to another. Furthermore, DeS architectures enhance the robustness and resilience of the system by decentralizing data collection and processing. In the event of a failure in one part of the system, the decentralized nature of DeS means that other sections can continue to function without disruption, minimizing downtime and ensuring continuous operation. By facilitating real-time adjustments and localized control, DeS architectures improve the overall efficiency of the plant, reduce response times to changing conditions, and allow for more precise management of resources such as energy and chemicals. The choice of control architecture is essential for the successful implementation of control systems in wastewater treatment plants. Distributed control (DeS) architectures have proven to be more effective compared to centralized control systems, as they allow for greater flexibility and local autonomy. These distributed systems facilitate data collection and the implementation of control loops that can be adjusted based on the specific operating conditions of each section of the plant (Katebi et al., 1998).

The implementation of real-time control systems, particularly those based on fuzzy logic, has revolutionized the way wastewater treatment plants respond to unexpected disturbances in the influent. These disturbances can arise from a variety of sources, such as sudden influxes of organic matter, chemical spills, or industrial discharges that introduce toxic compounds into the system. Traditional control methods often struggle to react quickly and appropriately to these dynamic changes, which can compromise the efficiency and safety of the treatment process. However, fuzzy logic systems offer a more flexible and adaptive approach. By mimicking human decision-making processes, fuzzy logic allows for more nuanced control of the treatment plant's operations. These systems can continuously monitor key parameters, such as pH, biological oxygen demand (BOD), and chemical oxygen demand (COD), and apply predefined fuzzy rules to adjust the process in real time. For instance, when faced with an organic overload or the presence of toxic substances, the system can modulate aeration rates, chemical dosing, or sludge recirculation to stabilize the plant and prevent any dangerous imbalances. The ability to detect and address these issues as they arise ensures that the system remains in a safe operating state, minimizing the risk of process failures. This level of responsiveness not only enhances the reliability of the treatment plant but also ensures that it can maintain compliance with environmental regulations, even under challenging conditions. The use of real-time control systems, such as those based on fuzzy logic, can improve the plant's response to disturbances in the influent. These systems can detect dangerous conditions and adjust process flows to keep the system in a safe state. The implementation of fuzzy rules allows to handle organic overloads and toxic compounds more effectively, resulting in greater process reliability and stability (Müller et al., 1997).

The evaluation and comparison of control strategies are essential for enhancing the performance and efficiency of wastewater treatment plants. These facilities face a wide range of challenges due to the variability of influent water quality, which can fluctuate significantly based on factors such as rainfall, industrial discharge, and seasonal changes. Additionally, the complex biological and biochemical processes involved in wastewater treatment, such as microbial digestion and chemical reactions, add further layers of difficulty to the task of maintaining consistent treatment quality. Given these complexities, the selection and implementation of effective control strategies are critical. However, one of the key challenges has been the lack of standardized criteria for evaluating the effectiveness of different control strategies, making it difficult to compare their performance across different plants or under varying conditions. This gap in evaluation frameworks has historically slowed the adoption of innovative technologies, as plant operators struggled to measure the true impact of new approaches. Fortunately, the development of benchmarking protocols has begun to address this issue by providing a more structured and rigorous method for assessing control strategies. These protocols enable treatment plants to systematically evaluate their performance, not only in terms of regulatory compliance but also in terms of energy consumption, chemical use, and overall operational efficiency. By offering a clearer picture of the strengths and weaknesses of various strategies, benchmarking helps facilitate the adoption of cutting-edge technologies that can improve the reliability and sustainability of wastewater treatment operations. The evaluation and comparison of control strategies is crucial to improving wastewater treatment plant operations. However, this task can be challenging due to the variability of the influent and the complexity of the biological and biochemical phenomena involved. The lack of standard evaluation criteria has been a significant obstacle, but the development of benchmarking protocols has allowed for more rigorous evaluation of control strategies, facilitating the adoption of innovative technologies (Gernaey et al., 2014).

Monitoring and fault detection systems are integral to maintaining the operational efficiency and reliability of wastewater treatment plants. These systems serve as the backbone of process control, ensuring that the plant can consistently meet environmental standards while optimizing resource use. The ability to monitor critical variables in real-time, such as temperature, pH, and dissolved oxygen levels, is crucial for detecting any abnormalities that could compromise the treatment process. These variables directly impact the biological and chemical reactions that occur within the treatment plant, influencing everything from microbial activity to the effectiveness of aeration and chemical dosing. By continuously tracking these parameters, monitoring systems can identify potential issues—such as a drop in oxygen levels or a spike in pH—before they escalate into more serious problems that could affect effluent quality. Additionally, fault detection systems work in tandem with these monitoring tools to diagnose equipment malfunctions, sensor errors, or process inefficiencies, enabling rapid intervention. This capability not only minimizes downtime but also reduces the risk of regulatory violations or environmental harm due to untreated or improperly treated wastewater. With the data provided by these systems, plant operators can make informed, real-time adjustments to maintain optimal performance. In essence, these systems are vital for ensuring that the plant operates efficiently and sustainably, with minimal disruptions to the treatment process. Monitoring and fault detection systems are essential components of process control in wastewater treatment plants. These systems ensure that the data used to make decisions is reliable and accurate, which is vital for efficient plant operation. Monitoring critical variables such as temperature, pH, and dissolved oxygen levels allows for rapid detection and correction of problems before they significantly affect effluent quality (Marais et al., 2020).

The introduction of intelligent control systems in wastewater treatment plants has marked a significant advancement in the field of process control, offering more sophisticated and adaptable solutions than traditional methods. One of the most impactful developments in this area is the use of knowledge-based techniques, such as fuzzy logic controllers and neural networks. These systems are designed to mimic human decision-making processes by incorporating linguistic rules derived from the expertise of plant operators and engineers. By doing so, they can interpret complex, non-linear data and respond to varying conditions in a more nuanced manner than conventional control systems. For instance, fuzzy logic controllers excel at handling the uncertainty and variability inherent in wastewater treatment processes, such as fluctuating influent quality and biological activity. Neural networks, on the other hand, can learn from historical data to predict optimal operating conditions and make real-time adjustments that enhance the efficiency of treatment processes. A key advantage of these intelligent systems is their integration into commercial programmable logic controllers (PLCs), which are already widely used in the industry. This integration provides a robust, scalable, and flexible solution that enhances the overall control architecture of treatment plants. The ability to seamlessly incorporate advanced control strategies into existing PLC frameworks means that treatment plants can achieve greater operational stability, improve resource management, and reduce operational costs without overhauling their current infrastructure. As a result, intelligent control has not only revolutionized the way wastewater treatment plants operate but also paved the way for more resilient and sustainable systems in the future. Intelligent control of wastewater treatment plants has revolutionized the field of advanced control. Knowledge-based techniques, such as fuzzy and neural controllers, use linguistic rules extracted from human experts to improve process control. These systems can be integrated into commercial programmable logic controllers, providing a robust and flexible solution for treatment plant control (Manesis et al., 1998).

The introduction of Internet of Things (IoT)-based systems has enabled real-time monitoring and control of wastewater treatment plants. These systems collect sensor data in real time and upload it to the cloud, where it can be analyzed to detect anomalies and adjust processes automatically. This improves operational efficiency and ensures that the plant can handle variations in the influent effectively (Salem et al., 2022).

Finally, model-based control has proven to be a powerful tool for wastewater treatment plant optimization. Using concepts of model-based predictive control (MPC) and $H\infty$ control theory, these systems can manage multiple variables and operational constraints, optimizing the treatment process to achieve the best possible results in terms of effluent quality and operating costs (Menzi & Steiner, 1995).

In summary, process control in wastewater treatment plants is essential to ensure operational efficiency, comply with environmental regulations, and optimize resource use. The implementation of advanced technologies and intelligent control strategies allows the complexity and variability of these systems to be managed, improving their performance and long-term sustainability (Agarwal et al., 2016).

3.2 Recent Algorithms and Methods

3.2.1 Deep Neural Networks (DNN)

Deep neural networks (DNN) have become an essential tool in wastewater treatment due to their ability to handle and analyze large volumes of complex data. These networks, consisting of multiple layers of interconnected nodes, allow modeling nonlinear relationships and extracting meaningful patterns from historical and real-time data. In the field of wastewater treatment, DNNs are used to predict critical parameters such as chemical oxygen demand (COD) and biological oxygen demand (BOD), which is critical to optimize treatment processes and ensure effluent quality. A study on the application of DNNs for effluent quality prediction in treatment plants showed that these networks can achieve high prediction accuracy, outperforming traditional models in terms of efficiency and accuracy (Jafar et al., 2022).

The use of DNNs in wastewater treatment not only improves the prediction of water quality parameters but also facilitates real-time monitoring and control of processes. Deep convolutional neural networks (CNNs), for example, are effective in analyzing complex spatial and temporal data, which is essential for continuous monitoring of treatment plants. A recent study implemented a CNN-based model for monitoring and prediction of wastewater treatment data, achieving high accuracy in predicting key contaminants such as nitrogen and phosphorus. This approach allowed operators to adjust processes in real time, thereby improving plant operational efficiency and stability (Zhang & Gu, 2022).

DNNs have also proven useful for fault detection and diagnosis in wastewater treatment plants. High-order statistical information-based recurrent neural network (RNN) models (HSI-DRN) are used to identify anomalous patterns and predict potential failures before they occur, allowing proactive intervention and reducing downtime. These models are able to handle non-Gaussian and temporal features in the data, providing a more accurate and robust representation of treatment processes. A study applying HSI-DRN in fault monitoring at treatment plants demonstrated high detection capability with significantly low false alarm and omission rates (Peng et al., 2020).

Figure 4. Process control in wastewater treatment plants using AI.

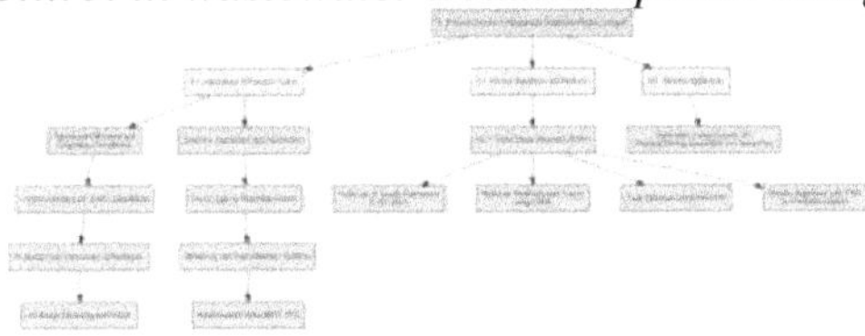

Another innovative approach is the integration of genetic algorithms with deep neural networks to improve the accuracy and reliability of predictive models. One study used a combination of genetic algorithms and deep belief networks (GA-DBN) to predict effluent quality in a wastewater treatment process of the paper industry. This approach allowed reducing the dimensionality of the input variables and improving the network structure, achieving significantly higher prediction accuracy compared to other models. The results indicated that the GA-DBN model can be structured with fewer variables or samples, while maintaining high accuracy and efficiency (Niu et al., 2020) (See Figure 4).

3.5 Specific Application

Development of an automatic control system for chemical dosing using recurrent neural networks (RNN) and TensorFlow, thereby optimizing chemical usage and improving treatment efficiency.

The code implements a Recurrent Neural Network (RNN) model using TensorFlow to predict chemical dosing in a wastewater treatment plant. First, a dataset is loaded and preprocessed, scaling the features and splitting the dataset into training and testing sets. Then, an RNN model is built and trained, which handles temporal sequences of data to predict the amount of chemicals needed for water treatment. After training the model, its performance is evaluated using Mean Absolute Error (MAE). Finally, the model's predictions are used to suggest adjustments in chemical dosing, thereby optimizing the water treatment process in real-time (See Figures 5 and 6).

Figure 5. Specific application flow.

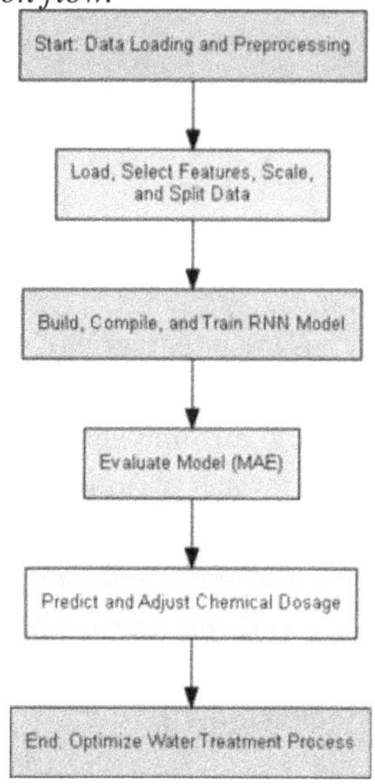

Figure 6. In the QR code you have access to the script.

The development of an automatic control system for chemical dosing in wastewater treatment plants using recurrent neural networks (RNN) and TensorFlow offers an advanced and efficient solution to optimize chemical usage and improve treatment efficiency. The implementation of RNN allows handling temporal sequences of data, which is crucial to adjust chemical dosing based on variations in influent water conditions. A study applying RNN for process control in a biological treatment plant demonstrated that this approach could keep the recycled biomass concentration proportional to the inlet flow, thus improving process stability against periodic disturbances (Baruch et al., 2005).

4. OPTIMIZATION OF KEY PROCESSES IN WASTEWATER TREATMENT

4.1 Optimization of Chemical Dosing

Optimizing chemical dosing in wastewater treatment is crucial to improve operational efficiency and reduce costs. Using intelligent dosing algorithms allows adjusting the amount of chemicals used based on variable influent characteristics such as turbidity, pH, and conductivity. A recent study proposed a dosing model based on LightGBM, which uses inlet water parameters to predict the optimal amount of coagulant needed. This approach proved to be more accurate than traditional methods, resulting in more efficient and economical dosing (Fang et al., 2022).

An innovative approach to optimizing coagulant dosing in industrial wastewater treatment involves the combination of genetic algorithms (GA) and particle swarm optimization (PSO) with regression models. This hybrid technique leverages the strengths of each method to enhance both the accuracy and efficiency of the dosing process. Genetic algorithms, inspired by natural selection, are particularly effective in navigating large and complex optimization spaces, while particle swarm optimization, modeled after the social behavior of animals, efficiently converges on optimal solutions by refining parameters based on the performance of individual particles in a population. When combined with regression models, which offer predictive insights based on historical data, this approach allows for more precise and dynamic adjustments to coagulant dosing. The result is a significant improvement in the treatment process, as the optimal amount of coagulant is applied based on real-time influent conditions, leading to better contaminant removal. One of the major benefits of this method is its ability to reduce the generation of sludge, which is a common byproduct of wastewater treatment and often incurs additional disposal costs. Additionally, by optimizing chemical usage, the overall operational costs are reduced, contributing to a more sustainable and cost-effective treatment process. A study applied this technique in a wastewater treatment plant of a printed circuit board factory, where it achieved remarkable results, including a 27% improvement in copper removal and a 10% reduction in operating costs. This demonstrates the potential of GA and PSO-based optimization methods to revolutionize industrial wastewater treatment by enhancing both environmental and economic outcomes. Another effective approach to optimize coagulant dosing in industrial wastewater treatment combines genetic algorithms and particle swarm optimization with regression models. This method not only improves effluent quality, but also reduces the amount of sludge generated and operating costs. A study implemented this technique in a wastewater treatment plant of a printed circuit board factory, achieving a 27% improvement in copper removal and a 10% reduction in operating costs (Wang et al., 2021).

Response Surface Methodology (RSM) has proven to be a powerful tool in optimizing chemical dosing for wastewater treatment, as it enables a systematic evaluation of the effects and interactions of multiple experimental variables. This technique is particularly valuable in complex treatment processes where several factors, such as reagent dosage, temperature, pH, and influent quality, can influence the outcome. By using RSM, operators can model the relationship between these variables and the treatment efficiency, allowing them to pinpoint the optimal conditions that maximize contaminant removal while minimizing resource use. For instance, RSM can determine the precise amount of coagulant needed based on real-time data, which not only improves the quality of the effluent but also reduces the overall cost of the treatment process. This methodology allows treatment plants to move beyond a trial-and-error approach, offering a more data-driven and efficient way to manage chemical inputs. A practical application of RSM was demonstrated in a study focusing on chemical precipitation in flue gas desulfurization wastewater. The study identified the optimal reagent dosage, leading to significant cost savings and enhanced treatment efficiency compared to conventional methods. This illustrates how RSM can play a crucial role in modern wastewater treatment by improving operational performance, reducing chemical waste, and making the overall process more sustainable. Response surface methodology (RSM) is another technique used to optimize chemical dosing in wastewater treatment. This methodology allows the effects and interactions of multiple experimental variables, such as reagent dosage, temperature, and pH, to be evaluated and optimized. In a study that applied RSM to optimize chemical precipitation in flue gas desulfurization wastewater, the optimal reagent dosage was determined, reducing costs and improving treatment efficiency compared to traditional methods (Kang et al., 2017).

Real-time control of chemical dosing through the use of feedback and feedforward algorithms has proven to be a highly effective method for enhancing the efficiency and precision of wastewater treatment operations. These algorithms work by continuously monitoring the influent water quality and adjusting chemical dosing in real time based on the data collected. Feedback control involves making adjustments to chemical dosing in response to deviations from desired effluent quality, while feedforward control anticipates potential changes in influent conditions and preemptively alters dosing to maintain optimal treatment performance. The integration of these two approaches enables wastewater treatment plants to respond swiftly and accurately to variations in water composition, ensuring that the appropriate amount of chemicals is used at all times. This not only improves the overall quality of the effluent by better targeting the removal of contaminants such as total phosphorus and suspended solids but also significantly reduces chemical usage, leading to substantial cost savings. In one study, the implementation of these control algorithms in a municipal wastewater treatment plant resulted in a 70% reduction in coagulant usage and a 36% decrease in polymer dosage, all while enhancing the effluent quality by reducing key pollutants. This demonstrates the potential of real-time control systems to optimize chemical dosing, improve environmental outcomes, and lower operational costs in wastewater treatment. Real-time control of chemical dosing using feedback and feedforward algorithms has also shown promising results. A study implemented these algorithms in a municipal wastewater treatment plant, achieving a reduction in coagulant dosage by 70% and polymer dosage by 36%, while improving effluent quality by reducing total phosphorus and total suspended solids concentrations (Väänänen et al., 2023).

4.2 Aeration Control

Aeration control in wastewater treatment plants is a critical aspect due to its direct impact on process efficiency and energy consumption. Aeration is used to supply oxygen necessary for biological processes that break down organic contaminants present in wastewater. The amount of dissolved oxygen must be carefully controlled to optimize microbial activity and ensure efficient system operation. A recent study proposes an aeration control method that uses dissolved oxygen meters and the ASM1 activated sludge model, improving system stability and reducing energy consumption (Kawail et al., 2006).

Aeration control optimization not only focuses on the amount of dissolved oxygen, but also on energy efficiency. Aeration is one of the most costly activities in terms of energy consumption in a wastewater treatment plant. Therefore, implementing efficient control strategies can result in significant savings. A study on optimizing the aeration process through a data mining-based approach succeeded in reducing energy consumption without sacrificing treated water quality. The developed models allowed for a clear relationship between input and output variables, demonstrating that large energy savings can be achieved while maintaining water quality within allowable limits (Asadi et al., 2017).

The implementation of model-based predictive control (MPC) systems has also proven effective in optimizing aeration in wastewater treatment plants. These systems use mathematical models to predict system behavior and adjust aeration operations accordingly. One study implemented predictive control based on stochastic differential equations and genetic optimization, prioritizing aeration during periods with low electricity prices. This approach not only reduced energy costs, but also promoted the use of greener electricity, demonstrating the viability of this method with real plant data and electricity market prices (Stentoft et al., 2019).

Another effective strategy for aeration control is the combination of dissolved oxygen (DO) and ammonium (NH4)-based control. This dual control strategy allows for more precise adjustment of aeration, ensuring that oxygen requirements for biological oxidation of pollutants are met without exceeding necessary levels. A study evaluated different control strategies in wastewater treatment plants, including cascade DO control, ammonium-based control (ABAC), and a combination of ABAC with nitrate control and sludge recirculation. The results showed that these strategies not only improved nutrient removal, but also achieved savings of up to 43% in aeration costs (Várhelyi et al., 2018).

4.3 Sedimentation and Clarification

Sedimentation and clarification are essential processes in wastewater treatment, as they allow the separation of suspended solids and the clarification of treated water. These processes are crucial to ensure that the effluent complies with environmental regulations and to improve the efficiency of subsequent treatment stages. Sedimentation takes place in primary sedimentation tanks, where solids settle by gravity. A recent study showed that circular clarifiers are more efficient in removing suspended solids compared to rectangular clarifiers, highlighting the importance of clarifier design in process efficiency (Patel et al., 2021).

Optimizing the design and operation of clarifiers can significantly improve wastewater treatment efficiency. The implementation of computational fluid dynamics (CFD) models has made it possible to study and improve the performance of industrial clarifiers. A study used a CFD model to identify design issues in a clarifier and proposed geometric modifications, such as incorporating a submerged skirt and internal baffles, which were successful in reducing the concentration of suspended solids in the effluent by over 80% (Das et al., 2016).

The use of advanced clarification techniques, such as dissolved air flotation (DAF), has proven to be an effective alternative to conventional sedimentation. DAF uses small air bubbles to float suspended solids to the water surface, where they can be easily removed. This method has shown superior efficiency in removing suspended solids and grease compared to traditional sedimentation, especially in cold climates. One study found that DAF can improve effluent quality and reduce operating costs by decreasing organic and solids loading in later stages of treatment (Collin et al., 2020).

Chemical clarification, which involves the addition of coagulants and flocculants to enhance sedimentation, has also been widely studied. The addition of aluminum sulphate as a coagulant in highly contaminated wastewater has been shown to significantly improve the removal of suspended solids and other contaminants. This method not only improves effluent quality, but also optimizes the efficiency of clarification equipment, reducing the necessary clarifier area and, therefore, installation and operation costs (González et al., 2007).

Finally, the implementation of hybrid systems that combine sedimentation and advanced clarification, such as the use of sedimentation tanks with aeration elements, can improve the efficiency of urban wastewater treatment. One study developed and simulated a two-level sedimentation tank with aeration elements that increased the efficiency of sedimentation of suspended solids without the use of coagulants or flocculants. This system proved to be more efficient in clarifying wastewater, reducing the required tank area and improving effluent quality (Novikov et al., 2021).

4.4 AI Applications for Optimizing These Processes

The application of artificial intelligence (AI) in the optimization of wastewater treatment processes is transforming the efficiency and effectiveness of these systems. One of the key areas where AI is making a significant difference is in the optimization of chemical dosing. The implementation of machine learning models, such as neural networks and genetic algorithms, allows dosing needs to be more accurately predicted based on influent characteristics, thereby reducing excessive chemical use and improving effluent quality. One study highlighted how the use of AI-based techniques, specifically artificial neural networks combined with genetic algorithms, optimized coagulant dosing in a wastewater treatment plant, achieving significant improvements in process efficiency (Cairone et al., 2023).

Furthermore, AI is used to improve aeration control in wastewater treatment plants. The use of predictive models based on recurrent neural networks (RNN) and adaptive controllers allows to dynamically adjust aeration levels, optimizing energy consumption and improving the efficiency of biological treatment. An example of this is the implementation of a non-linear predictive control (NMPC) system that uses a self-organizing radial basis function network (SORBFNN) to identify and adjust system states in real time. This approach has been shown to improve the accuracy and efficiency of aeration control, reducing energy consumption and improving effluent quality (Han & Qiao, 2014).

Regarding sedimentation and clarification, AI is applied to optimize the design and operation of clarifiers. By using computational fluid dynamics (CFD) and machine learning techniques, it is possible to improve the geometric design of clarifiers and adjust operating conditions to maximize the removal of suspended solids. A study used CFD in conjunction with optimization algorithms to improve the performance of industrial clarifiers, achieving a significant reduction in the concentration of solids in the effluent (Das et al., 2016).

The application of AI also enables the implementation of advanced control strategies for chemical clarification. By using machine learning algorithms to model and predict the efficiency of different coagulants and flocculants, treatment plants can optimize the dosing of these chemicals, reducing costs and improving effluent quality. One study highlighted the use of artificial intelligence techniques to optimize chemical clarification in an industrial wastewater treatment plant, achieving improved efficiency in contaminant removal and a reduction in chemical use (González et al., 2007).

Finally, AI facilitates the integration of real-time monitoring and control systems, which improves the overall efficiency of wastewater treatment. Implementation of AI-based control systems enables rapid response to variations in influent conditions, continuously optimizing plant operations. A recent study reviewed various applications of AI models in wastewater treatment, highlighting how these technologies have improved process efficiency and reduced operating costs through better monitoring and control (Wang et al., 2023).

4.5 Examples of Optimization Algorithms

4.5.1 Genetic Algorithms

Genetic algorithms (GAs) are a powerful tool for process optimization in wastewater treatment due to their ability to handle complex and multidimensional problems. GAs mimic the process of natural selection, using operators such as mutation, crossover, and selection to find optimal solutions to optimization problems. A pioneering study on the use of GAs for optimization of wastewater treatment systems showed that these algorithms can solve very complex optimization problems, improving both the cost and environmental performance of treatment systems. This approach allows finding optimal solutions that would be difficult to achieve using traditional methods (Boomgaard et al., 2001).

The application of genetic algorithms in the optimization of aeration in wastewater treatment plants has proven to be especially effective. Aeration is an expensive and energy-intensive process, and its optimization can result in significant cost savings. One study used GAs to optimize the aeration sequence in an intermittent treatment plant, finding that this strategy not only improved nitrogen removal efficiency, but also reduced energy consumption by 10% compared to traditional control strategies (Holenda et al., 2007).

4.5.2 Bayesian Optimization

Bayesian optimization is a powerful technique used to optimize objective functions that are expensive to evaluate, a common scenario in wastewater treatment. This technique relies on probabilistic models, such as Gaussian processes, to build a surrogate model of the objective function. This model allows predicting the performance of new configurations and selecting the most promising ones for actual evaluation. In the context of wastewater treatment, Bayesian optimization has been used to tune biological and chemical process parameters, improving both treatment efficiency and effectiveness. A study on the prediction and analysis of wastewater treatment systems using Bayesian networks demonstrated how this technique can handle uncertainty and temporal variations, providing a robust model for effluent quality prediction (Li et al., 2013).

Bayesian optimization has also been applied in the design and optimization of materials used in wastewater treatment. For example, in materials engineering, Bayesian optimization guides the selection of experiments to discover optimal designs with the fewest possible experiments. This approach has been adapted for wastewater treatment, where membrane and coagulant materials can be optimized. A study on Bayesian optimization for materials design highlighted how this technique can significantly accelerate the materials discovery process by optimizing both quantitative and qualitative design variables (Frazier & Wang, 2015).

In the field of process control, Bayesian optimization has been used to dynamically adjust the operating parameters of wastewater treatment plants. One study implemented a dynamic multi-objective predictive control scheme based on Bayesian optimization, achieving simultaneous optimization of multiple performance indices, such as dissolved oxygen concentration and nitrate level. This approach not only improved process efficiency, but also reduced energy consumption and improved effluent quality (Qiao & Zhang, 2018).

4.5.3 Particle Swarm Algorithms (PSO)

Particle swarm algorithms (PSO) are an optimization technique inspired by the collective behavior of animals, such as flocks of birds or schools of fish. In wastewater treatment, PSO has been used to identify optimal parameters and improve the efficiency of various processes. For example, in parameter identification for anaerobic processes, PSO has proven to be an effective tool for solving high-dimensional multimodal optimization problems. One study used PSO for parameter estimation in an anaerobic wastewater treatment process, achieving accurate and efficient identification of critical system parameters (Sendrescu, 2013).

Figure 7. Optimization of key processes in wastewater treatment.

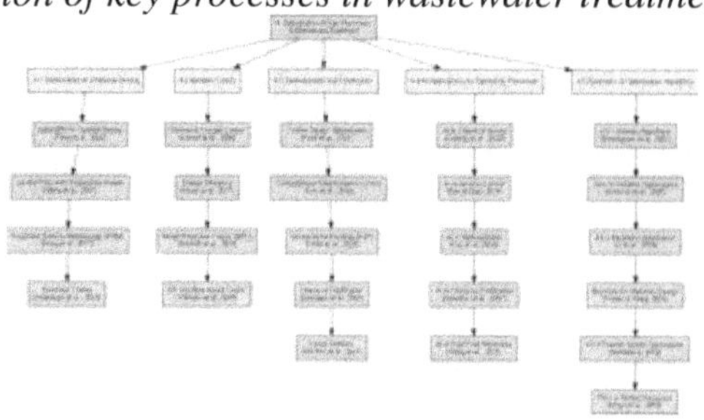

In wastewater treatment with activated sludge, PSO has also been applied for parameter identification of complex models. One study compared the performance of PSO to the Simplex search algorithm for parameter identification in an activated sludge treatment model with alternating phases. The results showed that PSO not only offered higher accuracy in parameter identification but also improved the operational efficiency of the system compared to traditional methods (Khoja et al., 2018).

In addition to parameter identification, PSO has been used for the optimal design of wastewater collection networks. This approach allows handling continuous and discrete variables, optimizing both the hydraulic and biological design of sewer networks. One study used PSO to optimize the design of a wastewater collection network, showing that PSO is a promising technique for solving optimal design problems and comparing favorably with traditional dynamic programming methods (Izquierdo et al., 2008) (See Figure 7).

Figure 8. Specific application flow.

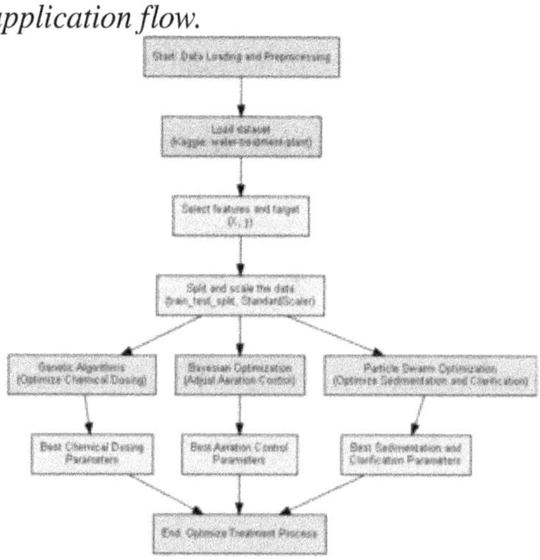

Figure 9. In the QR code you have access to the script.

4.6 Specific Application

Optimization of key processes in a wastewater treatment plant using Genetic Algorithms, Bayesian Optimization, and Particle Swarm Optimization (PSO). It enhances chemical dosing, aeration control, and sedimentation efficiency (See Figures 8 and 9).

The code implements a set of optimization techniques to enhance several key processes in a wastewater treatment plant using a relevant dataset. First, the dataset is loaded and preprocessed by scaling the data and splitting it into training and testing sets. Then, three optimization methods are applied: Genetic Algorithms to optimize chemical dosing, Bayesian Optimization to adjust aeration control parameters, and Particle Swarm Optimization (PSO) to optimize the sedimentation and clarification process by minimizing suspended solids concentration. Each of these methods aims to improve the efficiency of the treatment processes by optimizing critical parameters based on the available data.

5. CONCLUSIONS AND RECOMMENDATIONS

5.1. Main Conclusions

The application of artificial intelligence (AI) in wastewater treatment has demonstrated significant advancements in optimizing key processes such as chemical dosing, aeration control, and sedimentation. By integrating machine learning models like neural networks, genetic algorithms, and optimization techniques such as Particle Swarm Optimization (PSO) and Bayesian optimization, the efficiency of these processes has been significantly enhanced. These AI-driven methods allow for more precise adjustments, resulting in improved resource use, reduced operational costs, and better compliance with environmental regulations.

Furthermore, the incorporation of AI has revolutionized real-time monitoring and control within wastewater treatment plants. Techniques like deep learning and fuzzy logic have enabled the development of sophisticated models that can predict and adjust to varying conditions in real time. This not only improves the overall stability and reliability of treatment processes but also ensures that effluent quality consistently meets regulatory standards. The ability to make data-driven decisions rapidly has proven to be crucial, especially in managing the complexity and variability inherent in wastewater treatment.

Lastly, the advancements in AI technologies have fostered greater innovation and sustainability within the sector. By optimizing energy consumption, chemical usage, and other critical parameters, AI has contributed to making wastewater treatment more environmentally friendly and cost-effective. The ongoing evolution of AI tools is expected to further enhance the capabilities of wastewater treatment plants, leading to continued improvements in process efficiency and environmental impact.

5.2. Recommendations for Future Research

Future research in the application of artificial intelligence (AI) for wastewater treatment should focus on further enhancing the integration of AI with real-time control systems. This could involve the development of more sophisticated algorithms that can handle the dynamic and often unpredictable nature of wastewater influent. By improving the adaptability and responsiveness of AI systems, researchers can help ensure that treatment plants are better equipped to maintain consistent effluent quality, even under varying conditions. Additionally, exploring the potential of hybrid

AI models that combine different machine learning techniques could provide more robust and accurate predictions, leading to more effective process optimizations.

Another important area for future research is the exploration of AI-driven optimization techniques to minimize the environmental footprint of wastewater treatment processes. This includes optimizing energy consumption, reducing the use of harmful chemicals, and maximizing resource recovery, such as biogas production from sludge. By focusing on sustainability, future research can contribute to the development of greener and more efficient wastewater treatment solutions. Researchers should also investigate the scalability of these AI technologies to ensure they can be effectively implemented in treatment plants of varying sizes and capacities.

Lastly, future research should prioritize the development of standardized evaluation criteria and benchmarking protocols for AI applications in wastewater treatment. This would facilitate more rigorous comparisons of different AI approaches, enabling researchers and practitioners to identify the most effective methods for specific treatment challenges. Moreover, expanding the use of AI to monitor and predict emerging contaminants, such as pharmaceuticals and microplastics, could significantly enhance the capability of treatment plants to address new and evolving environmental threats. This research should also consider the integration of Internet of Things (IoT) technologies to enable more comprehensive data collection and real-time monitoring, thereby improving the overall management of wastewater treatment systems.

REFERENCES

Agarwal, M., Singh, K., Dohare, R., & Upadhyaya, S. (2016). Process control and optimization of wastewater treatment plants using simulation softwares: A review. *International Journal of Advanced Technology in Engineering and Exploration*, 3(2), 145–149. DOI: 10.19101/IJATEE.2016.322006

Asadi, A., Verma, A., Yang, K., & Mejabi, B. (2017). Wastewater treatment aeration process optimization: A data mining approach. *Journal of Environmental Management*, 203(Pt 2), 630–639. DOI: 10.1016/j.jenvman.2016.07.047 PMID: 27460213

Bao-lei, Y., De'an, Z., & Jun, Z. (2011). Prediction system of sewage outflow COD based on LS-SVM. *2011 2nd International Conference on Intelligent Control and Information Processing,* 1, 399–402. https://doi.org/DOI: 10.1109/ICICIP.2011.6008273

Baruch, I., Georgieva, P., Barrera-Cortés, J., & Azevedo, S. (2005). Adaptive recurrent neural network control of biological wastewater treatment. *International Journal of Intelligent Systems*, 20(2), 173–193. Advance online publication. DOI: 10.1002/int.20061

Boomgaard, M., Langeveld, J., & Clemens, F. (2001). Wastewater system optimization using genetic algorithms. *Proceedings of the World Environmental and Water Resources Congress,* 1–9. https://doi.org/DOI: 10.1061/40569(2001)498

Cairone, S., Hasan, S., Zarra, T., Belgiorno, V., & Naddeo, V. (2023). Optimization of wastewater treatment by integration of artificial intelligence techniques: Recent progress and future perspectives. *Global NEST International Conference on Environmental Science & Technology.*https://doi.org/DOI: 10.30955/gnc2023.00353

Choudhary, M., Peter, C., Shukla, S., Govender, P., Joshi, G., & Wang, R. (2019). Environmental issues: A challenge for wastewater treatment. *Green Materials for Wastewater Treatment,* 1–13. https://doi.org/DOI: 10.1016/j.scitotenv.2020.138415 PMID: 32348944

Collin, T., Cunningham, R., Asghar, M., Villa, R., MacAdam, J., & Jefferson, B. (2020). Assessing the potential of enhanced primary clarification to manage fats, oils, and grease (FOG) at wastewater treatment works. *The Science of the Total Environment*, 728, 138415. DOI: 10.1016/j.scitotenv.2020.138415 PMID: 32348944

Crini, G., & Lichtfouse, E. (2018). Wastewater treatment: An overview. *Environmental Chemistry for a Sustainable World*, 18, 1–21. DOI: 10.1007/978-3-319-92111-2_1

Das, S., Bai, H., Wu, C., Kao, J., Barney, B., Kidd, M., & Kuettel, M. (2016). Improving the performance of industrial clarifiers using three-dimensional computational fluid dynamics. *Engineering Applications of Computational Fluid Mechanics*, 10(1), 130–144. DOI: 10.1080/19942060.2015.1121518

Dwivedi, Y., Hughes, L., Ismagilova, E., Aarts, G., Coombs, C., & Crick, T.. (2019). Artificial intelligence (AI): Multidisciplinary perspectives on emerging challenges, opportunities, and agenda for research, practice, and policy. *International Journal of Information Management*. Advance online publication. DOI: 10.1016/j.ijinfomgt.2019.08.002

Fang, X., Zhai, Z., Zang, J., & Zhu, Y. (2022). An intelligent dosing algorithm model for wastewater treatment plants. *Journal of Physics: Conference Series*, 2224(1), 1–6. DOI: 10.1088/1742-6596/2224/1/012027

Fels, A., Mandi, L., Kammoun, A., Ouazzani, N., Monga, O., & Hbid, M. (2023). Artificial intelligence and wastewater treatment: A global scientific perspective through text mining. *Water (Basel)*, 15(19), 3487. DOI: 10.3390/w15193487

Gernaey, K., Jeppsson, U., Vanrolleghem, P., & Copp, J. (2014). Benchmarking of control strategies for wastewater treatment plants. https://doi.org/DOI: 10.3390/w15193487

González, T., Domínguez, J. R., Beltrán-Heredia, J., García, H. M., & Sanchez-Lavado, F. (2007). Aluminium sulfate as coagulant for highly polluted cork processing wastewater: Evaluation of settleability parameters and design of a clarifier-thickener unit. *Journal of Hazardous Materials*, 148(1–2), 6–14. DOI: 10.1016/j.jhazmat.2006.09.049 PMID: 17049726

Hamilton, R., Braun, B., Dare, R., Koopman, B., & Svoronos, S. (2006). Control issues and challenges in wastewater treatment plants. *IEEE Control Systems*, 26(4), 63–69. DOI: 10.1109/MCS.2006.1657877

Hasan, M. (2021). An emergent addition for the optimal systemization of wastewater utilization plants using artificial intelligence. *Water Science and Technology: A Journal of the International Association on Water Pollution Research, 84*(10–11), 2805–2817. https://doi.org/DOI: 10.3390/w15193487

Haupt, S. (2021). The history and practice of AI in the environmental sciences. *Bulletin of the American Meteorological Society*. Advance online publication. DOI: 10.1175/BAMS-D-20-0234.1

Jafar, R., Awad, A., Jafar, K., & Shahrour, I. (2022). Predicting effluent quality in full-scale wastewater treatment plants using shallow and deep artificial neural networks. *Sustainability (Basel)*, 14(23), 15598. Advance online publication. DOI: 10.3390/su142315598

Wang, Y., Cheng, Y., Liu, H., Guo, Q., Dai, C., Zhao, M., & Liu, D. (2023). A review on applications of artificial intelligence in wastewater treatment. *Sustainability (Basel)*, 15(18), 13557. Advance online publication. DOI: 10.3390/su151813557

West, D., & Mangiameli, P. (2000). Identifying process conditions in an urban wastewater treatment plant. *International Journal of Operations & Production Management*, 20(5), 573–590. DOI: 10.1108/01443570010318931

Zhao, L., Dai, T., Qiao, Z., Sun, P., Hao, J., & Yang, Y. (2020). Application of artificial intelligence to wastewater treatment: A bibliometric analysis and systematic review of technology, economy, management, and wastewater reuse. *Process Safety and Environmental Protection*, 133, 169–182. DOI: 10.1016/j.psep.2019.11.014

Chapter 11
An Optimal Method of Assessing the Quality of Water Using Integrated AI and Blockchain Methodology

R. Renugadevi
https://orcid.org/0000-0001-7462-9205
Saveetha Engineering College, India

ABSTRACT

Water quality deterioration has become increasingly severe recently due to various factors. Pollution, climate change, and population growth all contribute to this issue. Contaminants such as microorganisms, heavy metals, and excessive nitrogen and phosphorous disrupt water pH levels, posing significant health risks. Despite advancements in IoT for maintaining pH balance through chlorine and fluoride addition after disinfection, security threats like distributed denial of service attacks, data manipulation, and session hijacking affect water treatment plant operations. This compromises water safety, leading to health problems and reduced life expectancy. To address these challenges, we propose integrating AI and blockchain technology into water treatment plant management. Our method utilizes a standard dataset on water quality parameters such as pH and total hardness to classify water as potable or non-potable.

DOI: 10.4018/979-8-3693-8074-1.ch011

INTRODUCTION

Water quality degradation has become increasingly critical due to multiple factors in recent years. Pollution, climate change, and population growth all contribute to this issue. Contaminants like microorganisms, heavy metals, and excessive nitrogen and phosphorous disrupt water pH levels, posing significant health risks. Despite advancements in the Internet of Things (IoT) enabling pH balance through chlorine and fluoride addition after disinfection, various security concerns such as distributed denial of service attacks, data manipulation, and session hijacking affect water treatment plant operations. This compromises water safety, leading to health issues and reduced life expectancy. To tackle these challenges, we propose integrating artificial intelligence (AI) and blockchain technology into water treatment plant management. Our approach uses a standard water quality dataset featuring parameters like pH and total hardness for binary classification of water as potable or non-potable. AI classifiers such as stochastic gradient descent (SGDC), decision tree (DT), Naive Bayes (NB), K-nearest neighbors (KNN), and logistic regression (LR) are employed. Additionally, an InterPlanetary File System (IPFS)-based public blockchain is implemented to safeguard against data manipulation attacks, securely storing potable water samples in an immutable ledger. Evaluations include metrics like confusion matrix analysis, learning curve assessment, training accuracy, and blockchain scalability. Notably, the DT model performs best, achieving 99.41% accuracy and supporting 120 data transactions for scalability.

The motivation for our proposed approach is driven by several key factors:

1. **Streamlining Water Profiling:** Accurately profiling water is crucial for ensuring operational efficiency and delivering clean water. To simplify and enhance this process while maintaining safety, various approaches such as numerical modeling, IoT-based solutions, and AI-driven automation have been explored and implemented.
2. **Limitations of Numerical Modeling:** Previous studies have predominantly focused on numerical modeling for water treatment applications. However, these approaches often lack automation, which can lead to human errors and biases. Moreover, applying numerical optimization on resource-constrained smart sensors can degrade the performance of smart grid environments. This limitation motivated us to seek more accurate and automated methodologies, particularly leveraging AI.
3. **Machine Learning for Water Quality Forecasting:** Recent research has successfully applied machine learning techniques to forecast water quality with improved accuracy and reduced human biases. However, a significant drawback is the vulnerability of predicted data to tampering attacks. This gap prompted

us to combine machine learning with blockchain technology to ensure data integrity and immutability.

4. **Integration of AI and Blockchain:** Existing approaches often focus on either AI or blockchain individually to enhance security in Water Treatment Plants (WTPs). Our approach emphasizes the synergistic benefits of integrating AI and blockchain technologies. This integration not only enhances security but also improves overall operational efficiency and reliability.
5. **Incorporating IPFS for Enhanced Solutions:** Prior blockchain-based solutions in water treatment have not fully utilized the benefits of the InterPlanetary File System (IPFS). By incorporating IPFS, our solution aims to reduce latency and enhance scalability, crucial factors for deploying secure and efficient systems in WTPs.

In summary, our motivation lies in advancing the accuracy, automation, and security of water profiling and treatment processes by integrating AI and blockchain technologies, while leveraging the benefits of IPFS to achieve robust and scalable solutions.

Related Work

Water serves not only to quench thirst but also functions as a vital component of life. Approximately 60-70% of the human body consists of water, which plays crucial roles such as regulating body temperature and hydrating organ tissues. It facilitates nutrient and oxygen transport through blood circulation, reduces strain on organs like kidneys and liver by aiding in toxin removal, and supports overall bodily functions. However, the benefits of water are contingent upon its quality; impure water can lead to diseases, underscoring the critical importance of potable water in enhancing life expectancy. According to the Composite Water Management Index Report 2018 by the National Institution for Transforming India (NITI Aayog), inadequate access to safe water results in approximately two lakh deaths annually. The report predicts that by 2030, over 600 million people in India—nearly 40% of the population—will face water stress due to various factors, including climate change (Madhusoodhanan et al., 2016).

Addressing the challenges posed by climate change on India's water resources is complex due to interdependencies among energy, land use, water management, and climate factors. Effective policy development requires a holistic approach that recognizes these interconnected dynamics and distinguishes between local human activities and global climate impacts. Access to clean water, sanitation, and hygiene (WASH) is crucial for human health and well-being. Chemical pollution in water poses significant health risks, contributing to diseases such as diarrhoea, which annually causes over a million deaths globally due to poor WASH conditions. Improving WASH services can substantially reduce diarrheal deaths, highlighting the urgent need for interventions to address these public health challenges.

In areas like Thulamela, South Africa, studies reveal significant issues with drinking water quality. While water from its source (a dam) is safe, contamination occurs during storage and distribution, increasing the risk of waterborne diseases. Poor water treatment practices and shortages exacerbate these risks, emphasizing the necessity for stricter regulations and enhanced monitoring of water quality, particularly for trace metals like lead, which may pose cancer risks. Addressing water quality issues is paramount as contaminated water spreads diseases and poses serious health risks, jeopardizing individual well-being and life expectancy (Luvhimbi et al., 2022).

Water quality hinges on numerous factors, such as pH, mineral content (sodium, magnesium, calcium, chloride, potassium, carbonate, sulfate), total dissolved solids (TDS), electrical conductivity (EC), total hardness (TH), and the Water Quality Index (WQI). These elements are critical determinants of water's safety for drinking and its impact on human health. Imbalances in these factors can lead to digestive problems, kidney issues, cardiovascular diseases, and neurological disorders. Ensuring high water quality is essential for public health and life expectancy, as access to clean and safe drinking water is paramount. However, inaccuracies in water quality data reporting can result in consumption of contaminated water, causing gastrointestinal issues, dehydration, and severe long-term health consequences. This issue is particularly acute in areas with limited water resources, compounded by urbanization and pollution challenges in growing cities.

To mitigate these risks, accurate monitoring of water quality and implementation of sustainable practices are crucial. This ensures universal access to clean drinking water, addressing disparities in water quality that affect communities worldwide. The presence of water pollutants poses significant threats to human well-being, natural ecosystems, and biodiversity. Recent studies have focused on microplastics, such as polyethylene particles, exploring their behavior in water systems through energy loss, residence time distribution, and mixing models. Understanding how these pollutants move through urban water systems is vital for developing effective mitigation strategies. Furthermore, comprehensive studies are examining emerging contaminants like carbon nanotubes and graphite oxides, evaluating not only their original forms but also their byproducts and associated risks. This holistic approach aims to identify, measure, and mitigate the impact of these contaminants on water quality and human health.

The study detailed in reference (Rathi et al., 2021) investigated the behavior of microplastics within sewer systems, particularly focusing on manholes, which serve as critical junctions in these networks. Polyethylene particles were used as a representative type of microplastic, and their movement was analyzed in comparison to dissolved substances. The study employed energy loss, residence time distribution, and mixing models to understand how microplastics traverse city water systems. This research is crucial for mitigating contamination risks and enhancing the design and management of wastewater infrastructure. The findings from reference (Noori et al., 2022) indicated that the majority of plastic particles move through the sewer system similar to dissolved substances. However, a portion of these particles becomes trapped in manholes with elevated water levels. The longer these particles remain trapped, the higher the potential for contamination to occur. These insights underscore the importance of addressing how microplastics behave in urban water systems to minimize environmental and health risks effectively.

In their study, researchers (Guzmán-Avalos et al., 2023) have developed an enhanced turbine design tailored for Micro Hydroelectric Plants, specifically targeting energy extraction from Wastewater Treatment Plants (WWTPs). These plants typically feature significant unused low-energy water flow. The innovative design incorporates numerical modeling and a specialized algorithm to optimize the shape of turbine blades and other components. As a result, the propeller turbine can efficiently capture up to 76% of the available energy, presenting a promising avenue for generating renewable energy from WWTPs. In a separate investigation, study (Goodarzi et al., 2020) utilized numerical modeling to explore the impact of water temperature on disinfection processes in chlorine-utilizing tanks. The simulation accounts for substance movement, chlorine degradation, and the influence of temperature on water density and viscosity. Findings from study (Goodarzi et al., 2020) underscore that even minor fluctuations in temperature can markedly affect chlorine levels and the efficacy of disinfection procedures, thereby potentially influencing water treatment operations. The model's accuracy was rigorously validated against real-world data, affirming its reliability in optimizing temperature conditions to enhance water treatment efficiency and mitigate the formation of disinfection byproducts.

Additionally, research (Qiu et al., 2024) delves into the physical-chemical hydrodynamics of activated sludge reactors and their role in nitrogen oxide (N2O) emissions. This study aims to fill knowledge gaps concerning the design of these reactors by scrutinizing variables that impact aeration accuracy and N2O emission forecasts in surface-aerated oxidation ditch-type reactors using computational fluid dynamics simulations. Emphasizing the significance of design, operational conditions, and biokinetic factors, study (Qiu et al., 2024) compares laboratory findings with observations from an actual water treatment facility to better understand and predict N2O emissions. These studies collectively highlight advancements in water treatment technology and renewable energy generation, underscoring the interdisciplinary approaches necessary for optimizing efficiency and sustainability in water resource management.

The authors of reference (Sugumar et al., 2021) introduced an IoT and machine learning-driven solution aimed at automating and cost-effectively managing Water Treatment Plants (WTPs). Their proposed system utilizes the ESP8266 Wi-Fi module and ThingSpeak platform for real-time data monitoring. Biochemical oxygen and chemical oxygen demand sensors are integrated to ensure accurate water quality assessment. Machine learning algorithms enhance the precision of sensor data analysis, optimizing operational efficiency. In a related study, reference (Granata et al., 2017) applied machine learning techniques, including Support Vector Regression (SVR) and Regression Trees, to predict wastewater quality indicators from urban catchment data. SVR outperformed Regression Trees in forecasting total suspended solids, total dissolved solids, and chemical oxygen demand, while both models showed similar accuracy in predicting biochemical oxygen demand. These findings underscore the potential of machine learning in enhancing wastewater treatment plant planning and real-time management, addressing environmental challenges associated with wastewater.

However, both studies (Sugumar et al., 2021) and (Granata et al., 2017) highlight significant security concerns associated with their IoT and machine learning-based approaches. Given the potential vulnerabilities to security attacks in such systems, particularly those handling sensitive water quality data, there is a critical need for robust security measures. To address these security challenges, a proposed approach integrates artificial intelligence (AI) and blockchain technology. This method aims to enhance data integrity and security in WTP operations. The study employs various machine learning algorithms such as Stochastic Gradient Descent classifier (SGDC), Decision Tree (DT), Naive Bayes (NB), K Nearest Neighbours (KNN), and Logistic Regression (LR) to classify water samples as drinkable or non-drinkable based on their quality. Harnessing the capabilities of the Internet of Things (IoT) and machine learning for addressing environmental challenges has indeed gained significant attention in recent years (Kalpana et al., 2024; Ramalingam, 2024; Renugadevi, 2024; Devi et al., 2023; Renugadevi et al., 2023). By leveraging AI and blockchain, this approach not only enhances the reliability and security of data handling but also ensures the safety and integrity of water quality assessments in WTPs. This integrated framework represents a proactive step towards mitigating security risks and ensuring the sustainable management of water resources.

In reference (Jalal & Ezzedine, 2019), the authors introduced a water quality monitoring system that employs machine learning techniques, specifically Decision Trees (DT) and Support Vector Machines (SVM). Their study evaluated the performance of different SVM kernels (linear, fine tree, and medium tree) for predicting water quality parameters. It was observed that linear SVM generally provided comparable precision to decision trees (fine tree and medium tree) across the entire dataset. However, when focusing on a subset (1/4) of the samples, linear SVM exhibited superior precision.

In contrast, Wongburi and Park (Wongburi & Park, 2022) proposed an intelligent scheme for forecasting sludge volume index (SVI) in wastewater treatment plants using recurrent neural networks (RNNs) and explainable AI techniques. Their approach involved analyzing time series data spanning from 1996 to 2020. After assembling and training appropriate datasets, the prediction model achieved a root mean square error (RMSE) of 4.161 and a mean absolute error (MAE) of 3.284 for the first dataset (1996-2000). For the subsequent dataset (2001-2020), which includes data from 2001 to 2020, the model achieved an RMSE of 3.360 and an MAE of 2.156, similar to the third dataset (2010-2020). These studies exemplify the application of advanced machine learning techniques in water quality monitoring and wastewater treatment optimization. Reference (Jalal & Ezzedine, 2019) focuses on real-time monitoring using SVM and DT, highlighting the effectiveness of SVM in specific sample subsets. Conversely, reference (Wongburi & Park, 2022) demonstrates the capability of RNNs and explainable AI in predicting SVI over extended time periods, showcasing improvements in predictive accuracy over successive datasets. These approaches underscore the potential of AI-driven solutions in enhancing operational efficiency and sustainability in water management practices.

The ANN-ELM model, along with the chemical oxygen demand and biological oxygen demand models, achieved accuracies of 82.17%, 85.68%, and 88.31%, respectively, while the deep convolutional neural network with sine cosine algorithm achieved 90.11% accuracy using a dataset of 100 data points. In contrast, the ANN-ELM model demonstrated the highest accuracy at 93.14%. For a dataset of 300 data points, the accuracies were 83.77%, 86.82%, 89.24%, and 91.77% for the ANN-ELM, chemical oxygen demand, biological oxygen demand, and deep convolutional neural network with sine cosine algorithm models, respectively. The ANN-ELM model also exhibited the lowest root mean square error at 30.66% for a dataset of 600 data points, compared to 44.82%, 40.74%, 37.18%, and 32.85% for the ANN-ELM, chemical oxygen demand, biological oxygen demand, and deep convolutional neural network with sine cosine algorithm models, respectively.

The authors of (Banerjee et al., 2022) introduced a machine learning methodology to forecast water contamination, taking into account factors such as the geographical location of water bodies, latitude, longitude, and elevation. They applied multivariable linear regression, support vector regression, decision tree regression, and lasso regression. For temperature prediction, multivariable linear regression achieved the highest performance with an R^2 score of 40.94 during training and 52.51 during testing. Support vector regression excelled in predicting pH, achieving R^2 scores of 92.09 in training and 46.97 in testing. Multivariable linear regression proved most effective for biological oxygen demand, with perfect

R^2 scores of 99.99 in both training and testing phases. Predicting dissolved oxygen, multivariable linear regression yielded the best results with R^2 scores of 69.58 in training and 54.83 in testing. The correlation among data for turbidity, hardness, and alkalinity was too weak to determine the superior model. For chlorides, support vector regression performed the best with R^2 scores of 33.53 in training and 71.188 in testing. In the case of chemical oxygen demand, multivariable linear regression outperformed other models with R^2 scores of 12.01 in training and 9.02 in testing. In another study referenced as (Narendar et al., 2022), artificial neural networks (ANNs) were utilized to optimize wastewater treatment plant efficiency. The ANN models accurately predicted both influential and effluent chemical oxygen demand, achieving accuracies of 92.67% and 89.23% across different models, respectively.

Novelty of the Approach

The novelty of this proposed approach stems from addressing critical gaps in existing literature concerning the security of Water Treatment Plants (WTPs). Current literature has primarily utilized either basic AI modules or blockchain technology separately to enhance security in WTP operations. However, these approaches have not integrated both AI and blockchain technologies to provide a comprehensive and resilient security framework for WTP profiling systems.

In this novel methodology, AI models are employed for solving a binary classification problem to distinguish between potable and non-potable water data. This distinction is crucial because attackers may manipulate smart sensors to introduce hazardous chemicals or pollutants into the water supply, endangering public health. Consumption of such contaminated water can significantly impact human life expectancy. This AI model effectively classifies water potability based on predefined thresholds for each feature. However, there remains a risk of data manipulation attacks on the predicted data from AI models, which raises concerns about data integrity. To mitigate this risk, we incorporate InterPlanetary File System (IPFS)-based blockchain technology. This blockchain securely stores the predicted data in an immutable ledger, ensuring data integrity and resistance against tampering.

Furthermore, integrating IPFS with blockchain enhances the solution's response time and scalability. This improvement is crucial for managing a large volume of data transactions effectively within WTP operations. Our evaluation shows that the Decision Tree (DT) model performs exceptionally well, achieving an accuracy of 99.41% in classifying water as potable or non-potable. The blockchain component of our solution supports scalability up to 35 with 120 data transactions, demonstrating its robustness and suitability for real-world deployment. In contrast, existing approaches have not fully explored the synergistic benefits of integrating AI and blockchain technologies in WTP security. As a result, their solutions may not provide comprehensive end-to-end security or fully exploit the potential to safeguard WTPs from adversaries intent on compromising operational efficiency and public health with contaminated water.

PROPOSED BLOCKCHAIN AND AI-BASED WATER PROFILING FRAMEWORK FOR ENHANCING HUMAN LIFE EXPECTANCY

This proposed approach in figure 1 represents a significant advancement by combining AI for accurate classification and blockchain for secure data storage, thereby offering a robust defense against threats to WTPs and ensuring safe drinking water for the population.

Figure 1. Proposed model architecture

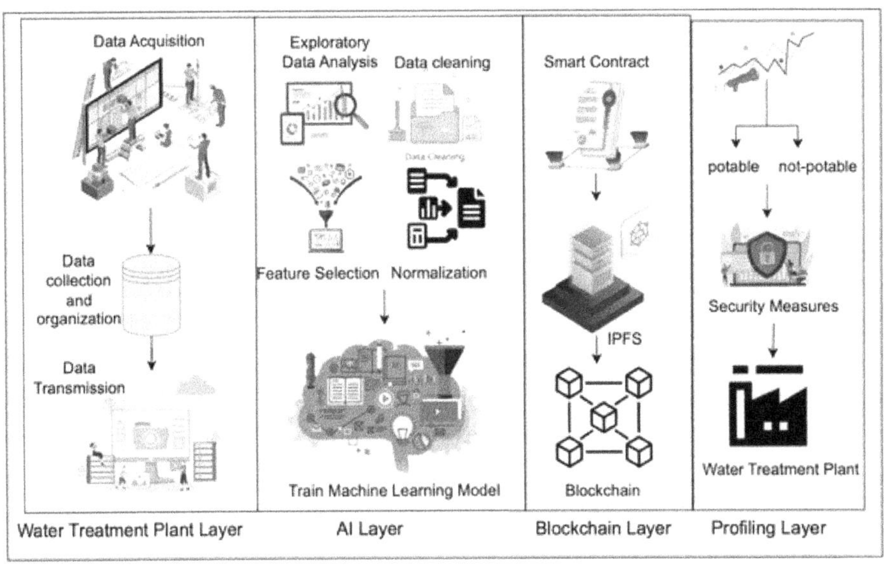

Water Treatment Plant Layer

In this operational layer, we implement an AI-driven classification algorithm aimed at enhancing human life expectancy by accurately predicting water potability within Water Treatment Plants (WTPs). The data is sourced from diverse sensors specifically designed to measure the nutritional composition of water. Each WTP is equipped with a unique array of sensors, denoted as $\{s1, s2, s3, ..., sn\} \in S$, where each sensor si captures water-related data {δ1, δ2, δ3, ..., δn} ∈ I. These data encompass a wide range of parameters crucial for assessing water quality, including pH levels, conductivity, oxidation-reduction potential, dissolved oxygen levels, and other pertinent metrics. Furthermore, this layer integrates blockchain technology to uphold the security and integrity of the collected data. Blockchain ensures that the data remains tamper-proof and immutable throughout its lifecycle, safeguarding against unauthorized modifications or breaches. This combination of AI-driven analysis and blockchain technology establishes a robust framework for ensuring the reliability and safety of water quality assessments within WTPs.

AI Layer

Dataset Description

This process involves merging data from various sensors to form a comprehensive dataset detailing water characteristics. Each entry in the dataset corresponds to specific data points, organized into columns representing different aspects of water quality. This structured CSV format facilitates further analysis, serving as the basis for predicting water potability using various machine learning techniques. Following the Exploratory Data Analysis (EDA), the subsequent step involved training the dataset. Data is partitioned into training and testing sets with an 80:20 ratio, respectively. Subsequently, we applied several AI algorithms to the training data, including Logistic Regression (LR), Bayesian Network (BN), K-Nearest Neighbors (KNN), Stochastic Gradient Descent Classifier (SGDC), and Decision Tree (DT). For each algorithm, we utilized grid-search to fine-tune hyperparameters systematically. Grid search involves exhaustively testing predefined sets of hyperparameter values to identify combinations that optimize specific performance metrics, such as accuracy. For instance, in Logistic Regression, hyperparameters like regularization parameter (C), penalty, random state, and maximum iterations (max_iter) were adjusted. Different sets of parameters were tailored for each algorithm to enhance their effectiveness in classifying water potability. This methodological approach was instrumental in maximizing the predictive performance of the models by selecting optimal hyperparameter configurations tailored to the task of water potability classification.

Data Preprocessing

Exploratory Data Analysis (EDA) techniques are utilized to prepare collected data and gain insights into its characteristics. EDA involves employing various methods to comprehend the structure, patterns, and relationships within the dataset. In the current study, EDA encompasses statistical analysis, data visualization, and feature engineering. Statistical analysis plays a crucial role in understanding the characteristics of the collected data, particularly in predicting water potability. Given the classification nature of the problem, a chi-square test for independence was initially conducted to assess significant associations between variables. This test provided chi-squared statistics and corresponding p-values. Additionally, Cramer's V was computed to gauge the strength of these associations more comprehensively. Cramer's V considers the chi-squared statistic, the total number of observations, and the dimension of the contingency table. Notably, the attribute showing the highest association with potability was the water quality index (WQI), approaching a value close to 1. Data visualization techniques were employed to gain visual insights into the dataset. Histograms and density plots were utilized to visualize the distribution, shape, and variability of data attributes. Scatter plots were also employed to explore relationships between numerical variables, providing a visual understanding of correlations and trends within the dataset.

Feature interactions were investigated through various methods, including creating combinations of features such as pH multiplied by conductivity, to capture potential interactions that could influence water potability. Additionally, domain-specific features such as water hardness and alkalinity were computed based on established relationships in the field. These feature engineering steps were crucial in enabling our model to utilize attributes effectively, resulting in enhanced performance and more accurate predictions. By incorporating these engineered features, the model was able to better discern complex relationships within the dataset, thereby improving its overall predictive capability for water potability. The entire AI layer architecture representing the proposed system is shown in figure 2.

Figure 2. AI layer implementation

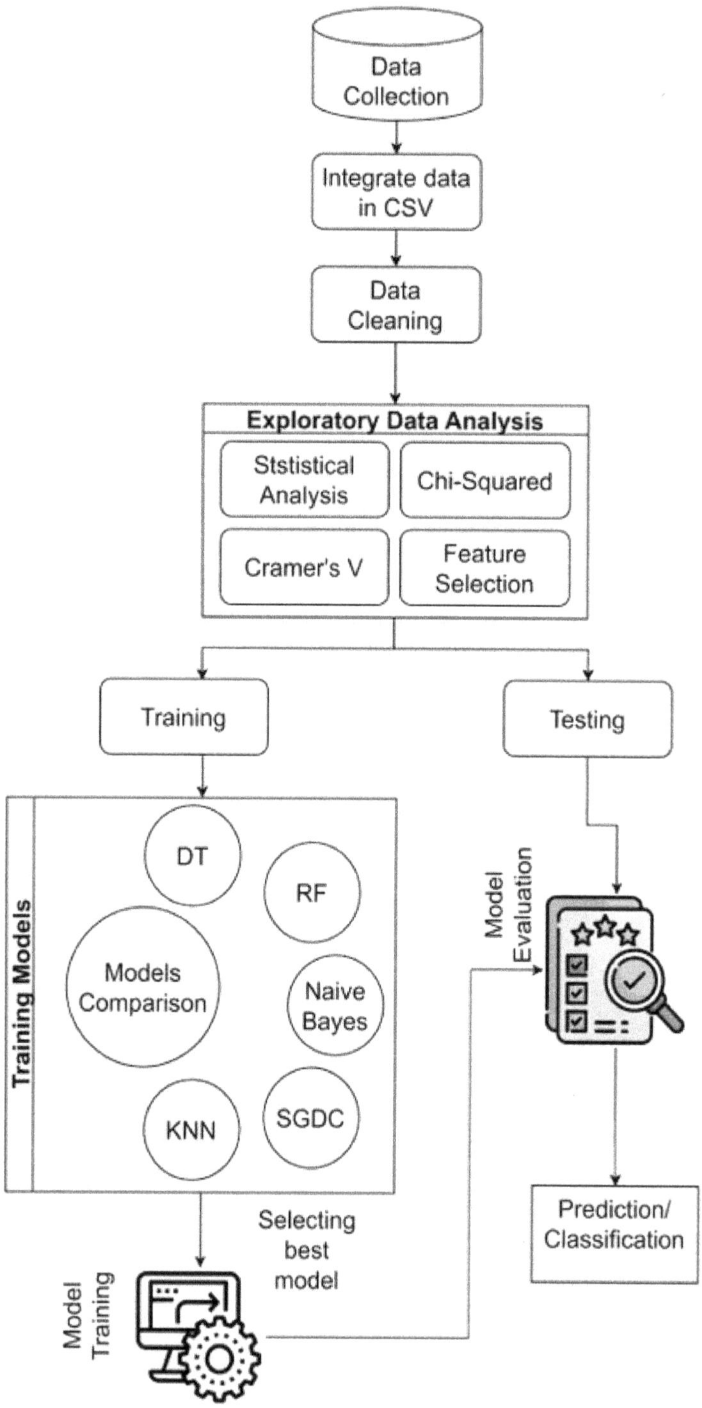

Model Training

After completing the Exploratory Data Analysis (EDA), the next step involved training the dataset, splitting it into a ratio of 80:20 for training and testing respectively. Various AI algorithms were applied to the training data, including LR (Logistic Regression), BN (Bayesian Networks), KNN (K-Nearest Neighbors), SGDC (Stochastic Gradient Descent Classifier), and DT (Decision Tree). Each algorithm's hyperparameters were fine-tuned using grid-search, a methodical approach that systematically explores a predefined set of hyperparameter values to optimize model performance metrics such as accuracy. For example, in LR, hyperparameters like regularization parameter (c), penalty, random state, and max iterations were adjusted accordingly. Among these algorithms, Decision Tree (DT) exhibited the highest accuracy, surpassing 99.44%. DT's ability to create hierarchical decision rules based on attribute thresholds proved particularly effective in accurately predicting water potability. Learning curves and ROC curves were plotted to further analyze model performance over the training period, providing insights into the models' capabilities and helping determine optimal classification thresholds.

The combination of EDA, statistical analysis, feature engineering, and AI algorithm application formed a comprehensive framework for addressing water potability prediction challenges. Insights gained from this process contributed to a deeper understanding of water quality dynamics and facilitated informed decision-making to enhance public health and life expectancy. Specifically, in the Decision Tree model, nodes in the tree corresponded to decision points based on feature tests. The tree traverses through these nodes, evaluating features at each step until reaching a leaf node that determines the final predicted class. In this study, the Water Quality Index (WQI) was identified as the feature responsible for the initial split in the decision tree due to its lowest Gini index. A threshold value of 49.87 derived from the decision tree indicated that water with a WQI above this threshold was classified as non-potable.

Blockchain Layer

Blockchain technology comprises a series of blocks containing records that store transactions across various entities within the framework. It provides a digital, decentralized, and immutable ledger that enhances the security of data stored within these blocks. In the context of our system, data from the AI layer is transmitted to the blockchain layer. Here, we deploy a digital smart contract designed to validate incoming predicted data. This smart contract plays a crucial role in ensuring trust and reliability by eliminating third-party intermediaries from the framework. By doing so, it enhances the operational efficiency and performance of Water Treatment Plants (WTPs). The integration of blockchain technology in this layer serves to:

Secure Data: Blockchain's decentralized nature and cryptographic mechanisms ensure that data stored in blocks remains tamper-proof and transparent.
Trust and Reliability: The use of smart contracts facilitates automated validation of predicted data, thereby reducing the dependency on external verification processes and enhancing reliability.
Performance Improvement: By streamlining transaction validation and data integrity checks, blockchain technology contributes to improving the overall performance and operational efficiency of WTPs.

In essence, the blockchain layer adds a layer of security, transparency, and efficiency to our framework, crucially supporting the reliable prediction and management of water potability within Water Treatment Plants. In the proposed system, validated data undergoes the following process: it is first sent to IPFS (on-site storage), where a hash is computed and linked to a unique content identifier. This hash is then transmitted to the blockchain's immutable ledger, which enhances the blockchain network's response time. The rationale is that instead of storing raw Water Treatment Plant (WTP) data directly in the immutable ledger, only the hash is stored. Blockchain members can access the original data by querying the content identifier through the IPFS web console. Storing only the hash, which is smaller in size compared to the raw data, significantly boosts the blockchain's efficiency in the proposed framework. Moreover, the blockchain layer restricts access to authorized users such as water suppliers, government entities, and other relevant stakeholders involved in using treated water from WTPs. It functions as a unified platform facilitating secure communication between government authorities and water suppliers. This collaborative approach is essential for predicting the impact of contaminated water on public health and life expectancy.

Profiling Layer

Within the profiling layer, collected data is securely distributed across multiple Water Treatment Plants (WTPs) to enhance overall water treatment effectiveness. This secure sharing ensures precise chemical application to non-potable water sources. By employing advanced profiling techniques, the system gains comprehensive insights into water quality characteristics. This knowledge enables WTPs to optimize their water treatment procedures, thereby positively impacting water quality and potentially extending human life expectancy.

RESULTS AND DISCUSSION

Experimental Setup

The proposed framework is deployed within the Google Colab integrated development environment (IDE), utilizing Python 3.10.5 as the primary programming language. To facilitate development in Python, essential libraries such as NumPy, Pandas, Scikit-learn, Seaborn, and Matplotlib were employed. Pandas was utilized for structured data manipulation and analysis, particularly handling CSV files. NumPy supported numerical calculations and efficient matrix operations essential for scientific computing. Scikit-learn played a critical role in model selection, evaluation, classification tasks, and data preprocessing. Seaborn was employed for advanced data visualization, offering capabilities for plotting graphs and customizing styles. Matplotlib complemented these functionalities by providing additional tools for visualizing data through graphs and curves.

Following data processing and analysis in Google Colab, the processed data from the AI layer is transmitted to an Ethereum-based public blockchain. This blockchain implementation is conducted in the Remix IDE (version 0.39.1), an online platform tailored for developing, testing, and deploying smart contracts on the Ethereum network. Remix IDE supports the Solidity programming language (version 0.8.23), enabling developers to write, compile, deploy, and debug smart contracts efficiently within a web-based environment. This integration ensures secure and transparent data management and transaction validation through blockchain technology.

The framework incorporates smart contracts developed using various functions to manage Water Treatment Plants (WTPs) effectively. Functions such as *\`startPlant()\`* initiate specific WTP operations, allowing subsequent actions like adding or releasing water using *\`addWater()\`* and *\`releaseWater()\`* functions. While the plant operates, *\`addSample()\`* and *\`getSample()\`* functions handle water profiling, crucial for categorizing samples into potable and non-potable categories. Once operations are completed, the *\`stopPlant()\`* function halts the WTP. Upon execution, authenticated data is securely stored in the IPFS-based blockchain ledger. Performance metrics including precision, recall, and F1-score are calculated based on the confusion matrix to evaluate model effectiveness. Learning curves visualize the Machine Learning model's learning rate and performance across training and validation datasets. Additionally, the Receiver Operating Characteristic (ROC) curve assesses the binary classification model comprehensively, aiding in understanding its discrimination ability between classes.

AI-Based Results

In this section, a thorough analysis of various AI-based algorithms is employed to classify Water Treatment Plants (WTPs) as either potable or non-potable. To accurately assess the performance and interpret the outcomes, key metrics such as precision, recall, and F1 score are essential. These metrics provide insights into the model's reliability and effectiveness. Confusion matrix visually represents the model's predictions by categorizing them into True Positives (TP), False Positives (FP), True Negatives (TN), and False Negatives (FN). In this context, the model identifies potable samples as positive (correctly classified) and non-potable samples as negative (correctly classified). True Positives (TP) indicate instances where potable water samples are correctly identified. False Positives (FP) occur when non-potable water samples are incorrectly classified as potable. Conversely, False Negatives (FN) represent potable water samples misclassified as non-potable, while True Negatives (TN) denote correctly classified non-potable water samples. These components of the confusion matrix are instrumental in computing precision and recall, pivotal for evaluating how well the model distinguishes between potable and non-potable water samples.

The learning curve provides a graphical representation of how a model's accuracy evolves as it learns from varying amounts of training data. It illustrates the percentage improvement in accuracy over the training process, offering insights into how the model's performance stabilizes with more data. A steeper slope indicates faster learning and better model performance. It is evident that the model gradually improves its accuracy as more data is trained, ultimately achieving a final accuracy of 99.44%. Figure 3 presents a bar graph comparison of the accuracy achieved by different learning algorithms. The analysis reveals that the Decision Tree (DT) algorithm outperforms others like Logistic Regression (LR), Naive Bayes (NB), Stochastic Gradient Descent Classifier (SGDC), and K-Nearest Neighbors (KNN), achieving the highest accuracy of 99.44%. This superior performance of the DT model can be attributed to its selection of the Water Quality Index (WQI) as a pivotal feature for classifying data points.

Figure 3. Accuracy comparison of ML models

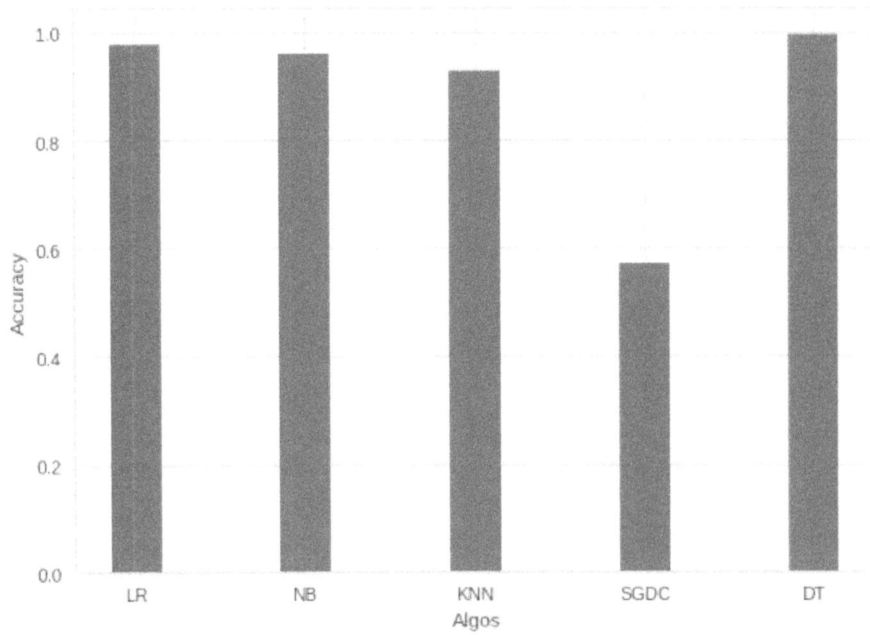

Precision, which measures the accuracy of positive predictions, and recall, which assesses the model's ability to correctly identify positive instances, are crucial metrics evaluated here. Precision is calculated as the ratio of True Positives (TP) to the sum of TP and False Positives (FP), while recall is the ratio of TP to the sum of TP and False Negatives (FN). The F1 score, which balances precision and recall, is also computed to gauge overall model performance. The DT model demonstrates exemplary precision with a value of 1, indicating that all positive predictions are correct. Its recall score of 0.990291 and F1 score of 0.995122 further underscore its effectiveness in accurately classifying water samples based on potability.

A Receiver Operating Characteristic (ROC) curve illustrates the performance of classification models by plotting the True Positive Rate (TPR) against the False Positive Rate (FPR). The x-axis represents the FPR values, while the y-axis represents the TPR values. A desirable ROC curve typically exhibits high TPR and low FPR, ideally covering the top-left corner of the graph, indicative of superior model performance. In this context, the ROC curve serves to compare the efficacy of different classifiers. Upon comparison, it is evident that the Decision Tree (DT) classifier produces the most accurate results, approaching a near-perfect ROC curve.

Additionally, the evaluation includes a log-loss curve, which assesses the proximity of predicted probabilities to actual values. The DT classifier notably achieves a log-loss of 0.2, underscoring its exceptional predictive accuracy and reliability. Among the models evaluated, the Stochastic Gradient Descent Classifier (SGDC) exhibits the lowest log-loss, indicating strong predictive performance, while the K-Nearest Neighbors (KNN) model demonstrates the highest log-loss. Evaluating these metrics alongside training times is crucial for practical implementation, as training times can significantly influence operational efficiency and deployment feasibility of the models.

Blockchain-Based Results

The data storage and operational control of the water treatment plant (WTP) are managed through a smart contract deployed on the blockchain network using Remix IDE. This smart contract includes several functions and variables essential for WTP operations. Functions like *startPlant()* initiate the WTP, while *addWater()* and *releaseWater()* allow controlled water addition and release during operation. When the WTP is active, *addSample()* and *getSample()* functions manage water profiling by collecting and retrieving sample details necessary for classifying water as potable or non-potable. The *stopPlant()* function halts WTP operations by setting the *isOperating* variable to false, disabling further water management functions. Additionally, variables such as *waterLevel* track the current water level, adjusting as water is added or released, and *sampleCount* monitors the number of samples processed.

Figure 4. Scalability comparison with proposed blockchain with IPFS and conventional blockchain

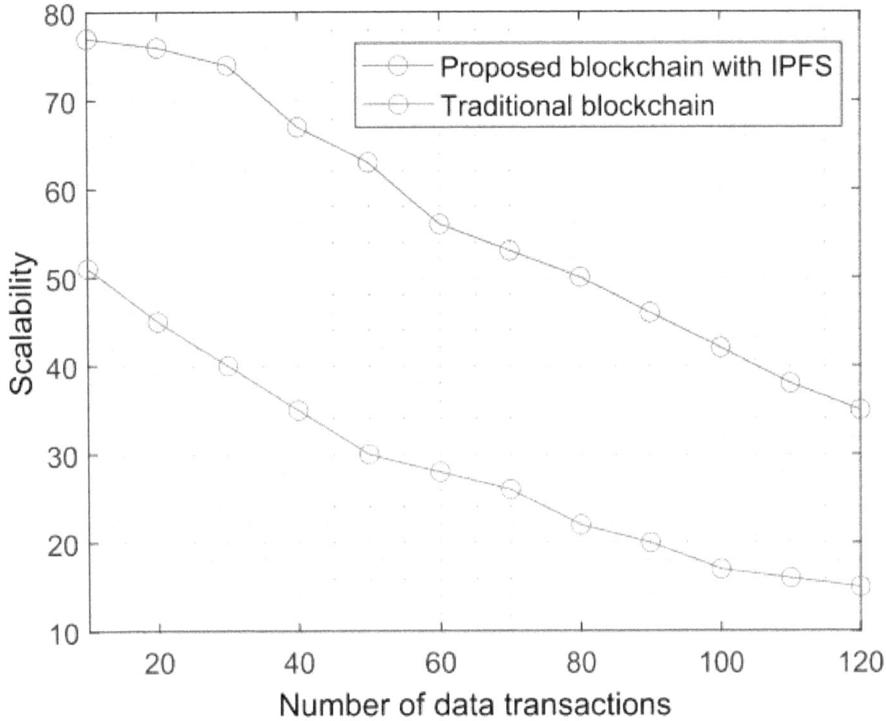

Figure 5. Scalability comparison with blockchain's response time

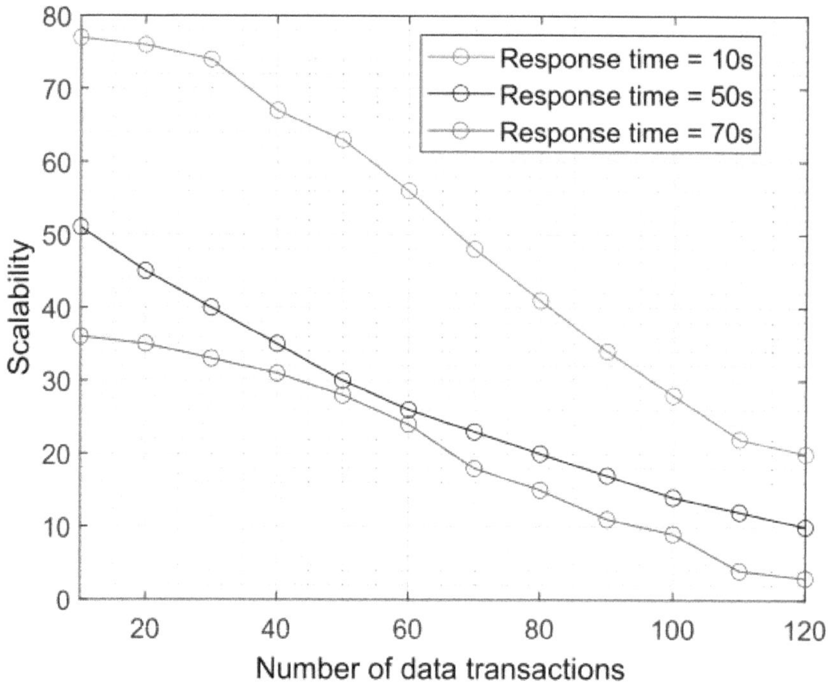

Within the blockchain network, each user operates a node that communicates with others for data exchange. The proposed framework utilizes the InterPlanetary File System (IPFS), which computes hashes for validated data from the smart contract. These hashed data are then stored in the blockchain's immutable ledger, enhancing efficiency in storage and retrieval. By storing only hashes instead of raw data, the framework minimizes processing delays and boosts scalability by enabling blockchain nodes to handle more data transactions. Figure 4 illustrates a scalability comparison between traditional blockchain methods (without IPFS) and the proposed framework (with IPFS), demonstrating superior scalability with IPFS integration. This observation in figure 5 underscores the inverse relationship between response time and blockchain scalability, highlighting the critical impact of timely transaction processing on overall system performance.

CONCLUSION

Assessing water potability poses a significant challenge due to various external factors influencing water quality. Traditional methods rely on subjective analysis and limited attributes, prompting the adoption of machine learning and deep learning algorithms in recent years. Our study focuses on evaluating water potability using machine learning models. We explored several algorithms including Decision Trees (DT), K-Nearest Neighbors (KNN), Naive Bayes (NB), Stochastic Gradient Descent Classifier (SGDC), and Random Forest (RF). Among these, decision tree algorithms proved most effective in accurately determining water potability. Leveraging decision tree models enhances the efficiency of water quality assessment, providing reliable classifications that support informed decision-making in water treatment and resource allocation. This contributes significantly to enhancing public health and safety measures. Recognizing the complexity of water potability prediction influenced by external factors, our future research aims to advance hybrid models to further improve prediction accuracy. Upon analysis, it is observed that the K-Nearest Neighbors (KNN), Naive Bayes (NB), and Decision Tree (DT) algorithms exhibit shorter training times, specifically 0.0046, 0.0047, and 0.0054 seconds, respectively. In contrast, the Logistic Regression (LR) and Stochastic Gradient Descent Classifier (SGDC) require more time for training, with durations of 0.028 and 0.0182 seconds, respectively. Further comparisons assess the blockchain's scalability across different response times (10s, 50s, and 70s). The findings indicate that higher response times correlate with reduced scalability, exemplified by a response time of 70s yielding a scalability value of 5 at transaction 120. This underscores the efficiency gains of incorporating IPFS into blockchain architecture, enhancing scalability and operational performance in managing WTP data and operations securely.

REFERENCES

Banerjee, K., Bali, V., Nawaz, N., Bali, S., Mathur, S., Mishra, R. K., & Rani, S. (2022, February). A machine-learning approach for prediction of water contamination using latitude, longitude, and elevation. *Water (Basel)*, 14(5), 728. DOI: 10.3390/w14050728

Devi, R. R., Dharshini, P., Hemala, R., & Swetha, D. "Heart Disease Prediction using Random Forest Classifier," *2023 International Conference on Data Science, Agents & Artificial Intelligence (ICDSAAI)*, Chennai, India, 2023, pp. 1-8, DOI: 10.1109/ICDSAAI59313.2023.10452459

Goodarzi, D., Abolfathi, S., & Borzooei, S. (2020). Modelling solute transport in water disinfection systems: Effects of temperature gradient on the hydraulic and disinfection efficiency of serpentine chlorine contact tanks. *Journal of Water Process Engineering*, 37(Oct), 101411. DOI: 10.1016/j.jwpe.2020.101411

Granata, F., Papirio, S., Esposito, G., Gargano, R., & De Marinis, G. (2017, February). Machine learning algorithms for the forecasting of wastewater quality indicators. *Water (Basel)*, 9(2), 105. DOI: 10.3390/w9020105

Guzmán-Avalos, P., Molinero-Hernández, D., Galván-González, S., Herrera-Sandoval, N., Solorio-Díaz, G., & Rubio-Maya, C. (2023, January). Numerical design and optimization of a hydraulic micro-turbine adapted to a wastewater treatment plant. *Alexandria Engineering Journal*, 62, 555–565. DOI: 10.1016/j.aej.2022.07.004

Jalal and T. Ezzedine. ''Performance analysis of machine learning algorithms for water quality monitoring system,'' in Proc. Int. Conf. Internet Things, Embedded Syst. Commun. (IINTEC), Dec. 2019, pp. 86–89.

Kalpana, A. V., Chandrasekar, T., Ramalingam, R., Tamizhselvi, S. P., Ramesh, T., & Chitra, M. (2024). Enhancing Healthcare Integration With IoT for Seamless and Responsive Patient Care. In Khade, S., & Mishra, R. (Eds.), *Future of AI in Biomedicine and Biotechnology* (pp. 147–181). IGI Global., DOI: 10.4018/979-8-3693-3629-8.ch008

Luvhimbi, N., Tshitangano, T. G., Mabunda, J. T., Olaniyi, F. C., & Edokpayi, J. N. (2022, April). Water quality assessment and evaluation of human health risk of drinking water from source to point of use at thulamela municipality, limpopo province. *Scientific Reports*, 12(1), 6059. DOI: 10.1038/s41598-022-10092-4 PMID: 35411067

Madhusoodhanan, C. G., Sreeja, K. G., & Eldho, T. I. (2016, October). Climate change impact assessments on the water resources of India under extensive human interventions. *Ambio*, 45(6), 725–741. DOI: 10.1007/s13280-016-0784-7 PMID: 27170012

Narendar, S. D., Murugamani, C., Kshirsagar, P. R., Tirth, V., Islam, S., Qaiyum, S., Suneela, B., Al Duhayyim, M., & Waji, Y. A. (2022, February). IoT based smart wastewater treatment model for Industry 4.0 using artificial intelligence. *Scientific Programming*, 2022, 1–11.

Noori, R., Farahani, F., Jun, C., Aradpour, S., Bateni, S. M., Ghazban, F., Hosseinzadeh, M., Maghrebi, M., Naseh, M. R. V., & Abolfathi, S. (2022). A nonthreshold model to estimate carcinogenic risk of nitrate-nitrite in drinking water. *Journal of Cleaner Production*, 363(Aug), 132432. DOI: 10.1016/j.jclepro.2022.132432

Qiu, Y., Ekström, S., Valverde-Pérez, B., Smets, B. F., Climent, J., Domingo-Félez, C., Cuenca, R. M., & Plósz, B. G. (2024). Numerical modelling of surface aeration and N2O emission in biological water resource recovery. *Water Research*, 255(May), 121398. DOI: 10.1016/j.watres.2024.121398 PMID: 38503179

Ramalingam, R. (2024). An Innovative Investigation on Predicting Forest Fire Using Machine Learning Approach. In Ouaissa, M., Ouaissa, M., Boulouard, Z., Iwendi, C., & Krichen, M. (Eds.), *AI and IoT for Proactive Disaster Management* (pp. 61–77). IGI Global., DOI: 10.4018/979-8-3693-3896-4.ch004

Rathi, B. S., Kumar, P. S., & Vo, D.-V.-N. (2021). Critical review on hazardous pollutants in water environment: Occurrence, monitoring, fate, removal technologies and risk assessment. *The Science of the Total Environment*, 797(Nov), 149134. DOI: 10.1016/j.scitotenv.2021.149134 PMID: 34346357

Renugadevi, R. (2024). Disease Monitoring and Management: Healthcare Information Systems (HIS) in Smart Cities. In Baral, S., & Goel, R. (Eds.), *Analyzing Current Digital Healthcare Trends Using Social Networks* (pp. 178–192). IGI Global., DOI: 10.4018/979-8-3693-1934-5.ch010

Renugadevi, R., & Rajeswari, M. A. R, R. Rajalakshmi and G. Ramkumar, "A Distinctive Explainable Deep Learning Framework to Predict Pulmonary Function at Chest CT Scan," *2023 First International Conference on Advances in Electrical, Electronics and Computational Intelligence (ICAEECI)*, Tiruchengode, India, 2023, pp. 1-7, DOI: 10.1109/ICAEECI58247.2023.10370935

Sugumar, S. J., Sahana, R., Phadke, S., Prasad, S., & Srilakshmi, G. R. ''Real time water treatment plant monitoring system using IoT and machine learning approach,'' in Proc. Int. Conf. Design Innov. 3Cs Compute Communicate Control (ICDI3C), Jun. 2021, pp. 286–289. DOI: 10.1109/ICDI3C53598.2021.00064

Wongburi, P., & Park, J. K. (2022, May). Prediction of sludge volume index in a wastewater treatment plant using recurrent neural network. *Sustainability (Basel)*, 14(10), 6276. DOI: 10.3390/su14106276

Chapter 12
Artificial Intelligence and Applications in Drinking Water Management

Ricardo A. Barrera-Cámara
https://orcid.org/0000-0002-3170-4671
Universidad Autónoma del Carmen, Mexico

Ana A. Canepa-Sáenz
https://orcid.org/0000-0003-0583-439X
Universidad Autónoma del Carmen, Mexico

ABSTRACT

The increase in the urban population and climate change have driven the development of Smart Water systems, which integrate artificial intelligence to improve drinking water management. AI optimizes water distribution, monitors water quality in real-time, detects leaks, and manages demand efficiently, thus addressing current challenges in water resource management. The aim of this research is to analyse the applications of AI in drinking water management within Smart Water systems. The method used includes a literature review, case studies and an analysis of the data obtained. The results show that AI improves the management of drinking water through continuous monitoring of its quality, accurate leak detection, optimization of the distribution and efficient use of water resources, demand prediction and management, and predictive maintenance. In addition, it reduces energy consumption in water treatment and distribution. However, there are technical, economic, and regulatory challenges that need to be addressed in order to achieve effective and sustainable implementation.

DOI: 10.4018/979-8-3693-8074-1.ch012

INTRODUCTION

Water management includes activities such as planning, efficient distribution, and optimal use of water resources so that it can meet current and future needs (Jenny et al., 2020), this under clearly defined policies and regulations (Manawadu et al., 2021). Traditional water management systems use a variety of techniques, methods, and technology to manage water resources. These activities allow for the measurement of consumption, inspection of the infrastructure and limited data availability. Measurements and fault detection are carried out manually and physically. In traditional methods, consumption measurements and fault detection are carried out manually and physically, the data generated is not sufficient to be analyzed, thus limiting the achievement of a efficient and sustainable water management.

Properly managing water is essential in a situation of climate change and water scarcity. Intelligent water management aims to monitor and manage the aquatic environment, allocate and manage water resources, monitor the drainage network and water supply, as well as achieve the effectiveness and efficiency of flood monitoring, for this artificial intelligence and new information technologies must be used (This is the 2022 edition).

In the context of the Sustainable Development Goals (SDGs), SDG 6, called "Clean water and sanitation", seeks to ensure the availability of water and its sustainable management for all. On the other hand, SDG 11, "Sustainable cities and communities", aims to make cities and human settlements inclusive, safe, resilient and sustainable. These two objectives are closely linked to smart water management.

Organizations dedicated to water supply, by making use of data analysis and artificial intelligence, lead to make better decisions (water demand, leak detection, etc.), reduce costs, increase efficiency and improve services (Jenny et al., 2020)

BACKGROUND

Related Works

In (Krishnan et al., 2022) Water management techniques are analyzed, including water resource collection, recycling, and conservation. It uses Artificial Intelligence, deep and machine learning, and addresses challenges in deploying an efficient water management system, and proposes future directions.

Various works of water-oriented smart applications using Internet of Things technologies were reviewed, in the analyzed works it was identified that 5G technologies are not used for communication, they incorporate 3D printing and solar energy in the applications, and smart technologies for the management of water resources were identified (Around & Kabaso, 2024).

Artificial intelligence (AI)-based techniques, specifically machine learning (ML) models, are applied in water treatment processes to improve decision-making, control, process optimization, and cost management. This paper reviews the use of AI in wastewater treatment plants (WWTPs), evaluating its development, applications, limitations, and areas for improvement (Chaudhary & Karanfil, 2024). AI techniques have improved predictions of coagulant doses, membrane flow, disinfection byproduct (DBP) formation, and contaminant removal.

Artificial intelligence has a large number of applications that benefit companies dedicated to carrying out operations with water, however many of these companies consider using them, but it is not clear if AI is adopted in their operations. The study focuses on identifying the adoption, form and effect of artificial intelligence in companies linked to the water industry (Rapp Alyson et al., 2023)

Traditional water resources management faces problems of data acquisition, data analysis, and decision-making. The applications of Artificial Intelligence and Big Data Analytics as applicable technologies for the management of water resources are analyzed. These apply to water quality monitoring, water demand allocation and forecasting (Kamyab et al., 2023).

In (Nova, 2023), explores how artificial intelligence (AI) can help conserve water in five key areas: water management and distribution, precision agriculture, water quality monitoring, drought prediction, and smart water grids. Artificial intelligence can improve resource management, make water use in agriculture more efficient, ensure drinking water quality, predict droughts accurately, and optimize water distribution, thereby reducing loss and waste. All of this is crucial to address water scarcity and droughts, important challenges for sustainable development.

SMART WATER MANAGEMENT SYSTEMS

Smart water management aims to optimize water use to promote efficiency, sufficiency, and sustainability. To this end, management integrates new technologies such as sensors (Internet of Things), intelligent water metering (Artificial Intelligence), information systems, data acquisition and decision support systems (Big Data) (BABLE, 2024).

Technologies associated with smart water management stand out (Tosin Michael et al., 2024):

– **IoT Sensors and Devices:** Smart water management relies on IoT sensors and devices, which are deployed in all water systems to collect data on various parameters, such as flow rates, pressure, water quality, and leaks. These sensors allow real-time monitoring and control of the water infrastructure, facilitating early detection of leaks, predictive maintenance and optimization of water distribution.

– **Data Analytics and Artificial Intelligence (AI):** Data collected from IoT sensors and devices is analyzed using advanced data analytics and AI algorithms to extract useful insights. Machine learning models can predict water demand, identify patterns in water use, and optimize treatment processes. AI can also improve decision-making by predicting system failures and recommending preventative measures.

– **Cloud Computing:** It allows you to store, process and manage the large amounts of data generated by intelligent water management systems. It enables scalable and flexible access to computing resources, facilitating data sharing and collaboration between different stakeholders involved in water management. Cloud platforms also support the deployment of data analytics and artificial intelligence applications, making it easier to implement sophisticated water management solutions.

– **Remote Sensing:** This technology involves the use of satellites and drones to collect comprehensive data on bodies of water, land use, and environmental conditions. Remote sensing is invaluable for large-scale water resource monitoring, as it offers insights into changes in water availability, land-use patterns that affect water cycles, and the health of aquatic ecosystems.

For the management and measurement of water quality, systems are also called cyber-physical systems which are controlled and composed of SCADA systems that feature sensors, actuators, programmable logic controllers (PLCs), remote terminal units (RTUs) and field devices, geographic information systems, and software (Balta et al., 2023).

- **SCADA (Supervisor and Control and Data Acquisition) Systems:** Fundamental tools in intelligent water management. They enable remote control and monitoring of processes related to the water cycle, such as drinking water treatment and distribution, sewerage networks, and wastewater treatment plants. These systems acquire data from sensors and devices, process it and display it on computer equipment for analysis and decision-making. Its main functions include monitoring, controlling, acquiring and analysing data. In addition, SCADA systems improve efficiency, reduce costs, and optimize energy and water resource consumption.

- **Geographic Information Systems:** Geographic Information Systems (GIS) combine maps and data to help manage water intelligently. With the help of artificial intelligence (AI), GIS can analyze large amounts of information to predict problems, optimize water use, and improve real-time decision-making.
- **Data Analytics Platforms:** Data Analytics Platforms for smart water management with AI are systems that collect, process, and analyze large volumes of data related to the water cycle. AI allows you to identify patterns, predict demands, and detect anomalies. These platforms integrate data from various sources, such as IoT sensors, SCADA, and GIS systems, providing valuable insights in real-time. This allows for more informed decision-making, optimizes the use of water resources, improves operational efficiency, and reduces costs.

METHODOLOGY

It consists of the analysis of various works and approaches related to artificial intelligence for water management. In this sense, techniques and strategies used to collect and analyze non-numerical data were applied. The perceptions, experiences and contexts identified in various works were also explored.

APPLICATIONS OF ARTIFICIAL INTELLIGENCE IN WATER MANAGEMENT

There is a diversity of applications of artificial intelligence in various sectors and areas, specifically those related to water management are presented below

Monitoring, Detection of Leaks and Water Loss: Supports the identification and location of leaks in drinking water, sewer, or other water supply networks efficiently and accurately. Also to optimize the management of water resources, reduce water losses, improve operational efficiency and protect the environment.

This approach is based on three key elements: a) Smart sensor networks: The implementation of sensor networks that collect real-time data on water pressure, flow, and quality makes it possible to identify anomalous patterns that could signal the presence of leaks. b) Advanced data analysis: Machine learning algorithms process sensor data to detect leaks accurately and quickly, even in complex distribution networks. c) Exact location of leaks: AI-powered geographic information systems (GIS) make it possible to accurately identify the location of leaks, significantly reducing repair time and costs. Below are some examples and situations related to the identification and monitoring of water leaks.

Traditional water leak detection, based on manual auscultation, has been replaced by a technological solution at the Metropolitan Waterworks Authority of Thailand. Using artificial intelligence and information management in the cloud, this system has developed a predictive model capable of identifying leaks from audios (Vanijjirattikhan et al., 2022). Operators, through a mobile application, receive precise indications to locate leaks, thus optimizing water management and reducing unbilled losses.

In China, the combination of the Internet of Things and artificial intelligence algorithms make possible an intelligent remote monitoring system of the water supply network (Li & Chen, 2023). This system allows the water supply pipe network to be monitored, detecting and locating leaks more accurately by using piping pressure data.

To address the problem of water loss in distribution networks in the African and Asian continents, where there is population growth and water scarcity, the implementation of an Internet of Things (IoT) system that uses acoustic sensors and deep learning algorithms to detect leaks in pipes in real time is proposed (Boujelben et al., 2023). This system has a high accuracy in the identification of leaks and a high accuracy in the location of their position.

In (Ebisi et al., 2023), to reduce water losses in a supply network, sensors were introduced IoT and AI algorithms, such as Isolation Forest, SVC, and RNN-LSTM. The results demonstrated the effectiveness of these methods in detecting leaks.

The integration of infrared thermography, ground penetrating radars, drones and artificial intelligence (convolutional neural networks) allows the detection of water leaks in irrigation systems (Massaro et al., 2021). This solution allows for more accurate location and rapid repair of leaks, optimizing water management.

Prediction or Forecasting and Management of Water Demand: AI-based prediction algorithms estimate future water demand with high accuracy, allowing utilities to optimize production and distribution.

In the face of increasing water scarcity, accurate prediction of short-term water demand becomes crucial. In Hubli City, India, deep and machine learning models were deployed to improve the accuracy of these predictions. By combining historical consumption data with climatic and calendar variables, these models made it possible to optimize water distribution, reducing operating costs and ensuring a more efficient supply (Kavya et al., 2023).

To mitigate the effects of the drought, an early warning system based on artificial intelligence was developed. This system uses deep neural network (DNN) and long-term memory (LSTM) models to predict water demand at the Gurye intake station, South Korea (Kim et al., 2023). The models were trained with historical data and achieved high accuracy in predicting demand. In addition, decision tree and random forest models were used to classify the severity of drought at different stages, considering factors such as the previous day's water demand, rainfall, and temperature. These models allow us to anticipate the evolution of drought and take preventive measures more effectively, contributing to a more efficient management of water resources in drought-prone regions.

The study (Filho et al., 2024), compares four machine learning models to predict daily water demand in the Brazilian coastal city of Guaratuba. Using historical consumption, weather, and calendar data, the models were trained and evaluated. The results indicate that nonlinear models, such as the multilayer perceptron, are better suited to predict water demand in this type of complex systems, more historical data, and better accuracy in predictions.

Predictive and Preventive Maintenance: Analyze sensor data to predict equipment failures and schedule preventative maintenance, avoiding service interruptions and urgent repair costs.

The integration of artificial intelligence and smart pipes represents an important support in water management. By analyzing the data collected by sensors installed in the pipes, wear patterns can be identified and failures can be accurately predicted (Rebahi et al., 2023). This makes it possible to optimise maintenance tasks, reduce water losses and protect the environment, contributing to more sustainable water management.

Efficient management of drinking water networks benefits from predictive maintenance, combining statistics and machine learning to predict pipe failures (Rahbaralam et al., 2020). In Barcelona, models were trained with data on the age, material and soil conditions of the pipes, managing to predict failures and optimize repairs. This allows for more informed decisions, optimizing resources and extending the life of the infrastructure.

An artificial intelligence model, based on recurrent neural networks, has been developed to optimize the predictive maintenance of local purification systems (LSR) in industrial environments. By analyzing historical data, the model allows to anticipate failures and schedule interventions, maximizing efficiency and reducing operating costs (Chang et al., 2020).

Water Quality Monitoring and Analysis. AI is used to monitor and analyze water quality in real-time. Smart sensors and AI algorithms detect contaminants in water quickly and accurately, enabling timely actions to protect public health and the environment.

Artificial intelligence, especially machine learning, makes it possible to assess water quality by analyzing spectral data efficiently. These technologies identify contaminants and predict changes in water quality more accurately. They require large data sets to function optimally, improving early detection of contaminants and optimizing water treatment processes, which reduces costs and promotes more sustainable practices(Pérez-Beltrán et al., 2024).

On the campus of the University of Lille, they developed an AI system to improve the safety of drinking water. This system, which overcomes the limitations of traditional models, uses vector support machines and neural networks to detect in real time any alteration in water parameters, offering an early warning of possible contamination (Tinelli & Juran, 2019). To test the feasibility of an intelligent monitoring system, an alert system was developed to detect contaminants, this by simulating the spread of bio-contamination risk under real conditions in the water distribution system at the University of Lille. This system supports water quality control by detecting contamination, having preventive risk management, and ensuring water supply (Pérez-Beltrán et al., 2024)

Neural networks are applied for the assessment, monitoring and estimation and management of surface water quality (Rana et al., 2023). To predict and classify water quality, neural networks and long-term memory (LSTM) were used, the data that were analyzed are temperature, pH, oxygen, conductivity, dissolved solids, turbidity and chlorides, in this sense LSTM models are the most accurate in water prediction (Nasir et al., 2022).

Artificial intelligence is used in conjunction with data on contaminants in water, such as total dissolved solids, potassium, sodium, and other metals, to classify water quality with high accuracy (Nallakaruppan et al., 2024). In order to assess whether water is drinkable, explainable artificial intelligence (XAI) is used to identify and understand the factors that have the greatest impact on water drinkability.

In order to monitor water quality and ensure the supply of safe drinking water, artificial intelligence models (ANFIS and FFNN) were developed that analyze various water parameters. ANFIS is effective in predicting a general water quality index, while the FFNN model stands out for its extreme accuracy in classifying water as safe or unsafe for human consumption (Hmoud Al-Adhaileh & Waselallah Alsaade, 2021).

Artificial intelligence is a tool to tackle growing water pollution, especially in the treatment of difficult contaminants such as antibiotics(Ganthavee & Trzciński, 2024). By analyzing large amounts of water quality data, AI optimizes treatment processes, reduces costs, and improves water quality, thereby helping to address the global water crisis and ensure a safe supply for all

Efficient Management of Water Resources. It refers to the use of artificial intelligence technologies to optimize the use and distribution of water. To this end, it combines the collection and analysis of advanced data, hydrological modelling and management optimisation, water geographic information systems (SIGA) with AI allow for a more efficient, sustainable and resilient use of water.

Artificial intelligence algorithms are used to optimize and reduce water losses due to poor water distribution (Danu, 2018). These algorithms, by analyzing large volumes of data in real time and considering factors such as flow, pressure, and water quality, allow for improved accuracy of distribution models.

On the other hand(Ferrari et al., 2023), the application of AI as a means to optimize water distribution efficiently and without compromising the quality of service. To do this, machine learning is used to identify patterns in historical data that generate decision rules for system optimization.

Drought Management and Prediction. AI helps manage droughts by implementing water-efficient measures, smart consumption restrictions, and finding new sources of supply.

Drought, a slow-evolving and high-impact phenomenon, is complex to define and mitigate. With the use of Artificial Intelligence, drought prediction is significantly improved, optimizing water management and reducing its consequences(Singh et al., 2022). In Turkey (Sabamehr & Ekin, 2023), an AI-based drought management system was developed that, by analyzing data in real time, allows crop monitoring and drought predictions more accurately. This optimizes agricultural management and a more efficient use of water.

Using neural networks and a large volume of climate data, droughts in Munich and Lisbon have been predicted a month in advance (Felsche & Ludwig, 2021). The North Atlantic Oscillation Index and atmospheric pressure, along with seasonality, are key factors for accurate prediction, improving the management of risks associated with drought.

Artificial intelligence, by analysing real-time data on consumption and weather conditions, makes it possible to optimise water management, ensuring its availability even during periods of drought (Nova, 2023).

Using artificial intelligence, the causes of flash droughts in China were identified and analyzed, demonstrating that machine learning models can predict and monitor these events with high accuracy, allowing for more effective management of water resources (Zhang et al., 2022).

Reduced Energy Consumption. AI identifies and eliminates unnecessary energy consumption in treatment plants, optimizing the use of resources and reducing operating costs.

Neural networks within the field of artificial intelligence are used to analyze and reduce the consumption of electrical energy in wastewater treatment plants (Esteves Francisco et al., 2023).

Using neural networks and fuzzy logic, wastewater treatment plants (WWTPs) optimize their processes and minimize energy consumption, mainly in aeration(-Monday et al., 2024). This allows for real-time adjustment of operating parameters, adapting to changing conditions and ensuring optimal wastewater treatment.

To improve the energy efficiency of the Melbourne WWTP, deep learning models such as the LSTM and BiGRU neural networks were used (Harrou et al., 2023) These models, trained on large amounts of data, were able to predict the plant's energy consumption with high accuracy.

To optimize and save energy consumption in wastewater treatment plants, an intelligent system based on genetic algorithms was implemented. This system analyzes daily plant operating data, such as inlet flow and nitrogen levels, to identify the factors that have the greatest impact on energy consumption(Wang et al., 2024). With its application, it was possible to reduce electricity costs without the need to install additional equipment, thus offering an economical and efficient solution.

Smart Metering of Consumption. AI-powered smart meters record water consumption in real-time, allowing customers to monitor their usage, detect leaks, and take steps to save water.

In Morocco (Naim et al., 2021), a system based on artificial intelligence and image recognition of water meters was designed, this system automates the reading of data, reducing errors and improving accuracy in billing. In addition, it detects leaks and visualises consumption, thus optimising water management.

An AI-based framework is proposed for water meter recognition that detects and recognizes numbers in meter images (Ktari et al., 2022). Through an app, water consumption data is read and recorded on the meters.

In (Duc et al., 2021), uses AI to automatically read water meter numbers through a camera. LoRa technology ensures accurate and continuous monitoring of consumption, transmitting data over long distances.

Intelligent Water Pressure Control. Adjust water pressure in real-time based on demand, reducing energy consumption, infrastructure wear and tear, and the risk of breakage.

Artificial intelligence is used to optimize pressure management in urban water networks. Through machine learning algorithms, pipe valves are adjusted to reduce average pressure and prevent leaks, resulting in a more efficient and sustainable system (Negm et al., 2023). In tests, it was possible to reduce the average pressure by 2% by adjusting only two valves.

An AI-powered water control system regulates pressure using an intelligent controller, pressure control valve, and sensors (Wang et al., 2024). The controller employs machine learning algorithms to analyze sensor data and adjust valve opening in real time, reducing consumption and improving the efficiency of the distribution system.

Another application is the use of smart models to simulate interstitial water filtration and pressure in dams, providing detailed information on the effectiveness and validity of various modeling methods in this context (Beiranvand & Rajaee, 2022).

Improved Customer Service. The systems and other smart applications provide customers with personalized information about their water consumption, electronic bills, outage alerts and efficient communication channels, improving the customer experience.

To solve the communication problems with customers that generate complaints about drinking water and sewerage services in Peru, a chatbot was implemented (Salcedo Huarcaya et al., 2020). This chatbot streamlines service, provides quick responses, and improves communication with customers.

In (JAPAMA, 2024), a chatbot is used that allows users to report water leaks, sewage, potholes, various problems and pay bills and report employees, thus providing better service to users.

Aigües de Barcelona (Aigües de Barcelona, 2024), incorporates the chatbot called *chatbot* of Aigües, this as a service for customer service that improves communication and attention to possible incidents related to the water service.

Anglian Water (Anglian Water Services, 2023), incorporated a complementary customer service through a chatbot, which facilitates the resolution of doubts, the preparation of reports, payments and account consultations, with the aim of improving user satisfaction with the service.

DEWA Emirates (Rammas), offers support to customers of the Dubai Electricity and Water Authority (DEWA) through a chatbot (DEWA, 2024). This chatbot helps to solve doubts, manage services, make payments and access information on invoices quickly and efficiently, using digital channels such as WhatsApp, DEWA's mobile app and its website.

Water Saving in Agriculture: Crop irrigation is optimized by precise dosing of water according to the specific needs of each plant and climatic conditions, reducing water consumption and improving productivity.

Artificial intelligence, using LSTM neural networks, optimizes the irrigation of date palms in arid regions. By analyzing soil and climate data, these models accurately predict the exact amount of water needed, maximizing crop yields and reducing water waste(Mohammed et al., 2023). This allows farmers to automate and efficiently adjust irrigation, promoting more sustainable and resilient agriculture in the face of climate change.

The smart irrigation system, powered by artificial intelligence and IoT, optimizes the use of water in agriculture. It analyzes factors such as climate and soil moisture to determine the exact amount of water each plant needs and the right time to water it(R et al., 2023). By avoiding over- or under-irrigation, this system contributes significantly to water savings and promotes more efficient and sustainable agriculture.

The use of artificial intelligence in precision agriculture makes it possible to optimize and increase the efficiency in the use of water for irrigation, which entails both economic and environmental benefits (Del-Coco et al., 2024). In addition, a generic structure model for smart agriculture based on machine learning is proposed.

Smart Water Treatment in Treatment Plants. It consists of the application of artificial intelligence technology and automated systems to optimize and automate processes in water treatment plants.

Machine learning is applied in wastewater treatment by analyzing data and comparing models, optimizing processes and improving water quality(El Alaoui El Fels et al., 2023).

The integration of the Internet of Things (IoT) and artificial intelligence in water treatment plants uses sensors and cloud computing, thereby developing automated systems that monitor and control treatment processes more efficiently (Alprol et al., 2024). In addition, they make it possible to predict water quality, optimize the use of resources, and make informed decisions to ensure a clean water supply.

The use of artificial neural networks in wastewater treatment plants is highly effective in accurately predicting critical parameters such as chemical oxygen demand(D et al., 2022). This technology optimizes treatment processes, reduces the discharge of pharmaceutical pollutants into the environment, and contributes to the protection of aquatic ecosystems.

Energy Optimisation in the Distribution or Supply of Water. Detection or reduction of energy consumption in water treatment or distribution.

In (Ferrari et al., 2023), also proposes to optimize energy consumption in a water distribution network through AI and rules-based control. The proposed method optimizes the control of the pumps in real time.

Artificial intelligence can support the optimization of energy consumption in water treatment plants. Based on the analysis of data from various plants in Taiwan, a machine learning model was developed that predicts energy consumption(Lee et al., 2024). This model serves as a reference to compare performance between plants and detect opportunities for reduction.

In (Dada et al., 2024), water treatment processes are optimized by adjusting chemical doses and energy consumption according to water demand and conditions, while adaptive learning algorithms allow the system to continuously improve its performance.

STANDARDS AND FRAMEWORKS RELATED TO WATER MANAGEMENT

There are various national and international standards, frameworks or regulations for water management, each with a specific focus and objectives.

- **ISO 24510:** It establishes guidelines for the evaluation and improvement of the management and operation of drinking water and sanitation services. Its scope includes aspects of service quality, customer satisfaction and operational efficiency. However, it does not cover specific technical aspects of the design and construction of water and sanitation infrastructures (ISO, 2024a)
- **ISO 24516:2016:** Provides guidelines for the management of drinking water system assets. Its scope includes technical aspects, tools and best practices to maintain the value of drinking water distribution networks, and is applicable to organizations of any size and type that operate these systems, and to various roles and functions within asset management. However, it does not apply to the management of waterworks assets such as catchment, treatment, pumping and storage, or to specific infrastructure design and construction details (ISO, 2016).
- **ISO 24591:** This standard is Part 1 and Part 2 provides guidelines for the efficient management of drinking water system assets, including technical aspects, tools and best practices. Part 1 focuses on the management of water distribution networks (ISO, 2024b), while Part 2 addresses the management of capture, treatment, pumping and storage infrastructures(ISO, 2024c). Both parties exclude specific details about the design and construction of these infrastructures.
- **AWS International Standard for Sustainable Water Management 2.0:** A framework for organizations to manage water resources in an equitable, sustainable, inclusive way, with economic benefits, and also promotes continuous improvement. It consists of the following steps: collect and understand, engage and plan, implement, evaluate, and communicate and disseminate (AWS, 2019).
- **European Water Stewardship Standard:** Comprehensive framework for sustainable water management, designed to continuously improve water efficiency. Its key principles are: the protection of surface and groundwater, the assurance of sustainable water use, the promotion of water quality and the preservation of the natural environment. The EWS process includes several steps: initial assessment, planning and goal setting, implementation of sustainable water management practices, and ongoing monitoring and reporting(EWS, 2011). Water Framework Directive (Directive 2000/60/EC).

– **European Union:** Legal framework for the sustainable protection and management of water resources in Europe. It focuses on achieving the "good status" of all European waters, both surface and groundwater, where the EU member is responsible for setting the hydrological plans (Council, 2020).
– **National Water Program 2020-2024:** Document that establishes strategies for sustainable water management in Mexico, focusing on the conservation of water resources, improvement of infrastructure, and guarantee of equitable access to drinking water and sanitation. Through this program, it seeks to promote efficiency in the use of water, the protection of aquatic ecosystems and resilience to climate change, addressing both technical and social and economic aspects (Ministry of Environment and Natural Resources, 2020).
– **ISO 46001:** It specifies requirements for an efficient water use management system, applicable to any organization that aims to reduce its water consumption and improve its water performance. It includes the planning, implementation, monitoring and improvement of water management practices, it does not consider specific technical aspects of the design or construction of hydraulic infrastructures (ISO, 2019). The ISO 46001:2019/Amd 1:2024 amendment (ISO, 2024d), integrates changes into climate action.

The Economic Commission for Latin America and the Caribbean (ECLAC) is responsible for carrying out important activities in water management in the region. It publishes and promotes water resources policies, promotes integrated water management and water governance. It considers the importance of community-based management of water and sanitation services, the integration of the economy, legislation and administration into water management to ensure the availability of and equitable access to water in Latin America and the Caribbean. Some of the work he does are regional water dialogues, the Latin American and Caribbean water observatory, and the water governance program (ECLAC, 2024).

The 2023 UN Water Conference addressed the global water crisis, where leaders and experts met to discuss solutions and policies to ensure universal access to safe drinking water and sanitation (United Nations, 2023). The importance of international cooperation and concrete actions to meet the water-related Sustainable Development Goals is highlighted.

The Africa Water Vision for 2025 seeks to ensure equitable and sustainable access to safe drinking water, improve water resources management, and strengthen transboundary cooperation to combat scarcity and promote economic development and public health. This vision is based on Strengthening Water Governance, Improving Water Wisdom, Meeting Urgent Water Needs, and Strengthening the Financial Foundation for the Future of Water (Economic Commission for Africa, 2003)

CONCLUSION

The implementation of AI in drinking water management within Smart Water systems offers multiple benefits, including greater operational efficiency, reduced losses, improved water quality and optimization of resource use. However, there are significant challenges, such as systems integration, implementation costs, and regulatory barriers. To maximize the benefits of these technologies, it is crucial that public policies support research and development, promote public-private collaboration, and establish appropriate regulatory frameworks. Ultimately, the adoption of AI in drinking water management has the potential to contribute significantly to the sustainability and resilience of water supply systems, securing a vital resource for future generations.

CONFLICTS OF INTEREST

The author(s) declare that there is no conflict of interest with respect to the publication of this chapter

REFERENCES

Aigües de Barcelona. (2024). *Aigues Chatbot*. Retrieved July 12, 2024 from https://www.aiguesdebarcelona.cat/ca/aigues-de-barcelona

Alprol, A. E., Mansour, A. T., Ibrahim, M. E., & Ashour, M. (2024). Artificial Intelligence Technologies Revolutionizing Wastewater Treatment: Current Trends and Future Prospective. *Water (Basel)*, 16(2), 314. DOI: 10.3390/w16020314

Anglian Water Services. (2023). *Anglian Water chatbot*. Retrieved July 10, 2024 from https://www.anglianwater.co.uk/

BABLE. (2024). *Smart Water Management*. Retrieved July 2, 2024 from https://www.bable-smartcities.eu/explore/solutions/solution/smart-water-management.html

Balta, D., Balta Kaç, S., Balta, M., & Eken, S. (2023). Water Quality Estimation and Anomaly Detection: A Review. *EAI Endorsed Transactions on Internet of Things*, 9(4), e2. DOI: 10.4108/eetiot.v9i4.3660

Beiranvand, B., & Rajaee, T. (2022). Application of artificial intelligence-based single and hybrid models in predicting seepage and pore water pressure of dams: A state-of-the-art review. *Advances in Engineering Software*, 173, 103268. DOI: 10.1016/j.advengsoft.2022.103268

Boujelben, M., Benmessaoud, Z., Abid, M., & Elleuchi, M. (2023). An efficient system for water leak detection and localization based on IoT and lightweight deep learning. *Internet of Things : Engineering Cyber Physical Human Systems*, 24, 100995. DOI: 10.1016/j.iot.2023.100995

CEPAL. (2024). *Water resources*. Retrieved July 17, 2024 from https://www.cepal.org/en/subtopics/water-resources

Chang, T.-Y., Cho, W.-T., Tseng, S.-Y., Ouyang, Y., & Lai, C.-F. (2020). *Predictive Maintenance of Water Purification Unit for Smart Factories*. Cognitive Cities.

Chowdhury, S., & Karanfil, T. (2024). Applications of artificial intelligence (AI) in drinking water treatment processes: Possibilities. *Chemosphere*, 356, 141958. DOI: 10.1016/j.chemosphere.2024.141958 PMID: 38608775

Council, E. P. a. o. t. (2024). *Directive 2000/60/EC* Retrieved July 12, 2024 from https://eur-lex.europa.eu/legal-content/EN/TXT/PDF/?uri=CELEX:02000L0060-20141120

Dada, M., Majemite, M., Obaigbena, A., Daraojimba, O., Oliha, J., & Nwokediegwu, Z. (2024). Review of smart water management: IoT and AI in water and wastewater treatment. *World Journal of Advanced Research and Reviews*, 21(1), 1373–1382. DOI: 10.30574/wjarr.2024.21.1.0171

Danu, R. (2018). Optimization of water distribution algorithms using AI. *International Journal of Civil Engineering and Technology*, 9(13), 2130–2146. https://doi.org/https://doi.org/10.17605/OSF.IO/FU2PD

Del-Coco, M., Leo, M., & Carcagnì, P. (2024). Machine Learning for Smart Irrigation in Agriculture: How Far along Are We? *Information (Basel)*, 15(6), 306. DOI: 10.3390/info15060306

DEWA. (2024). *RAMMAS Chatbot*. Retrieved July 14, 2024 from https://www.dewa.gov.ae/en/rammas

El Alaoui El Fels, A., Mandi, L., Kammoun, A., Ouazzani, N., Monga, O., & Hbid, M. L. (2023). Artificial Intelligence and Wastewater Treatment: A Global Scientific Perspective through Text Mining. *Water (Basel)*, 15(19), 3487. DOI: 10.3390/w15193487

Esteves Francisco, A., Cardoso, J., Leitão, S., & Pires, E. (2023). Impact of artificial intelligence in the reduction of electrical consumption in wastewater treatment plants: A review. *Journal of Information Systems Engineering & Management*, 8(3). https://doi.org/https://doi.org/10.55267/iadt.07.13623

EWS. (2011). *European Water Stewardship (EWS) Standard*. European Water Partnership. Retrieved July 3, 2024 from https://ews.info/media/1007/document1.pdf

Felsche, E., & Ludwig, R. (2021). Applying machine learning for drought prediction in a perfect model framework using data from a large ensemble of climate simulations. *Natural Hazards and Earth System Sciences*, 21(12), 3679–3691. DOI: 10.5194/nhess-21-3679-2021

Ferrari, E., Verda, D., Pinna, N., & Muselli, M. (2023). Optimizing Water Distribution through Explainable AI and Rule-Based Control. *Computers*, 12(6), 123. DOI: 10.3390/computers12060123

Filho, J. V., Scortegagna, A., Vieira, A. P. S. D., & Jaskowiak, P. A. (2024). Machine learning for water demand forecasting: Case study in a Brazilian coastal city. *Water Practice & Technology*, 19(5), 1586–1602. DOI: 10.2166/wpt.2024.096

Ganthavee, V., & Trzcinski, A. P. (2024). Artificial intelligence and machine learning for the optimization of pharmaceutical wastewater treatment systems: A review. *Environmental Chemistry Letters*, 22(5), 2293–2318. Advance online publication. DOI: 10.1007/s10311-024-01748-w

Harrou, F., Dairi, A., Dorbane, A., & Sun, Y. (2023). Energy consumption prediction in water treatment plants using deep learning with data augmentation. *Results in Engineering*, 20, 101428. DOI: 10.1016/j.rineng.2023.101428

Hmoud Al-Adhaileh, M., & Waselallah Alsaade, F. (2021). Modelling and Prediction of Water Quality by Using Artificial Intelligence. *Sustainability (Basel)*, 13(8), 4259. DOI: 10.3390/su13084259

ISO. (2016, November). *ISO 24516-1:2016 Guidelines for the management of assets of water supply and wastewater systems Part 1: Drinking water distribution networks*. Retrieved July 3, 2024 from https://www.iso.org/standard/64117.html

ISO. (2019, July). *ISO 46001:2019 Water efficiency management systems — Requirements with guidance for use*. Retrieved July 5, 2024 from https://www.iso.org/standard/68286.html

ISO. (2024a, January). *ISO 24510:2024 Activities relating to drinking water and wastewater services — Guidelines for the assessment and for the improvement of the service to users*. Retrieved July 4, 2024 from https://www.iso.org/standard/81484.html

ISO. (2024b, January). *ISO 24591-1:2024 Smart water management Part 1: General guidelines and governance*. Internacional Organization for Standardization. Retrieved July 4, 2024 from https://www.iso.org/standard/79033.html

ISO. (2024c, Apri). *ISO 24591-2:2024 Smart water management Part 2: Data management guidelines*. Internacional Organization for Standardization. Retrieved July 4, 2024 from https://www.iso.org/standard/79034.html

ISO. (2024d, February). *ISO 46001:2019/Amd 1:2024 Water efficiency management systems — Requirements with guidance for use*. Retrieved July 5, 2024 from https://www.iso.org/standard/68286.html

JAPAMA. (2024, January 29). *JAPAMA's chatbot, the AI tool where you report leaks and even corrupt employees*. Retrieved July 1, 2024 from https://www.luznoticias.mx/2024-01-29/sinaloa/chatbot-de-japama-la-herramienta-con-ia-en-donde-reportas-fugas-y-hasta-empleados-corruptos/193696

Jenny, H., Wang, Y., Garcia Alonso, E., & Minguez, R. (2020). Using Artificial Intelligence for Smart Water Management Systems. *ADB BRIEFS*(143). http://dx.doi.org/10.22617/BRF200191-2

Kamyab, H., Khademi, T., & Chelliapan, S. (2023). The latest innovative avenues for the utilization of artificial Intelligence and big data analytics in water resource management. *Results in Engineering, 20*, 101566. https://doi.org/10.1016/j.rineng.2023.101566

Kavya, M., Mathew, A., Shekar, P. R., & P, S. (2023). Short term water demand forecast modelling using artificial intelligence for smart water management. *Sustainable Cities and Society, 95*, 104610. https://doi.org/10.1016/j.scs.2023.104610

Kim, D., Choi, S., Kang, S., & Noh, H. (2023). A Study on Developing an AI-Based Water Demand Prediction and Classification Model for Gurye Intake Station. *Water (Basel)*, 15(23), 4160. DOI: 10.3390/w15234160

Krishnan, S. R., Nallakaruppan, M. K., Chengoden, R., Koppu, S., Iyapparaja, M., Sadhasivam, J., & Sethuraman, S. (2022). Smart Water Resource Management Using Artificial Intelligence—A Review. *Sustainability (Basel)*, 14(20), 13384. DOI: 10.3390/su142013384

Lee, D.-S., Fu, S.-K., & Lai, C.-W. (2024). Establishment and analysis of online energy baseline for water purification plants: A case study in Kaohsiung city, Taiwan. *Heliyon*, 10(14), e33981. DOI: 10.1016/j.heliyon.2024.e33981 PMID: 39092258

Li, L., & Chen, H. (2023). Artificial Intelligence and Internet of Things-Based Leak Detection Method for the Water Supply Network. *International Transactions on Electrical Energy Systems*, 2023(1), 3443047. DOI: 10.1155/2023/3443047

Manawadu, L., Gunathilaka, M. D. K. L., Wijerathne, V. P. I. S., & Gunathilake, K. L. W. I. (2021). How and Why the Land Resources Changed: A Chronological Change of Land Use and Land Cover in Sri Lanka. In Hasnat, G. N. T., & Hossain, M. K. (Eds.), *Examining International Land Use Policies, Changes, and Conflicts* (pp. 88–110). IGI Global. DOI: 10.4018/978-1-7998-4372-6.ch005

Massaro, A., Panarese, A., & Galiano, A. (2021, June 7-9). Technological Platform for Hydrogeological Risk Computation and Water Leakage Detection based on a Convolutional Neural Network. *Agronomy (Basel)*, 13(4).

Ministry of Environment and Natural Resources. (2020, December 30). *National Water Program 2020-2024*. Retrieved July 5, 20224 from https://www.gob.mx/cms/uploads/attachment/file/642632/PNH_2020-2024__ptimo.pdf

Monday, C., Zaghloul, M. S., Krishnamurthy, D., & Achari, G. (2024). A Review of AI-Driven Control Strategies in the Activated Sludge Process with Emphasis on Aeration Control. *Water (Basel)*, 16(2), 305. DOI: 10.3390/w16020305

Naim, A., Aaroud, A., Akodadi, K., & El Hachimi, C. (2021). A fully AI-based system to automate water meter data collection in Morocco country. *Array (New York, N.Y.)*, 10, 100056. DOI: 10.1016/j.array.2021.100056

Nasir, N., Kansal, A., Alshaltone, O., Barneih, F., Sameer, M., Shanableh, A., & Al-Shamma'a, A. (2022). Water quality classification using machine learning algorithms. *Journal of Water Process Engineering*, 48, 102920. DOI: 10.1016/j.jwpe.2022.102920

Negm, A., Ma, X., & Aggidis, G. (2023). *Water Pressure Optimisation for Leakage Management Using Q Learning*. DOI: 10.1109/CAI54212.2023.00120

Ni, J., & Zhenxiang, X. (2022). Using Artificial Intelligence for Smart Water Management *Water*(Special Issue). https://www.mdpi.com/journal/water/special_issues/Artificial_Intelligence_Water

Nova, K. (2023). AI-Enabled Water Management Systems: An Analysis of System Components and Interdependencies for Water Conservation. *Eigenpub Review of Science and Technology, 7*(1), 105-124. https://studies.eigenpub.com/index.php/erst/article/view/12

Okoli, N. J., & Kabaso, B. (2024). Building a Smart Water City: IoT Smart Water Technologies, Applications, and Future Directions. *Water (Basel)*, 16(4), 557. DOI: 10.3390/w16040557

Pérez-Beltrán, C. H., Robles, A. D., Rodriguez, N. A., Ortega-Gavilán, F., & Jiménez-Carvelo, A. M. (2024). Artificial intelligence and water quality: From drinking water to wastewater. *Trends in Analytical Chemistry*, 172, 117597. DOI: 10.1016/j.trac.2024.117597

Rahbaralam, M., Modesto, D., Card'us, J., Abdollahi, A., & Cucchietti, F. M. (2020). Predictive Analytics for Water Asset Management: Machine Learning and Survival Analysis. *ArXiv, abs/2007.03744*.

Rana, R., Kalia, A., Boora, A., Alfaisal, F. M., Alharbi, R. S., Berwal, P., Alam, S., Khan, M. A., & Qamar, O. (2023). Artificial Intelligence for Surface Water Quality Evaluation, Monitoring and Assessment. *Water (Basel)*, 15(22), 3919. DOI: 10.3390/w15223919

Rapp Alyson, H., Capener Annelise, M., & Sowby Robert, B. (2023). Adoption of Artificial Intelligence in Drinking Water Operations: A Survey of Progress in the United States. *Journal of Water Resources Planning and Management*, 149(7), 06023002. DOI: 10.1061/JWRMD5.WRENG-5870

Rebahi, Y., Hilliger, B., Lowin, P., Zheng, B., Da Bormida, G., & Ladjeri, K. (2023). Towards AI-Based Condition Monitoring and Predictive Maintenance for Water Smart Pipes: The SANDMAN Approach. *Artificial Intelligence and Applications (Commerce, Calif.)*, 2(2), 100–110. DOI: 10.47852/bonviewAIA32021513

Sabamehr, M., & Ekin, C. Ç. (2023). AI-Driven Drought Management System: A Turkish Case Study. 2023 4th International Informatics and Software Engineering Conference (IISEC), DOI: 10.46908/rict.v3i1.67

Singh, V. P., Khedikar, S., Mathew, J., & Garg, T. (2022). Harnessing Artificial Intelligence for Drought Management. In Gupta, R. K., Jain, A., Wang, J., Singh, V. P., & Bharti, S. (Eds.), *Artificial Intelligence of Things for Weather Forecasting and Climatic Behavioral Analysis* (pp. 130–143). IGI Global. DOI: 10.4018/978-1-6684-3981-4.ch009

Tinelli, S., & Juran, I. (2019). Artificial intelligence-based monitoring system of water quality parameters for early detection of non-specific bio-contamination in water distribution systems. *Water Science and Technology: Water Supply*, 19(6), 1785–1792. DOI: 10.2166/ws.2019.057

Tosin Michael, O., Fatai Adeshina, A., & Zamathula Queen, S. (2024). A review of smart water management systems from Africa and the United States. *Engineering Science &Technology Journal*, 5(4), 1231–1242. DOI: 10.51594/estj.v5i4.1014

United Nations. (2023). *UN 2023 Water Conference* UN 2023 Water Conference, New York. https://sdgs.un.org/conferences/water2023

Vanijjirattikhan, R., Khomsay, S., Kitbutrawat, N., Khomsay, K., Supakchukul, U., Udomsuk, S., Suwatthikul, J., Oumtrakul, N., & Anusart, K. (2022). AI-based acoustic leak detection in water distribution systems. *Results in Engineering*, 15, 100557. DOI: 10.1016/j.rineng.2022.100557

Wang, Z., Zhou, X., Wang, H., Huang, Z., Ji, J., Peng, Z., & Jiang, K. (2024). XGB-SEGA coupled energy saving method for wastewater treatment plants. *Applied Water Science*, 14(2), 29. DOI: 10.1007/s13201-023-02081-3

Zhang, L., Liu, Y., Ren, L., Teuling, A. J., Zhu, Y., Wei, L., Zhang, L., Jiang, S., Yang, X., Fang, X., & Yin, H. (2022). Analysis of flash droughts in China using machine learning. *Hydrology and Earth System Sciences*, 26(12), 3241–3261. DOI: 10.5194/hess-26-3241-2022

KEY TERMS AND DEFINITIONS

Drinking Water Management: It refers to ensuring a safe and quality supply to protect public health.

ODS: They are global goals to combat poverty, inequality, climate change, environmental degradation, peace and justice, to achieve a more sustainable and equitable future.

Chapter 13
AI Applications in Drinking-Water Management

P. Selvakumar
https://orcid.org/0000-0002-3650-4548
Nehru Institute of Technology, India

Ratan Rajan Srivastava
https://orcid.org/0009-0000-4565-9667
Shri Ramswaroop Memorial College of Engineering and Management, India

Sumanta Bhattacharya
https://orcid.org/0000-0003-2563-2787
Maulana Abul Kalam Azad University of Technology, India

Abhijeet Das
https://orcid.org/0000-0003-4599-5462
C.V. Raman Global University, Bhubaneswar, India

T. C. Manjunath
https://orcid.org/0000-0003-2545-9160
Rajarajeswari College of Engineering, India

Sandeep Gupta
https://orcid.org/0000-0002-3734-3723
Graphic Era University, Dehradun, India

ABSTRACT

This introduction explores the transformative potential of AI technologies in addressing complex challenges facing drinking water systems, while also examining the ethical considerations and technical hurdles that must be navigated for responsible and effective deployment. Drinking water is a fundamental resource essential for human health, economic prosperity, and ecosystem integrity. However, managing water quality and distribution systems presents significant challenges, exacerbated by population growth, urbanization, climate change impacts, aging infrastructure, and emerging contaminants. Traditional methods of water quality monitoring and management rely on periodic sampling, laboratory analysis, and manual intervention, which are often time-consuming, resource-intensive, and may not provide real-time

DOI: 10.4018/979-8-3693-8074-1.ch013

insights needed to prevent waterborne diseases or respond swiftly to contamination events. Firstly, AI enables real-time detection of water quality deviations and potential contaminants through advanced sensor networks and predictive analytics.

INTRODUCTION TO AI IN DRINKING WATER MANAGEMENT

This introduction explores the transformative potential of AI technologies in addressing complex challenges facing drinking water systems, while also examining the ethical considerations and technical hurdles that must be navigated for responsible and effective deployment. Drinking water is a fundamental resource essential for human health, economic prosperity, and ecosystem integrity. However, managing water quality and distribution systems presents significant challenges, exacerbated by population growth, urbanization, climate change impacts, aging infrastructure, and emerging contaminants. Traditional methods of water quality monitoring and management rely on periodic sampling, laboratory analysis, and manual intervention, which are often time-consuming, resource-intensive, and may not provide real-time insights needed to prevent waterborne diseases or respond swiftly to contamination events. The key opportunities presented by AI in drinking water management are manifold. Firstly, AI enables real-time detection of water quality deviations and potential contaminants through advanced sensor networks and predictive analytics. By analyzing historical data and sensor readings, AI algorithms can forecast water quality trends, identify early warning signs of contamination, and trigger alerts or automated responses to mitigate risks before they escalate. This proactive approach not only enhances public health protection but also minimizes operational costs associated with reactive measures and infrastructure maintenance.

Secondly, AI facilitates predictive modeling and scenario analysis to optimize water treatment processes, resource allocation, and infrastructure planning. ML algorithms predict water demand patterns based on historical consumption data and environmental variables, enabling utilities to adjust pumping schedules, optimize chemical dosing, and manage reservoir levels efficiently. Deep learning models analyze complex datasets from satellite imagery and remote sensing technologies to monitor watershed dynamics, assess water availability, and inform drought management strategies in water-stressed regions (Antunes A, et al., 2018). Furthermore, AI-driven decision support systems enhance operational efficiencies and resilience in response to dynamic environmental conditions and regulatory requirements. Autonomous systems can autonomously adjust treatment parameters, optimize energy consumption, and prioritize maintenance activities based on real-time data and predictive insights. This adaptive approach reduces energy consumption, minimizes carbon footprints, and enhances the overall sustainability of drinking water treatment facilities. Despite these opportunities, integrating AI into drinking water management poses several challenges that warrant careful consideration. One of the primary challenges is ensuring data quality, reliability, and interoperability across heterogeneous data sources. AI algorithms rely on accurate and diverse datasets to generate reliable predictions and actionable insights. Therefore, harmonizing data standards, improving sensor accuracy, and addressing data biases are critical for enhancing the robustness and reliability of AI-driven decision-making in water quality management.

AI-POWERED WATER QUALITY PREDICTION AND MONITORING: REAL-TIME DETECTION OF CONTAMINANTS

AI-powered water quality prediction and monitoring systems represent a cutting-edge approach to ensuring the safety and sustainability of water resources. This capability is particularly valuable in dynamic environments where factors influencing water quality, such as weather patterns or land use changes, can vary unpredictably. Satellite technology plays a crucial role in AI-powered water quality monitoring by providing spatially extensive data on water bodies. Satellites equipped with remote sensing instruments can capture images of water surfaces and analyze spectral signatures to assess parameters like algae blooms or sedimentation. These data are then integrated into AI models to enhance spatial understanding and provide insights into regional water quality trends (Abhilasha A., et al., 2022). By combining satellite data with ground-based sensor networks, AI-powered systems offer a holistic view of water quality across different scales, from local water bodies to entire watersheds. The integration of AI into water quality monitoring also addresses challenges related to data management and interpretation. sensor networks, the volume of data generated from water monitoring has increased exponentially. AI algorithms can sift through this data, identify relevant information, and prioritize actionable insights for decision-makers. In addition to real-time monitoring and predictive capabilities, AI-powered systems facilitate adaptive management strategies for water quality improvement. By continuously analyzing feedback loops between management interventions and water quality outcomes, these systems can optimize resource allocation and intervention strategies over time (Balduzzi D, et al., 2017). For example, AI models can recommend optimal locations for installing green infrastructure or suggest adjustments to wastewater treatment processes based on current environmental conditions.

Moreover, AI-powered water quality monitoring has the potential to enhance public engagement and transparency in water management practices. By making real-time water quality data accessible to the public through interactive platforms and mobile applications, communities can become more actively involved in monitoring efforts and advocating for sustainable water use practices. This transparency fosters trust between stakeholders and decision-makers, encouraging collaborative efforts to address water quality challenges effectively (Cutore P, et al., 2008). Despite these advancements, AI-powered water quality monitoring systems also face certain limitations and challenges. One critical consideration is the need for reliable data sources and calibration of sensor technologies to ensure the accuracy and reliability of AI predictions. Additionally, ethical concerns related to data privacy, algorithm bias, and the equitable distribution of monitoring resources must be carefully addressed to uphold fairness and accountability in water management practices. Looking ahead, the ongoing evolution of AI technologies holds promise for further enhancing the capabilities of water quality prediction and monitoring systems. Future advancements may include the integration of advanced AI techniques, such as reinforcement learning and anomaly detection, to improve system robustness and adaptability. Furthermore, partnerships between researchers, technology developers, and water managers will be essential to leverage AI innovations effectively and address emerging challenges in water quality management. In conclusion, AI-powered water quality prediction and monitoring systems represent a transformative approach to safeguarding water resources in an increasingly complex environment (Drogkoula M, et al., 2023). By harnessing the power of AI, these systems enable real-time detection of contaminants, predictive insights into water quality trends, and adaptive management strategies to promote sustainable water use practices. While challenges remain, the continued advancement and integration of AI technologies hold the potential to revolutionize how we monitor, manage, and protect water resources for future generations.

Machine Learning for Water Treatment Process Optimization: Efficiency and Cost Savings

Machine learning (ML) has emerged as a powerful tool in optimizing water treatment processes, offering substantial benefits in terms of efficiency improvements and cost savings. Water treatment is critical for ensuring the availability of clean and safe drinking water, as well as for industrial processes and environmental protection (Ehret DL, et al., 2011). One of the primary applications of ML in water treatment is in predictive modeling and process optimization. By learning from past performance and real-time data inputs, these models can anticipate changes in water quality and adjust treatment processes accordingly. For example, predictive models can optimize chemical dosing schedules based on incoming water quality characteristics, thereby reducing chemical usage while maintaining water quality standards. ML techniques also contribute to improving the efficiency of filtration and disinfection processes in water treatment. By analyzing data on filter performance and microbial levels, ML algorithms can optimize backwashing schedules for filters to prevent clogging and ensure consistent water flow rates. Moreover, ML can enhance the effectiveness of disinfection processes by dynamically adjusting chlorine dosing rates based on real-time measurements of microbial contamination levels, thus reducing operational costs while meeting regulatory requirements (Firat M, et al., 2010). Furthermore, ML plays a crucial role in anomaly detection and fault diagnosis within water treatment systems. By continuously monitoring sensor data and operational parameters, ML algorithms can detect deviations from expected performance indicators that may indicate equipment malfunctions or process inefficiencies. Early detection of anomalies allows operators to take proactive measures, such as scheduling maintenance or adjusting operational parameters, to prevent costly breakdowns and ensure uninterrupted water supply.

In addition to operational efficiency improvements, ML-based optimization strategies contribute to significant cost savings for water treatment facilities. By minimizing chemical usage, energy consumption, and maintenance costs through more precise process control and predictive maintenance, ML algorithms help reduce overall operational expenditures. For instance, optimizing energy-intensive processes like pumping and aeration based on real-time demand and environmental conditions can lead to substantial energy savings over time. ML also supports sustainable water management practices by facilitating data-driven decision-making and resource allocation. By analyzing historical trends and environmental factors, ML algorithms can recommend strategies for water reuse and recycling, optimizing the allocation of water resources in regions prone to water scarcity (Gao X, et al., 2020). This capability is particularly valuable in enhancing resilience to climate change impacts and variability in water availability. Despite these benefits, deploying ML in water treatment optimization poses several challenges that must be addressed. These include the need for high-quality data for training and validation, as well as ensuring the robustness and interpretability of ML models in dynamic and complex water treatment environments. Moreover, the integration of ML into existing operational frameworks requires collaboration between water utilities, technology developers, and regulatory agencies to ensure compliance with safety standards and regulatory requirements. Looking ahead, the ongoing advancement of ML techniques holds promise for further enhancing water treatment process optimization. Future developments may include the application of advanced ML and adaptive control strategies (Gautam J, et al., 2020). Moreover, the continued expansion of data sources and digitalization initiatives in the water sector will provide opportunities for leveraging big data analytics and AI-driven insights to improve operational efficiencies and sustainability outcomes.

AI-Driven Leak Detection and Pipe Failure Prediction: Reducing Water Loss and Damage

AI-driven leak detection and pipe failure prediction systems represent a significant advancement in addressing the longstanding challenges of water loss and infrastructure damage in water distribution networks. These systems harness artificial intelligence (AI) algorithms to analyze data from sensors, flow meters, and other sources in real-time, enabling early detection of leaks and predicting potential failures before they occur. By leveraging patterns, these systems offer proactive solutions to mitigate water loss, minimize infrastructure damage, and improve overall operational efficiency (Huang H, et al., 2021). In contrast, AI algorithms can continuously analyze data from a network of sensors installed throughout the water distribution system. These sensors monitor parameters such as flow rates, pressure variations, acoustic signals, and water quality indicators. AI models trained on historical data can distinguish normal operating conditions from anomalies that indicate potential leaks, thereby triggering alerts to operators in real-time.

Moreover, AI facilitates predictive maintenance strategies by forecasting potential pipe failures based on data analysis and machine learning techniques. By integrating data from sensors, maintenance records, weather forecasts, and historical failure patterns, AI models can predict the likelihood and location of future pipe failures. This predictive capability enables utilities to prioritize maintenance activities, replace aging infrastructure proactively, and allocate resources efficiently to prevent costly repairs and service disruptions. AI-driven leak detection and failure prediction systems also contribute to reducing non-revenue water (NRW) losses, which encompass both physical losses from leaks and unauthorized consumption (Ioffe S, et al., 2015). NRW is a significant issue globally, with some regions experiencing high levels of water loss due to aging infrastructure, inadequate maintenance, and operational inefficiencies. By quickly identifying and repairing leaks, AI-powered systems help utilities minimize water wastage and improve revenue recovery, thereby supporting financial sustainability and ensuring reliable water supply to consumers. Furthermore, AI enhances the accuracy and reliability of leak detection by incorporating advanced data analytics techniques such as anomaly detection, pattern recognition, and statistical modeling. This adaptive learning process is particularly valuable in dynamic environments where factors influencing leak occurrence, such as seasonal variations or changes in water demand, can affect system performance.

By minimizing leaks and optimizing system efficiency, utilities can lower their carbon footprint and mitigate the environmental impact of water distribution operations. This aligns with global efforts to promote sustainable water management practices and achieve water conservation goals in the face of increasing water scarcity and climate change impacts. Moreover, the deployment of AI in leak detection and pipe failure prediction fosters innovation in water infrastructure management and smart city initiatives. As part of digital transformation efforts, utilities are increasingly adopting IoT devices, cloud computing platforms, and advanced sensor technologies to enhance data collection and analysis capabilities. AI serves as a cornerstone of these smart technologies, enabling utilities to harness real-time data insights for decision-making. Despite the benefits, the implementation of AI-driven leak detection systems poses several challenges that must be addressed (Kambalimath S, et al., 2020). Additionally, integrating AI technologies into existing water infrastructure requires investments in sensor deployment, network connectivity, and staff training to maximize system effectiveness and reliability.

Smart Water Grids: AI-Enabled Infrastructure for Efficient Water Distribution

Smart water grids leverage AI algorithms, advanced sensor technologies, and real-time data analytics to optimize operations, improve asset management, and respond effectively to changing conditions. At the heart of smart water grids is the integration of AI algorithms that analyze vast amounts of data from sensors deployed throughout the distribution network. These sensors monitor parameters such as water flow rates, pressure levels, water quality indicators, and operational statuses of pumps and valves. By continuously collecting and analyzing data in real-time. This proactive approach allows utilities to identify potential issues before they escalate, thereby minimizing service disruptions and reducing operational costs associated with reactive maintenance. One of the primary benefits of AI-enabled smart water grids is their ability to optimize water distribution efficiency. AI algorithms can dynamically adjust water flow and pressure levels based on real-time demand patterns and network conditions. For example, during periods of low demand, AI can reduce pumping rates to conserve energy and minimize leakage losses. Conversely, during peak demand periods, AI can optimize pumping schedules to ensure adequate water supply while maintaining system reliability. This adaptive control strategy not only improves water distribution efficiency but also reduces energy consumption and operational expenses for utilities.

Moreover, AI enhances asset management practices within smart water grids by enabling predictive maintenance strategies (Leitão J, et al., 2019). By analyzing historical maintenance records, sensor data, and environmental factors, AI models can forecast the remaining useful life of critical infrastructure components such as pipes, pumps, and valves. Predictive maintenance allows utilities to prioritize maintenance activities, schedule repairs proactively, and extend the lifespan of assets, thereby optimizing capital investments and reducing lifecycle costs. In addition to operational efficiency improvements, AI-driven smart water grids contribute to water conservation and environmental sustainability. By accurately monitoring and controlling water flows, detecting leaks promptly, and optimizing distribution networks, these systems help minimize water losses and promote efficient water use practices. This is particularly crucial in regions facing water scarcity challenges, where every drop of water saved through smart management practices contributes to long-term water resource sustainability. Furthermore, the deployment of AI in smart water grids supports resilience and reliability in water supply systems. AI algorithms can analyze historical data on weather patterns, demand fluctuations, and infrastructure performance to develop predictive models for managing droughts, floods, and other extreme weather events. By anticipating these challenges, utilities can implement adaptive strategies such as water rationing, demand management programs, and emergency response plans to ensure continuity of service and minimize impacts on consumers.

The integration of AI technologies into smart water grids also facilitates real-time decision-making and data-driven insights for water utility operators and stakeholders. Advanced data analytics platforms enable utilities to visualize network performance metrics, identify trends, and generate actionable insights to optimize operational workflows and strategic planning (Mahadev Madgule, et al.,2023). This transparency and accessibility of data empower utilities to make informed decisions, allocate resources efficiently, and enhance customer service levels through improved communication and responsiveness. Additionally, AI-enabled smart water grids support the evolution of smart cities and digital transformation initiatives by integrating with other urban infrastructure systems such as energy grids, transportation networks, and environmental monitoring systems. By sharing data and coordinating operations across interconnected infrastructure domains, smart cities can achieve synergies in resource management, enhance urban resilience, and improve overall quality of life for residents. Despite the transformative benefits, the adoption of AI-enabled smart water grids faces several challenges that require careful consideration and mitigation. Moreover, investments in workforce training and capacity-building are essential to equip personnel with the skills and knowledge needed to deploy, operate, and maintain AI-driven technologies effectively.

Looking ahead, the continued advancement of AI technologies holds promise for further enhancing the capabilities and scalability of smart water grids. Future developments may include the application of advanced AI techniques such as deep learning, reinforcement learning, and distributed AI for decentralized decision-making in water distribution networks (Oliveira L, et al., 2009). Furthermore, innovations in sensor technology, IoT devices, and cloud computing platforms will provide opportunities to expand real-time monitoring capabilities, optimize predictive modeling accuracy, and enable adaptive control strategies in response to evolving water management challenges.

AI-Assisted Water Quality Management: Regulatory Compliance and Risk Assessment

AI-assisted water quality management is revolutionizing how regulatory compliance and risk assessment are approached in the water industry. With the increasing complexity of environmental regulations and the growing challenges posed by pollution and climate change, water utilities and regulatory agencies are turning to artificial intelligence (AI) to enhance their monitoring, analysis, and decision-making capabilities. One of the primary applications of AI in water quality management is in monitoring and compliance with regulatory standards. For example, AI can identify exceedances in parameters such as turbidity levels, pH, dissolved oxygen, and concentrations of contaminants like heavy metals or pathogens. This early detection allows water utilities to take prompt corrective actions, such as adjusting treatment processes or implementing targeted interventions, to ensure compliance with regulatory requirements and protect public health (Pacchin E, et al., 2019). Moreover, AI enhances the accuracy and reliability of risk assessment in water quality management. By integrating data analytics and machine learning techniques, AI models can assess potential risks associated with sources of contamination, infrastructure vulnerabilities, and environmental factors affecting water quality. These models can predict the likelihood of pollution events, such as harmful algal blooms or industrial spills, based on historical patterns and real-time monitoring data. By forecasting these risks, water managers can implement preventive measures, develop contingency plans, and allocate resources strategically to minimize the impact on water resources and ecosystems.

AI-driven risk assessment also plays a crucial role in optimizing resource allocation and emergency response planning. By analyzing spatial data and modeling scenarios, AI algorithms can prioritize areas for monitoring and intervention based on the severity of risks identified. For example, in regions prone to water scarcity or pollution hotspots, AI can recommend targeted monitoring efforts and adaptive management strategies to mitigate risks and ensure sustainable water management practices (Richards CE, et al, 2023). In addition to regulatory compliance and risk assessment, AI-driven water quality management supports proactive decision-making and policy development. By providing insights into trends, correlations, and causal relationships within water quality data, AI enables evidence-based policymaking and regulatory reforms. These insights empower policymakers and regulatory agencies to set more informed standards, guidelines, and enforcement measures that promote sustainable water use, protect ecosystems, and safeguard public health. Moreover, AI technologies enhance transparency and accountability in water quality management by making data more accessible and understandable to stakeholders. Advanced data visualization tools and interactive dashboards allow policymakers, researchers, and the public to explore water quality trends, monitor compliance with regulatory standards, and track the effectiveness of management interventions. This transparency fosters trust, collaboration, and public engagement in decision-making processes, promoting shared responsibility for protecting water resources and ensuring equitable access to clean and safe drinking water.

Despite the transformative benefits of AI in water quality management, several challenges and considerations must be addressed to maximize its effectiveness and ethical use. These include the need for privacy protection, and security against cyber threats. Additionally, AI algorithms must be transparent, interpretable, and unbiased to facilitate trust among stakeholders and ensure equitable outcomes in decision-making processes (Ramakrishnan T., et al., 2022). Furthermore, investments in infrastructure, capacity-building, and interdisciplinary collaboration are essential to empower water utilities, regulatory agencies, and researchers to leverage AI technologies effectively and address emerging water quality challenges. Looking ahead, the continued evolution of AI technologies holds promise for advancing water quality management practices and addressing complex environmental issues. Future developments may include the integration of AI with emerging technologies such as blockchain for secure data sharing and decentralized decision-making in water governance. Moreover, advancements in sensor technology, satellite imagery, and IoT devices will provide opportunities to expand real-time monitoring capabilities, enhance predictive modeling accuracy, and support adaptive management strategies in response to evolving water quality dynamics.

Predictive Maintenance for Water Treatment Plants: AI-Driven Asset Management

Predictive maintenance powered by artificial intelligence (AI) is revolutionizing asset management strategies in water treatment plants, offering substantial benefits in efficiency, cost savings, and operational reliability (Selvakumar, P., et al., 2024). Water treatment plants are critical infrastructures tasked with ensuring the supply of clean and safe drinking water to communities while managing operational challenges such as aging infrastructure, stringent regulatory requirements, and increasing maintenance costs. The implementation of AI-driven predictive maintenance begins with the deployment of sensors and IoT devices throughout the water treatment plant. The data collected from these sensors, including temperature, pressure, vibration levels, flow rates, and energy consumption, are transmitted to AI platforms for real-time analysis. AI algorithms, including machine learning techniques such as regression, decision trees, and neural networks, process this data to identify patterns, anomalies, and early warning signs indicative of potential equipment failures.

One of the primary advantages of AI-driven predictive maintenance is its ability to detect subtle changes and degradation in equipment performance that may precede catastrophic failures. By analyzing historical data and learning from past maintenance records, AI models can develop predictive insights into the remaining useful life of assets. For example, algorithms can detect abnormal vibration patterns in pumps, indicating potential bearing wear or misalignment, and recommend preventive maintenance actions before these issues escalate to failure. Similarly, AI can predict the buildup of biofilm or scale in filtration systems based on water quality data, prompting timely cleaning or maintenance activities to maintain operational efficiency. Traditional preventive maintenance practices often involve scheduled inspections and component replacements based on predefined intervals, which may result in unnecessary maintenance or missed opportunities to address emerging issues (Selvakumar, P., et al., 2024). In contrast, AI algorithms analyze real-time data to dynamically adjust maintenance schedules based on the actual condition of equipment. This approach minimizes downtime, maximizes asset availability, and optimizes the allocation of maintenance personnel and resources.

Moreover, AI enhances asset performance optimization by facilitating condition-based monitoring and decision support systems. By integrating AI with digital twin technologies, water treatment plants can create virtual models of their physical assets to simulate operational scenarios, predict performance under different conditions, and optimize process parameters. For instance, digital twins can simulate the impact of varying water flow rates or chemical dosing strategies on equipment performance, allowing operators to identify optimal operating conditions and improve energy efficiency. In addition to operational benefits, AI-driven predictive maintenance contributes to sustainability and resilience in water treatment operations. By reducing unplanned downtime and improving asset reliability, utilities can minimize energy consumption, greenhouse gas emissions, and water wastage associated with emergency repairs or inefficient operations (Tang ZQ, et al., 2012). This supports environmental stewardship goals and enhances the overall sustainability of water treatment processes. Moreover, by extending the lifespan of assets through proactive maintenance strategies, utilities can defer capital expenditures for equipment replacement, thereby optimizing financial resources and improving long-term budget planning. The deployment of AI in predictive maintenance for water treatment plants also fosters a culture of innovation and continuous improvement within the water industry. By leveraging AI's capabilities to analyze big data, identify performance trends, and predict equipment failures, utilities can drive operational excellence and adaptability in response to evolving regulatory requirements, technological advancements, and customer expectations. This proactive approach enables utilities to stay ahead of challenges such as aging infrastructure, changing water quality standards, and population growth, ensuring reliable water supply and service delivery to communities.

AI-Driven Water Demand Forecasting: Optimizing Water Supply and Distribution

AI-driven water demand forecasting represents a critical advancement in optimizing water supply and distribution management, addressing challenges such as population growth, urbanization, climate variability, and aging infrastructure. Water utilities and authorities rely on accurate demand forecasts to efficiently allocate water resources, plan infrastructure investments, and ensure reliable service delivery to consumers (Tao M, et al., 2023). Traditional forecasting methods often rely on historical data trends and statistical models, which may not fully capture the complexity and dynamics of modern water systems. AI technologies, however, offer capabilities to analyze large datasets, integrate diverse sources of information, and predict future water demand patterns with greater accuracy and granularity. The foundation of AI-driven water demand forecasting lies in its ability to process and analyze diverse datasets from multiple sources. These datasets include historical water consumption records, demographic data, land use patterns, weather forecasts, economic indicators, and seasonal variations. By integrating these data streams, AI algorithms can identify correlations, trends, and relationships that influence water demand fluctuations over time. For example, AI can analyze historical consumption patterns in relation to factors such as population growth, weather conditions (e.g., temperature, precipitation), and socio-economic changes to develop predictive models.

Machine learning techniques play a crucial role in AI-driven water demand forecasting by enabling algorithms to learn from data, adapt to changing conditions, and improve forecast accuracy over time. Supervised learning algorithms, such as regression and time series analysis, are commonly used to model and predict water demand based on historical consumption patterns and relevant variables (Uhlenbrook S, et al., 2022). These models can capture seasonal variations, day-of-week effects, holidays, and special events that impact water usage behavior, allowing utilities to anticipate peak demand periods and adjust supply accordingly. Moreover, AI facilitates real-time demand forecasting by processing streaming data from smart meters, IoT devices, and sensor networks deployed throughout the water distribution network. These sensors monitor water flow rates, pressure levels, and operational parameters in real-time, providing continuous updates on system performance and consumer behavior. AI algorithms can analyze this real-time data to detect anomalies, identify emerging trends, and forecast short-term demand fluctuations, enabling utilities to optimize operational decisions and respond promptly to changing conditions.

One of the significant advantages of AI-driven water demand forecasting is its ability to support proactive water resource management strategies (Vijayakumar G, et al., 2024). By providing accurate predictions of future demand, utilities can optimize water supply planning, storage capacity, and distribution network design. For instance, AI forecasts can inform decisions on infrastructure investments, such as the construction of new reservoirs or pipelines, to meet projected demand growth and ensure resilience against climate-related uncertainties, such as droughts or floods. Furthermore, AI-driven forecasting enhances operational efficiency by optimizing water production and distribution processes. Utilities can use demand forecasts to adjust pumping schedules, manage reservoir levels, and optimize water treatment operations based on anticipated consumption patterns. This proactive approach helps utilities minimize energy consumption, reduce pumping costs, and optimize chemical usage while maintaining water quality standards and regulatory compliance. In addition to operational benefits, AI-driven water demand forecasting supports sustainability goals by promoting efficient water use practices and conservation efforts (Villarin MC, et al., 2019). Utilities can use demand forecasts to implement demand-side management programs, such as water conservation campaigns, leak detection initiatives, and tiered pricing structures, to encourage consumers to use water more responsibly. By aligning supply with actual demand and promoting water-saving behaviors, AI-driven forecasting contributes to reducing water wastage, preserving natural resources, and mitigating environmental impacts associated with water extraction and distribution. In conclusion, AI-driven water demand forecasting represents a transformative approach to optimizing water supply and distribution management, enhancing operational efficiency, resilience, and sustainability. By harnessing AI's analytical capabilities to predict demand patterns, utilities can make informed decisions, improve service delivery, and ensure reliable access to clean and safe water for communities worldwide (Wang J, et al., 2016). As these technologies continue to evolve, their integration into water management practices will play a crucial role in addressing global water challenges and advancing towards a more water-secure future.

Cybersecurity in AI-Powered Drinking Water Management: Threats and Countermeasures

Cybersecurity in AI-powered drinking water management systems is increasingly crucial as utilities leverage artificial intelligence (AI) to optimize operations, enhance efficiency, and ensure the safety and reliability of drinking water supplies. AI-driven technologies offer significant benefits in monitoring, managing, and controlling water treatment processes and distribution networks, but they also introduce new cybersecurity challenges and vulnerabilities that must be addressed to safeguard against potential threats. One of the primary cybersecurity concerns in AI-powered drinking water management systems is the protection of data integrity and confidentiality. These systems rely on vast amounts of sensitive data, including operational data from sensors, customer information, water quality metrics, and network configurations. Cyberattacks targeting these data can compromise system integrity, leading to unauthorized access, data manipulation, or loss of critical information (Wang C, et al., 2023). For instance, malicious actors could alter sensor readings to disrupt water treatment processes or manipulate flow rates in distribution networks, potentially compromising water quality and public health. Another critical cybersecurity risk is the threat of ransomware and denial-of-service (DoS) attacks against AI-powered systems. Ransomware attacks involve malicious software that encrypts data or systems, demanding ransom payments for decryption, while DoS attacks aim to overwhelm system resources, causing disruptions in service availability. These attacks can disrupt water treatment operations, affect control systems, and impair communication networks essential for monitoring and managing water infrastructure. Utilities must implement robust cybersecurity measures, such as network segmentation, encryption protocols, and intrusion detection systems, to mitigate these risks and ensure continuous service delivery.

In response to these cybersecurity challenges, water utilities and stakeholders are increasingly focusing on implementing proactive cybersecurity strategies and best practices. Collaboration with cybersecurity experts, industry partners, and regulatory agencies is essential to stay abreast of emerging threats, share best practices, and implement standardized security measures across the sector. Furthermore, regulatory frameworks and industry standards play a crucial role in guiding cybersecurity practices and ensuring compliance with legal requirements. Regulatory bodies may require utilities to adhere to cybersecurity standards, undergo audits, and report cyber incidents to mitigate risks and protect public safety (Xu TY, et al., 2015). Compliance with standards such as the NIST Cybersecurity Framework, ISO/IEC 27001, and industry-specific guidelines helps utilities build resilient cybersecurity defenses and demonstrate accountability in safeguarding drinking water infrastructure. Looking ahead, the future evolution of cybersecurity in AI-powered drinking water management will continue to be shaped by advancements in technology, threat landscapes, and regulatory frameworks. Emerging technologies such as quantum computing, AI-driven cybersecurity tools, and decentralized architectures offer opportunities to enhance detection capabilities, mitigate risks, and strengthen resilience against evolving cyber threats. Moreover, investments in workforce training, research, and development will be crucial to build cybersecurity expertise within the water sector and ensure sustainable protection of critical infrastructure.

AI-Enabled Customer Engagement: Personalized Water Usage Feedback and Education

AI-enabled customer engagement is transforming how water utilities interact with consumers by offering personalized water usage feedback and educational resources. Traditionally, water utilities have focused on delivering reliable water supply and managing infrastructure, with limited direct engagement with consumers regarding their water usage behaviors. However, advancements in artificial intelligence (AI) technologies are revolutionizing this approach by enabling utilities to analyze consumption patterns, provide real-time feedback, and empower consumers to make informed decisions about their water usage habits. At the core of AI-enabled customer engagement is the ability to leverage data analytics to understand and predict individual and community-level water consumption patterns (Yang L, et al., 2023). AI algorithms process data from smart meters, historical consumption records, weather conditions, demographic information, and behavioral data to identify trends and insights. By analyzing this data, utilities can create personalized profiles for consumers, highlighting their usage patterns, comparing them to similar households, and identifying opportunities for conservation. One of the key benefits of AI-enabled customer engagement is the delivery of personalized water usage feedback to consumers. Through mobile apps, web portals, or smart home devices, consumers can access real-time information about their water consumption, including daily usage trends, peak demand periods, and comparisons to previous periods or similar households. This transparency empowers consumers to monitor their water usage, identify inefficiencies, and take proactive steps to conserve water and reduce utility bills.

Moreover, AI facilitates targeted educational initiatives aimed at promoting water conservation and sustainability practices. Utilities can use AI insights to develop personalized conservation tips, behavior change nudges, and educational materials tailored to individual consumer profiles. For example, AI algorithms can recommend water-saving practices based on specific consumption patterns, such as adjusting irrigation schedules, fixing leaks, or upgrading to water-efficient appliances. By providing relevant and actionable information, utilities can enhance consumer awareness, foster behavioral change, and contribute to overall water resource sustainability.

AI-enabled customer engagement also enhances communication and interaction between utilities and consumers. By leveraging AI-powered chatbots or virtual assistants, utilities can provide personalized customer support, answer inquiries about billing, meter readings, conservation programs, and water quality concerns in real-time (Zubaidi SL, et al., 2020c). These interactive platforms improve customer satisfaction, streamline service delivery, and strengthen trust and transparency in utility-customer relationships. Furthermore, AI-driven customer engagement supports demand-side management strategies by encouraging participation in incentive programs, water efficiency rebates, and community initiatives. Utilities can use AI analytics to segment consumers based on their conservation potential or responsiveness to behavioral interventions, targeting outreach efforts more effectively. For example, AI can identify high-water users who may benefit from efficiency audits or personalized conservation coaching, thereby optimizing resource allocation and maximizing program effectiveness.

Looking ahead, the future evolution of AI-enabled customer engagement in water utilities will continue to be shaped by advancements in AI technologies, data analytics, and customer-centric strategies. Emerging technologies such as natural language processing (NLP), sentiment analysis, and predictive modeling offer opportunities to enhance personalized customer interactions, anticipate consumer needs, and deliver tailored services. Moreover, collaborations between utilities, technology providers, and research institutions will drive innovation in AI applications for water management, supporting sustainable development goals and enhancing water resource stewardship. In conclusion, AI-enabled customer engagement represents a transformative approach to promoting water conservation, enhancing consumer awareness, and improving operational efficiency in water utilities. By harnessing AI's analytical capabilities to deliver personalized usage feedback, educational resources, and responsive customer service, utilities can empower consumers to contribute actively to water sustainability efforts. As utilities continue to innovate and adapt to digital technologies, AI-enabled customer engagement will play a crucial role in shaping a resilient and water-secure future for communities worldwide.

Water Quality Monitoring with AI-Powered Sensors: Advanced Analytics and IoT Integration

Water quality monitoring is undergoing a significant transformation with the integration of AI-powered sensors, advanced analytics, and IoT (Internet of Things) technologies. Traditionally, water quality monitoring has relied on periodic sampling and laboratory analysis, which can be time-consuming, costly, and limited in spatial coverage. The emergence of AI-powered sensors and IoT devices offers real-time monitoring capabilities, enabling continuous data collection, rapid detection of water quality anomalies, and proactive management of water resources. Traditional monitoring methods may miss transient or localized pollution events that can impact water safety and ecosystem health. AI algorithms analyze continuous sensor data streams to detect subtle changes in water quality parameters, such as sudden spikes in pollutant levels or shifts in microbial populations, which may indicate potential contamination sources or environmental stressors. These data fusion techniques allow AI models to contextualize water quality data within broader environmental conditions and hydrological processes. For example, AI can correlate changes in water turbidity with precipitation events or land use activities upstream, providing insights into the sources and impacts of sedimentation or runoff pollution. Furthermore, IoT integration enhances the scalability and interoperability of AI-powered water quality monitoring systems. IoT devices, such as smart sensors and communication modules, facilitate data transmission, remote monitoring, and real-time connectivity across distributed networks. These devices enable utilities to monitor water quality in remote or hard-to-access locations, improve response times to emergencies, and optimize operational efficiency. For instance, IoT-enabled sensors can transmit real-time data to centralized monitoring platforms, where AI algorithms analyze trends, generate predictive models, and visualize insights for decision-makers.

In addition to real-time monitoring and data analytics, AI-powered sensors support predictive modeling and scenario analysis to enhance proactive management of water resources. By analyzing historical trends and forecasting future conditions, AI algorithms can predict potential water quality challenges, such as algal blooms, contamination events, or seasonal variations in pollutant levels. Utilities can use these predictions to implement preventive measures, develop contingency plans, and allocate resources strategically to mitigate risks and safeguard water supplies. Moreover, AI-driven water quality monitoring promotes sustainability by optimizing resource allocation, reducing operational costs, and supporting evidence-based decision-making. Utilities can prioritize investments in water treatment upgrades, infrastructure maintenance, and pollution control measures based on data-driven insights into water quality trends and risk assessments. This proactive approach helps utilities meet regulatory compliance requirements, enhance environmental stewardship, and ensure sustainable water management practices for future generations. Despite the transformative benefits of AI-powered sensors in water quality monitoring, several challenges and considerations must be addressed to maximize their effectiveness and adoption. These include the need for standardized protocols for sensor calibration, data validation, and quality assurance to ensure accuracy and reliability of sensor measurements. Additionally, interoperability standards and cybersecurity protocols are essential to protect sensor networks from cyber threats, data breaches, and unauthorized access.

Furthermore, investments in infrastructure, technology deployment, and workforce training are necessary to build capacity and expertise in AI-powered water quality monitoring. Collaboration between water utilities, technology developers, researchers, and regulatory agencies is crucial to advance research, innovation, and best practices in sensor technology and data analytics. By fostering partnerships and knowledge exchange, stakeholders can accelerate the adoption of AI-powered sensors and drive continuous improvement in water quality monitoring capabilities.

Looking ahead, the ongoing evolution of AI technologies holds promise for advancing water quality monitoring practices and addressing complex environmental challenges. Future developments may include the integration of AI with emerging technologies such as blockchain for secure data sharing and decentralized decision-making in water governance. Moreover, advancements in sensor technology, satellite imagery, and IoT devices will provide opportunities to expand real-time monitoring capabilities, enhance predictive modeling accuracy, and support adaptive management strategies in response to evolving water quality dynamics.

AI-Driven Water Supply Chain Management: Traceability and Transparency

AI-driven water supply chain management represents a transformative approach to enhancing traceability, transparency, and efficiency across the entire lifecycle of water resources—from sourcing and treatment to distribution and consumption. Traditionally, water supply chains have faced challenges such as aging infrastructure, resource constraints, regulatory compliance, and increasing demand due to population growth and urbanization. AI technologies offer utilities and stakeholders in the water sector powerful tools to optimize operations, improve decision-making, and ensure sustainable management of water resources. Moreover, AI-driven predictive analytics improves forecasting accuracy and resilience in water supply chain management. By analyzing historical data and real-time inputs, AI algorithms can predict water demand patterns, anticipate seasonal variations, and identify potential supply disruptions. Utilities can use these insights to optimize water storage, plan maintenance activities, and implement contingency measures to mitigate risks associated with climate variability, extreme weather events, or infrastructure failures. This proactive approach enhances operational resilience and ensures reliable water supply to meet growing demands. AI also enhances operational efficiency in water supply chain management by optimizing asset utilization and reducing operational costs. Utilities can use AI-driven optimization models to schedule maintenance activities, optimize pump operations, manage energy consumption, and streamline logistics for water treatment chemicals and equipment. For instance, AI algorithms can analyze real-time sensor data to adjust pumping rates based on demand fluctuations, minimize energy consumption, and extend the lifespan of equipment through predictive maintenance strategies. These efficiencies contribute to cost savings, resource conservation, and improved service delivery for consumers.

Furthermore, AI facilitates data-driven decision-making and strategic planning in water supply chain management. By integrating AI analytics with geographical information systems (GIS), utilities can visualize spatial data, identify geographic hotspots for water demand, and optimize infrastructure investments. AI-powered simulations and scenario analysis enable utilities to assess the impact of infrastructure upgrades, population growth, land use changes, and climate scenarios on water supply and demand dynamics. These insights inform long-term planning strategies, support sustainable development goals, and ensure equitable access to water resources for communities. In addition to operational benefits, AI-driven water supply chain management promotes sustainability by optimizing water use efficiency and reducing environmental impacts. Utilities can use AI to implement demand-side management programs, promote water conservation practices, and incentivize consumers to adopt water-efficient technologies. AI-powered analytics enable utilities to identify high-water-use sectors, implement targeted conservation initiatives, and monitor the effectiveness of water-saving measures over time. By promoting sustainable water management practices, AI contributes to environmental stewardship and resilience against water scarcity challenges. Looking ahead, the future evolution of AI technologies holds promise for advancing water supply chain management capabilities and addressing emerging challenges in the water sector. Innovations in AI algorithms, edge computing, IoT devices, and cloud-based analytics platforms will provide opportunities to enhance real-time monitoring, predictive modeling accuracy, and adaptive management strategies. Moreover, partnerships between utilities, technology providers, academia, and government agencies will drive innovation, research, and knowledge exchange to accelerate the adoption of AI-driven solutions for sustainable water management.

Resilience and Adaptation in AI-Powered Drinking Water Management: Climate Change and Disaster Response

Resilience and adaptation are becoming increasingly critical in the context of AI-powered drinking water management systems, especially in light of climate change impacts and the need for effective disaster response strategies. As climate change intensifies, water utilities are facing unprecedented challenges, including more frequent and severe droughts, floods, heatwaves, and extreme weather events that disrupt water supply, compromise water quality, and threaten infrastructure resilience. AI technologies offer innovative solutions to enhance the resilience of drinking water systems by enabling proactive monitoring, predictive modeling, adaptive management, and rapid response capabilities. AI-powered systems play a pivotal role in enhancing resilience by improving the monitoring and early detection of climate-related risks and water quality issues. Advanced AI algorithms analyze diverse datasets from sensors, satellites, weather forecasts, historical climate data, and operational parameters to identify trends, patterns, and anomalies indicative of changing environmental conditions. For example, AI can detect shifts in precipitation patterns or temperature extremes that may affect water availability and quality, enabling utilities to anticipate risks, optimize water allocation, and implement preemptive measures to mitigate potential impacts on water supply.

Moreover, AI-driven predictive modeling enhances the ability of water utilities to forecast and plan for climate-induced challenges. Machine learning algorithms analyze historical data to develop models that predict future water demand, anticipate changes in water availability, and simulate scenarios under different climate change projections. These predictive capabilities enable utilities to optimize reservoir operations, manage storage capacities, and adjust supply strategies to meet evolving demand patterns and environmental conditions. By integrating AI insights into decision-making processes, utilities can enhance adaptive capacity, improve resource efficiency, and ensure sustainable water management practices in the face of uncertainty. AI-powered systems also support adaptive management strategies that enable utilities to respond swiftly and effectively to climate-related emergencies and disasters. Real-time monitoring of sensor data and AI analytics provide utilities with early warning systems for detecting sudden changes in water quality, infrastructure failures, or operational disruptions caused by extreme weather events. For instance, AI can identify potential vulnerabilities in water distribution networks, prioritize response actions during emergencies, and optimize emergency preparedness plans to minimize service interruptions and protect public health.

Furthermore, AI technologies facilitate data-driven decision-making in disaster response and recovery efforts. By analyzing real-time data from IoT devices, satellite imagery, and social media platforms, AI algorithms can assess damage impacts, coordinate resource allocation, and optimize logistical operations for emergency response teams. Utilities can use AI-driven simulations to model flood scenarios, predict inundation areas, and implement flood mitigation measures to protect critical infrastructure and communities at risk. These proactive measures enhance resilience, reduce recovery costs, and accelerate restoration efforts in the aftermath of climate-related disasters. In addition to enhancing operational resilience, AI-powered drinking water management systems promote environmental sustainability by optimizing water use efficiency and reducing carbon footprints. Utilities can leverage AI analytics to identify opportunities for energy savings, optimize pumping schedules, and minimize greenhouse gas emissions associated with water treatment and distribution processes. AI-driven demand-side management programs promote water conservation behaviors among consumers, incentivize efficiency upgrades, and support sustainable urban water planning initiatives to mitigate climate impacts and enhance ecosystem resilience.

Looking ahead, the future evolution of AI technologies holds promise for advancing resilience and adaptation strategies in drinking water management systems. Innovations in AI algorithms, edge computing, decentralized AI applications, and blockchain technologies will provide opportunities to enhance real-time monitoring, predictive modeling accuracy, and adaptive management capabilities. Moreover, partnerships between utilities, technology providers, academia, and government agencies will drive innovation, research, and knowledge exchange to accelerate the adoption of AI-driven solutions for climate-resilient water management.

ETHICAL AND SOCIAL IMPLICATIONS OF AI IN DRINKING WATER MANAGEMENT: EQUITY AND ACCESS

The integration of AI (Artificial Intelligence) into drinking water management systems introduces a range of ethical and social implications, particularly concerning equity and access to safe drinking water. AI technologies have the potential to revolutionize water management by improving efficiency, optimizing resource allocation, and enhancing environmental sustainability. However, alongside these advancements come significant considerations related to fairness, transparency, accountability, and the equitable distribution of benefits and risks across communities. One of the primary ethical concerns surrounding AI in drinking water management is ensuring equitable access to safe and reliable water resources for all populations. Access to clean water is a fundamental human right, yet disparities in water quality and service levels persist, particularly in underserved or marginalized communities. AI-powered systems must address these disparities by prioritizing equitable access in resource allocation, infrastructure investments, and service delivery strategies. For example, AI analytics can identify areas with inadequate access to clean water and inform targeted interventions to improve water quality, reliability, and affordability in underserved communities. Moreover, the deployment of AI technologies in drinking water management raises concerns about data privacy, security, and algorithmic bias. AI algorithms rely on vast amounts of data, including sensitive information about water usage patterns, demographic characteristics, and infrastructure conditions. Safeguarding data privacy and protecting against unauthorized access or misuse of personal information are paramount to maintaining public trust and ensuring ethical AI deployment. Utilities must implement robust data protection measures, encryption protocols, and transparency mechanisms to mitigate risks and uphold ethical standards in data handling and AI-driven decision-making processes.

Furthermore, algorithmic bias poses a significant ethical challenge in AI applications for drinking water management. Biases embedded in AI algorithms, such as racial, socioeconomic, or geographic biases, can perpetuate inequalities in access to water services and exacerbate existing disparities. For instance, biased algorithms may inadvertently allocate resources unevenly, favoring certain populations over others or reinforcing historical inequities in infrastructure investment and service provision. To mitigate bias, utilities must adopt inclusive and diverse data sets, conduct regular audits of AI models, and incorporate fairness metrics to ensure equitable outcomes in decision-making processes. Another critical ethical consideration is transparency and accountability in AI-driven decision-making. As AI algorithms increasingly influence operational decisions, resource allocation, and policy formulation in water management, stakeholders must have transparency into how AI systems operate, the data they use, and the factors influencing their outputs. Utilities should establish clear governance frameworks, explainable AI techniques, and stakeholder engagement strategies to foster trust, accountability, and public understanding of AI applications in water management. Transparency enables stakeholders to scrutinize decision-making processes, challenge biases, and advocate for equitable and inclusive water policies. Additionally, the ethical implications of AI in drinking water management extend to broader societal impacts, including workforce displacement, community resilience, and environmental justice considerations. Automation of tasks through AI technologies may lead to workforce shifts, requiring upskilling and training programs to ensure a just transition for affected workers. Moreover, AI-driven resilience strategies can enhance community resilience to climate change impacts, such as droughts or floods, by improving water supply reliability and emergency response capabilities. Utilities must engage with communities to understand their needs, preferences, and concerns regarding AI technologies, ensuring that technological advancements contribute to equitable and sustainable outcomes for all residents.

Addressing the ethical and social implications of AI in drinking water management requires collaboration among utilities, policymakers, regulators, community organizations, and technology developers. Multidisciplinary approaches that integrate ethical principles, human rights frameworks, and community engagement strategies are essential to guide responsible AI deployment and mitigate potential harms. By prioritizing equity, transparency, accountability, and inclusivity in AI-driven water management initiatives, stakeholders can leverage technological innovations to enhance access to safe drinking water, promote environmental sustainability, and support resilient communities worldwide.

FUTURE DIRECTIONS IN AI-POWERED DRINKING WATER MANAGEMENT: EMERGING TECHNOLOGIES AND RESEARCH FRONTIERS

Future directions in AI-powered drinking water management are poised to leverage emerging technologies and explore new research frontiers to address complex challenges, enhance sustainability, and ensure resilient water systems. As technological advancements continue to evolve, the integration of AI (Artificial Intelligence) with other cutting-edge innovations holds promise for revolutionizing how drinking water is monitored, managed, and distributed. One of the key areas of advancement lies in the integration of AI with IoT (Internet of Things) devices and sensor networks. IoT-enabled sensors provide real-time data on water quality, infrastructure performance, and environmental conditions, which AI algorithms can analyze to detect anomalies, predict trends, and optimize operational efficiency. By leveraging AI's analytical capabilities with IoT's connectivity, utilities can achieve comprehensive monitoring across distributed networks, improve response times to water quality incidents, and enhance system resilience against climate variability and extreme weather events. Another frontier in AI-powered drinking water management is the development of decentralized AI applications and edge computing solutions. Decentralized AI leverages computing power and data processing capabilities at the network edge, closer to IoT devices and sensors, to enable real-time decision-making and reduce latency in data transmission. Edge AI technologies enhance system autonomy, improve responsiveness to local conditions, and support resilient operations in remote or resource-constrained environments where connectivity may be limited.

Furthermore, AI-driven optimization and control algorithms are advancing towards autonomous decision-making capabilities that enhance operational efficiency and adaptive management in water utilities. Autonomous systems can adjust water treatment processes, optimize pumping schedules, and manage energy consumption based on real-time sensor data and AI insights. These capabilities not only optimize resource use and reduce operational costs but also improve system reliability, resilience, and responsiveness to dynamic environmental conditions and consumer demands. Additionally, AI holds potential in revolutionizing water distribution systems through smart grid technologies and demand-side management strategies. AI algorithms can optimize water flow, pressure management, and leakage detection in distribution networks to minimize losses, improve water quality, and enhance service reliability. By integrating AI-powered smart grids with advanced metering infrastructure (AMI) and consumer engagement platforms, utilities can empower consumers to monitor their water usage, adopt conservation practices, and participate in demand response programs to achieve water efficiency goals.

CONCLUSION

In conclusion, AI applications in drinking water management represent a transformative approach to addressing complex challenges, enhancing operational efficiency, and ensuring sustainable water stewardship. The integration of AI technologies across various facets of water management—from source water protection and treatment to distribution, consumption, and regulatory compliance—offers significant opportunities to optimize resource use, improve water quality, and mitigate risks associated with climate change, population growth, and urbanization. Moreover, AI-driven advancements in predictive modeling and scenario analysis empower utilities to forecast future water demand, simulate climate change impacts, and develop resilience strategies to safeguard water resources and ensure reliable service delivery. These predictive capabilities enable utilities to optimize water allocation, manage storage capacities, and implement adaptive management practices that enhance system resilience against climate variability and extreme weather events. AI-powered technologies also facilitate proactive maintenance and asset management strategies by analyzing operational data, predicting equipment failures, and optimizing maintenance schedules to minimize downtime and extend asset lifespan. This approach not only improves operational efficiency but also reduces costs associated with infrastructure maintenance, enhances service reliability, and supports long-term sustainability goals. Furthermore, AI applications enhance regulatory compliance and risk management in drinking water management by automating monitoring, reporting, and compliance processes. AI algorithms can detect anomalies in water quality parameters, assess regulatory compliance, and generate real-time alerts to facilitate timely corrective actions and regulatory reporting. By streamlining regulatory compliance efforts, utilities can ensure adherence to water quality standards, maintain public health protections, and foster trust and transparency with regulatory agencies and stakeholders.

Additionally, AI-driven consumer engagement initiatives promote water conservation behaviors, empower consumers with real-time usage data, and enhance communication between utilities and customers. AI-powered analytics enable utilities to personalize customer interactions, offer tailored conservation tips, and incentivize efficient water use through behavioral nudges, educational campaigns, and smart metering technologies. By promoting consumer awareness and participation in water conservation efforts, utilities can reduce demand pressures, optimize water distribution, and support sustainable water management practices at the community level. Looking ahead, the future of AI applications in drinking water management holds promise for continued innovation, collaboration, and interdisciplinary research to address emerging challenges and opportunities. Advancements in AI algorithms, edge computing, IoT integration, and blockchain technologies will further enhance real-time monitoring capabilities, predictive modeling accuracy, and adaptive management strategies in water utilities. By fostering partnerships between utilities, technology providers, researchers, and policymakers, stakeholders can accelerate the development and adoption of AI-powered solutions that promote resilience, sustainability, and equitable access to safe drinking water worldwide. In conclusion, AI applications in drinking water management represent a paradigm shift towards data-driven decision-making, operational efficiency, and sustainable resource management. By leveraging AI's analytical capabilities, predictive insights, and technological innovations, utilities can enhance water quality, optimize infrastructure performance, and meet evolving demands in a rapidly changing world. As utilities embrace digital transformation and collaborative approaches, AI-powered solutions will play a pivotal role in shaping a resilient and water-secure future for communities, ensuring that safe drinking water remains accessible and sustainable for generations to come.

REFERENCES

Abhilasha, A., Sreenivasulu, A., & Manimozhi, T. Satheesh kumar P., Selvakumar P. and Singh P.,(2022). "The Model of Smart Sensing Device For Sensitive Nanoclusters Modification in Sensing Properties," 2022 2nd International Conference on Advance Computing and Innovative Technologies in Engineering (ICACITE), Greater Noida, India, 2022, pp. 1043-1047, DOI: 10.1109/ICACITE53722.2022.9823846

Antunes, A., Andrade-Campos, A., Sardinha-Lourenço, A., & Oliveira, M. S. (2018). Short-term water demand forecasting using machine learning techniques. *Journal of Hydroinformatics*, 20(6), 1343–1366. DOI: 10.2166/hydro.2018.163

Balduzzi, D., Frean, M., & Leary, L.. (2017). The shattered gradients problem: if resnets are the answer, then what is the question? In: Proceedings of the 34th international conference on machine learning. *Proceedings of Machine Learning Research*, 70, 342–350.

Cutore, P., Campisano, A., Kapelan, Z., Modica, C., & Savic, D. (2008). Probabilistic prediction of urban water consumption using the SCEM-UA algorithm. *Urban Water Journal*, 5(2), 125–132. DOI: 10.1080/15730620701754434

Drogkoula, M., Kokkinos, K., & Samaras, N. (2023). A comprehensive survey of machine learning methodologies with emphasis in water resources management. *Applied Sciences (Basel, Switzerland)*, 13(22), 12147. DOI: 10.3390/app132212147

Ehret, D. L., Hill, B. D., Helmer, T., & Edwards, D. R. (2011). Neural network modeling of greenhouse tomato yield, growth and water use from automated crop monitoring data. *Computers and Electronics in Agriculture*, 79(1), 82–89. DOI: 10.1016/j.compag.2011.07.013

Firat, M., Turan, M. E., & Yurdusev, M. A. (2010). Comparative analysis of neural network techniques for predicting water consumption time series. *Journal of Hydrology (Amsterdam)*, 384(1-2), 46–51. DOI: 10.1016/j.jhydrol.2010.01.005

Gao, X., Zeng, W., Shen, Y., Guo, Z., Yang, J., Cheng, X., Hua, Q., & Yu, K. (2020). Integrated deep neural networks-based complex system for urban water management. *Complexity*, 2020, 1–12. DOI: 10.1155/2020/8848324

Gautam, J., Chakrabarti, A., Agarwal, S., Singh, A., Gupta, S., & Singh, J. (2020). Monitoring and forecasting water consumption and detecting leakage using an IoT system. *Water Science and Technology: Water Supply*, 20(3), 1103–1113. DOI: 10.2166/ws.2020.035

Huang, H., Zhang, Z., & Song, F. (2021). An ensemble-learning-based method for short-term water demand forecasting. *Water Resources Management*, 35(6), 1757–1773. DOI: 10.1007/s11269-021-02808-4

Ioffe, S., & Szegedy, C. (2015). Batch normalization: accelerating deep network training by reducing internal covariate shift. In: Proceedings of the 32nd international conference on machine learning. *Proceedings of Machine Learning Research*, 37, 448–456.

Kambalimath, S., & Deka, P. C. (2020). A basic review of fuzzy logic applications in hydrology and water resources. *Applied Water Science*, 10(8), 191. DOI: 10.1007/s13201-020-01276-2

Leitão, J., Simões, N., Sá Marques, J. A., Gil, P., Ribeiro, B., & Cardoso, A. (2019). Detecting urban water consumption patterns: A time-series clustering approach. *Water Science and Technology: Water Supply*, 19(8), 2323–2329. DOI: 10.2166/ws.2019.113

Mahadev Madgule, N. (2023). Vinayaka, Yeshwant M. Sonkhaskar, Dhiren Ramanbhai Patel, R. Karthikeyan, P. Selvakumar,(2023). Mechanical properties and microstructure of activated TIG welded similar joints of Inconel alloys by desirability approaches. *Materials Today: Proceedings*, 77, 528–533. DOI: 10.1016/j.matpr.2022.12.250

Oliveira, L., Gonçalves, O., & Uchida, C. (2009). Performance evaluation of dual-flush WC cistern in a multifamily building in Brazil. *Building Services Engineering Research and Technology*, 30(1), 27–36. DOI: 10.1177/0143624408098445

Pacchin, E., Gagliardi, F., Alvisi, S., & Franchini, M. (2019). A comparison of short-term water demand forecasting models. *Water Resources Management*, 33(4), 1481–1497. DOI: 10.1007/s11269-019-02213-y

Ramakrishnan, T., Mohan Gift, M. D., Chitradevi, S., Jegan, R., & Subha Hency Jose, P. Nagaraja, Rajneesh Sharma H.N., Selvakumar P., Sintayehu Mekuria Hailegiorgis, (2022). "Study of Numerous Resins Used in Polymer Matrix Composite Materials", Advances in Materials Science and Engineering, vol. 2022, Article ID 1088926, 8 pages, 2022. DOI: 10.1155/2022/1088926

Richards, C. E., Tzachor, A., Avin, S., & Fenner, R. (2023). Rewards, risks and responsible deployment of artificial intelligence in water systems. *Nature Water*, 1(5), 422–432. DOI: 10.1038/s44221-023-00069-6

Selvakumar, P.. (2024). AI-Based Cameras to Detect a Leak of Chemicals or Gas in Real Time. In Satishkumar, D., Sivaraja, M., & Global, I. G. I. (Eds.), *Industry Applications of Thrust Manufacturing: Convergence with Real-Time Data and AI* (pp. 221–236)., DOI: 10.4018/979-8-3693-4276-3.ch009

Selvakumar, P., Muthusamy, S., Satishkumar, D., Vigneshkumar, P., Selvamurugan, C., & Satheesh Kumar, P. (2024). AI-Powered Tools. In Satishkumar, D., & Sivaraja, M. (Eds.), *Using Real-Time Data and AI for Thrust Manufacturing* (pp. 20–42). IGI Global., DOI: 10.4018/979-8-3693-2615-2.ch002

Tang, Z. Q., Liu, G., Xu, W. N., Xia, Z. Y., & Xiao, H. (2012). Water demand forecasting in Hubei province with BP neural network model. *Advanced Materials Research*, 599, 701–704. . DOI: 10.4028/www.scientific.net/AMR.599.701

Tao, M., Zhang, T., Xie, X., & Liang, X. (2023). Water footprint modeling and forecasting of cassava based on different artificial intelligence algorithms in Guangxi, China. *Journal of Cleaner Production*, 382, 135238. DOI: 10.1016/j.jclepro.2022.135238

Uhlenbrook, S., Yu, W., Schmitter, P., & Smith, D. M. (2022). Optimising the water we eat—Rethinking policy to enhance productive and sustainable use of water in agri-food systems across scales. *The Lancet. Planetary Health*, 6(1), e59–e65. DOI: 10.1016/S2542-5196(21)00264-3 PMID: 34998461

Vijayakumar, G., Rajkumar, M., Rajiv Chandar, N., Selvakumar, P., & Duraisamy, R. (2024). Muniyandi Rajkumar, Rajiv Chandar N, Selvakumar P and Ramesh Duraisamy, (2024). Environmentally friendly TDS removal from waste water by electrochemical ion exchange batch-type recirculation (EIR) technique. *Environmental Science. Water Research & Technology*, 10(4), 826–835. DOI: 10.1039/D3EW00793F

Villarin, M. C., & Rodriguez-Galiano, V. F. (2019). Machine learning for modeling water demand. *Journal of Water Resources Planning and Management*, 145(5), 04019017. DOI: 10.1061/(ASCE)WR.1943-5452.0001067

Wang, C., Li, Z., Ni, X., Shi, W., Zhang, J., Bian, J., & Liu, Y. (2023). Residential water and energy consumption prediction at hourly resolution based on a hybrid machine learning approach. *Water Research*, 246, 120733. DOI: 10.1016/j.watres.2023.120733 PMID: 37879283

Wang, J., & Liu, Y. (2016). RVM-BASED urban hourly water consumption forecasting model and its application. *Journal of Simulation*, 4, 37.

Wu, S., Han, H., Hou, B., & Diao, K. (2020). Hybrid model for short-term water demand forecasting based on error correction using chaotic time series. *Water (Basel)*, 12(6), 1683. DOI: 10.3390/w12061683

Xu, T. Y., & Qin, X. S. (2015). A sequential fuzzy model with general-shaped parameters for water supply-demand analysis. *Water Resources Management*, 29(5), 1431–1446. DOI: 10.1007/s11269-014-0884-8

Yang, L., Feng, Q., Lu, T., Adamowski, J. F., Yin, Z., Hatami, S., Zhu, M., & Wen, X. (2023). The response of agroecosystem water use efficiency to cropland change in northwest China's Hexi Corridor. *Agricultural Water Management*, 276, 108062. DOI: 10.1016/j.agwat.2022.108062

Zubaidi, S. L., Ortega-Martorell, S., Al-Bugharbee, H., Olier, I., Hashim, K. S., Gharghan, S. K., Kot, P., & Al-Khaddar, R. (2020c). Urban water demand prediction for a city that suffers from climate change and population growth: Gauteng province case study. *Water (Basel)*, 12(7), 1885. DOI: 10.3390/w12071885

Chapter 14
Integrating AI in Green and Blue Infrastructure for Sustainable Smart Cities

Josue Roman Martinez-Mireles
https://orcid.org/0000-0002-3167-5209
Polytechnic University of Pachuca, Mexico

Jazmin Rodriguez-Flores
https://orcid.org/0000-0002-3313-0325
Polytechnic University of Pachuca, Mexico

Rafael Alfonso Figueroa-Diaz
Technological Institute of Sonora, Mexico

Brenda Berenice Garcia-Escorza
Polytechnic University of Pachuca, Mexico

Marco Antonio Garcia-Marquez
https://orcid.org/0000-0002-0644-4839
Polytechnic University of Pachuca, Mexico

Arturo Austria-Cornejo
https://orcid.org/0009-0002-9291-4397
Polytechnic University of Pachuca, Mexico

ABSTRACT

The future of sustainable smart cities is based on effective collaboration between technology, policy and community engagement, with the integration of artificial intelligence (AI) into urban planning is rapidly transforming the way cities manage their environmental resources. This innovative technology has the potential to improve the functionality of green and blue infrastructure, offering sustainable solutions that address both ecological conservation and urbanization challenges. As cities strive for sustainability, the application of AI enables real-time monitoring and predictive analytics, facilitating more informed decision-making that balances human needs with ecological demands. In conclusion, while the integration of AI

DOI: 10.4018/979-8-3693-8074-1.ch014

holds great promise for transforming urban environments, it simultaneously requires a comprehensive framework that prioritizes sustainability, equity, and active citizen participation in shaping the cities of the future.

INTRODUCTION TO GREEN AND BLUE INFRASTRUCTURE

The need for innovative approaches to urban planning has led to an increasing focus on integrating natural ecosystems into urban environments. This concept, commonly referred to as green and blue infrastructure, emphasizes the importance of incorporating natural elements-such as parks, waterways, and green roofs-into city planning and development. These infrastructures serve multiple purposes, such as enhancing biodiversity, improving air and water quality, and providing recreational spaces for residents. Moreover, as urban areas face challenges such as climate change, high population density and resource depletion, the implementation of these infrastructures becomes crucial. Not only do they mitigate the urban heat island effect and manage stormwater runoff, but they also support public health by promoting physical activity and mental well-being, thus contributing to a holistic urban ecosystem that benefits both people and the environment.

The integration of green and blue infrastructure with advanced technologies, such as AI and big data, is transforming urban sustainability. AI optimizes the placement and maintenance of green spaces using predictive analytics, guiding planners to create parks tailored to demographic and environmental needs. Smart irrigation systems enhance water efficiency while minimizing waste. Studies show that incorporating innovative technologies into smart cities fosters cross-sector collaboration, strengthening urban interventions (Suresh et al., 2020). This synergy promotes resilient cities that prioritize environmental and public health. The COVID-19 pandemic underscored the interconnectedness of urban life, nature, and technology, prompting a shift in urban design principles (Mohammadian et al., 2020). Green and blue infrastructures are now central to adaptive, sustainable cities, enhancing livability, economic vitality, and environmental resilience through holistic, tech-enabled approaches.

Definition and Importance of Green Infrastructure

Green infrastructure integrates natural and built environments to deliver ecological, social, and economic benefits. By incorporating vegetation, soil, and natural processes, it enhances biodiversity, air and water quality, and flood mitigation while mimicking natural systems for resilience. Features like green roofs and permeable pavements serve functional purposes, regulating temperatures and managing stormwater. Blue-green infrastructure (BGI) further addresses complex urban challenges, aligning environmental benefits with stakeholder objectives (Lacroix et al., 2024). As cities face climate change, green infrastructure mitigates extreme weather impacts, such as heatwaves and flooding, while promoting urban equity. Practical implementations, like urban forests and community gardens, improve mental health, foster social cohesion, and empower marginalized communities, emphasizing its critical role in sustainable, inclusive urban development.

Integrating technology, including AI and big data analytics, into the framework of green infrastructure can further amplify its effectiveness and efficiency. Advanced modeling tools allow for the real-time assessment of green spaces' ecological impacts, facilitating informed decision-making regarding urban planning and resource management. AI-driven systems can optimize the placement of green facilities, predict maintenance needs, and simulate environmental conditions, thus enhancing both performance and sustainability. As cities evolve into smart environments, the incorporation of data-driven strategies will be essential for maximizing the benefits of green infrastructure, minimizing costs, and ensuring adaptability to changing ecological and social conditions. Therefore, recognizing green infrastructure as a dynamic and technology-augmented entity is vital for achieving holistic and sustainable urban development in the context of a rapidly urbanizing world.

Definition and Importance of Blue Infrastructure

Blue infrastructure is vital for promoting ecological resilience within urban development. It integrates natural and engineered systems, such as rivers, lakes, and wetlands, to manage stormwater, enhance biodiversity, and improve water quality while serving as a recreational amenity. This dual role supports sustainable urban ecosystems and balances built and natural environments, particularly as cities face climate challenges. Strategic planning and stakeholder engagement are key to optimizing site selection and maximizing the multifunctional benefits of blue infrastructure (Lacroix et al., 2024).

Blue infrastructure extends beyond environmental benefits to include significant socio-economic impacts, such as boosting recreation, property values, and local economies. Interdisciplinary approaches combining urban planning, ecology, and community engagement enhance its effectiveness. For example, integrating hydroponic systems into blue infrastructure supports local food production and improves water quality, showcasing its multifaceted role in sustainable communities. Advances in AI and machine learning further amplify the potential of blue infrastructure by enabling real-time stormwater management, resource optimization, and data-driven decision-making. This synergy between AI and blue infrastructure fosters resilient, adaptive smart cities, addressing future challenges while improving urban quality of life.

Successful Green and Blue Infrastructure Planning

Integrating AI and machine learning into urban planning can greatly enhance the effectiveness of green and blue infrastructure. For example, AI algorithms in railway maintenance optimize resource allocation by distinguishing between climatic and non-climatic failures, promoting resilience and sustainability (Soleimani-Chamkhorami et al., 2024). Similarly, data-driven approaches are essential for addressing challenges like flooding and urban heat, enabling adaptive systems that optimize the coexistence of green spaces and water bodies (Song et al., 2024). These systems must balance ecological health, socioeconomic benefits, and climate resilience to maximize the societal returns of green and blue infrastructure investments.

The emphasis on collaborative, intelligent systems reflect a shift from traditional approaches to a more integrated, evidence-based framework that prioritizes ecological integrity and human well-being. Furthermore, the successful implementation of these infrastructures requires ongoing evaluation and adaptation strategies grounded in data analytics. The use of AI not only facilitates initial planning but also enables continuous monitoring and reassessment of infrastructure performance against dynamic environmental conditions. This is particularly crucial in urban areas facing rapid demographic and climatic changes. Implementing robust feedback mechanisms that harness real-time data can guide urban planners and stakeholders in making informed decisions about resource allocation and maintenance, ensuring infrastructure remains effective over time. With an increasing focus on sustainability, strategies that incorporate advanced computational techniques -like those employed in the economic planning of hydrogen infrastructures- can identify optimal pathways for integrating green and blue solutions, ultimately leading cities toward more resilient and sustainable futures.

OVERVIEW OF ARTIFICIAL INTELLIGENCE IN URBAN PLANNING

AI is transforming urban planning by enabling data-driven decision-making and proactive problem-solving. Its ability to analyze large datasets on demographics, environmental conditions, and spatial trends allows planners to predict issues like traffic congestion or inadequate green spaces before they escalate. AI facilitates resource optimization, scenario simulations, and long-term impact assessments on green and blue infrastructure. Additionally, AI enhances community engagement by analyzing feedback and public sentiment, ensuring urban solutions are both technically sound and aligned with public needs.

AI-driven mapping and visualization tools make complex data accessible, fostering community engagement in sustainable urban design. These tools help integrate green and blue infrastructure while encouraging residents to connect with their environment. However, the adoption of AI in urban planning raises equity concerns, as disparities in access to technology may deepen existing inequalities. Planners must ensure inclusive practices, providing marginalized communities with equitable access and training to engage with AI tools. Addressing these challenges requires ethical policy frameworks to ensure AI benefits are shared across all urban demographics, balancing sustainability with social equity.

Benefits of AI in Enhancing Urban Infrastructure

The integration of AI in urban infrastructure is transforming how cities approach sustainability and resilience. AI technologies facilitate data-driven decision-making, enhancing the management of resources such as water, energy, and green spaces. For instance, smart water management systems can predict usage patterns and detect leaks in real time, thereby reducing waste and conserving vital resources. Through the application of big data analytics, urban planners can identify trends and develop proactive strategies for maintaining green and blue infrastructure. This aligns with the findings of (Lacroix et al., 2024), which highlight the need for objective-driven frameworks in enhancing urban spaces, ensuring that stakeholder values are prioritized in the planning processes.

The deployment of AI tools leads not only to improved efficiency but also to a stronger alignment between urban development and environmental preservation, ultimately fostering sustainable smart cities. Moreover, the benefits of AI extend into the realm of enhancing the social and psychological well-being of urban residents. AI systems can help optimize the design and accessibility of urban green spaces, crucial for community health and engagement. By analyzing factors such as naturalness and park layout, cities can create environments that promote physical activity and reduce stress, as evidenced by (Bressane et al., 2024). As cities grapple with mental health challenges exacerbated by urbanization, leveraging AI to improve the quality and distribution of recreational spaces becomes increasingly important. Facilitating stakeholder engagement through platforms powered by AI can also lead to more inclusive urban planning, where diverse community needs are considered. By fostering collaboration, AI not only capitalizes on expert insights but also enriches the collective vision for urban development, forming spaces that are beneficial for all citizens. AIs potential in predictive analytics further empowers municipalities to forecast infrastructure needs effectively, reducing maintenance costs and enhancing service delivery. By employing machine learning algorithms on historical data, cities can anticipate failures in infrastructure systems, allowing for timely interventions. This proactive approach optimizes operational efficiency and extends the lifespan of existing infrastructure.

Additionally, as (Lacroix et al., 2024) indicates, integrating value-focused thinking into the planning processes enables cities to align infrastructure development with community objectives. Effective integration of AI can lead to informed decision-making that weighs both economic and environmental benefits, ultimately paving the way for a holistic approach to urban sustainability. In recognition of these advantages, municipalities are increasingly adopting AI technologies to foster not just resilient infrastructure, but also vibrant communities capable of thriving amidst the challenges of urbanization.

Challenges and Limitations of AI Implementation

In the realm of smart cities, the incorporation of AI faces notable impediments linked to the quality and availability of data. A substantial amount of data emerges from diverse origins, and the variation in data types and formats constitutes a major obstacle. Additionally, the diverse nature of data often results in inconsistent outcomes during analysis by AI systems, which depend on inputs that are well-organized to produce precise outputs. This dilemma is further amplified in initiatives related to green and blue infrastructure, where it is crucial that data is diligently gathered and preserved to support effective decision-making. Poor data management practices add to these complications, often leading to incomplete datasets that can diminish the overall efficacy of AI applications (Abdulaziz et al, 2023). Tackling these data-related challenges is vital for those stakeholders who aspire to exploit AI technologies in sustainable urban planning aimed at improving environmental results. In conjunction with data issues, ethical implications and the prospect of bias in AI algorithms represent significant hurdles for implementation. Given that these algorithms are developed using historical data, there exists a risk they may inadvertently reinforce existing inequalities or biases embedded in that data. Such circumstances could yield skewed insights that have a disproportionate effect on marginalized populations, thus undermining the equitable foundations of sustainable development.

Moreover, AI systems may attract scrutiny regarding their transparency and accountability, as there is a growing demand from stakeholders for elucidation on the decision-making processes and the rationale behind those decisions. This insistence on transparency is especially crucial within the context of green and blue infrastructure projects, which have a profound impact on both the ecological and social dynamics of urban settings. Neglecting these ethical considerations can prompt public opposition and reduce trust in AI applications, ultimately obstructing the advancement of sustainable urban initiatives (Sanchez et al, 2024). Considering these challenges, the swift rate of technological progress underscores the urgent requirement for ongoing education and training among those involved in AI implementation. Elected officials, city planners, and community members must cultivate a certain proficiency in AI and its implications for urban development. This transfer of knowledge is critical for nurturing collaborative settings where diverse viewpoints contribute to decision-making processes. Furthermore, adopting interdisciplinary methods could bolster the resilience of AI systems against the limitations previously mentioned. For example, collaborations between technologists and environmental scientists might yield groundbreaking solutions that address data quality deficiencies while ensuring the ethical use of AI.

Ultimately, successfully navigating the complex challenges intrinsic to deploying AI in green and blue infrastructure initiatives necessitates a dedicated effort to align technology with community values, making certain that the benefits of smart city projects are equitably shared across all demographic groups (Hodson et al, 2023).

THE ROLE OF BIG DATA IN SUSTAINABLE SMART CITIES

A significant change in how cities are managed is brought on by the combination of big data technologies, which greatly improves the functioning efficiency of smart cities. The employment of real-time data analytics enables city planners and decision-makers to observe urban life patterns such as traffic movement, energy usage, and waste handling. By utilizing such data, local governments can pinpoint inefficiencies and allocate resources more wisely. For example, strategies based on data can forecast high energy demand periods, which helps in incorporating renewable energy sources into urban power systems and decreasing dependence on fossil fuels. This method aligns with the broader aim of creating sustainable urban areas that harmonize with ecological principles, thus laying the groundwork for smart urban planning that boosts quality of life while reducing environmental damage. Improved data analytics not only enhance the operational functionalities of cities but also promote engagement within communities and social fairness.

Big data has the potential to ensure that the advantages of smart city projects are shared fairly amongst diverse populations. By examining demographic statistics, city officials can customize services according to the distinct needs of different neighborhoods, making certain that underprivileged communities get necessary support and resources, this research backs up this claim by showing that smart city initiatives frequently falter because of inadequate inclusiveness and engagement from stakeholders, underlining the necessity for a data-centric strategy that values the input of residents (Ehwi, 2020).

The encouragement of participatory governance through the transparency and availability of data can, therefore, increase trust among citizens and foster cooperative urban growth, transforming the idea of sustainable smart cities into a joint effort that adheres to human-centered design principles. Moreover, beyond dealing with social issues, big data stands out as a critical element in promoting environmental sustainability within smart city contexts. By applying advanced analytics, urban planners can predict environmental outcomes and adjust their plans as needed, aiding in efforts to tackle climate change. For instance, insights drawn from data can optimize green infrastructure such as city parks and green rooftops, aiding in stormwater management and increasing biodiversity in urban spaces. This active use of big data strengthens the blend of sustainable practices by enabling cities to react effectively to ecological challenges, as highlighted in recent research regarding eco-cities and data-oriented smart solutions (Biscaya, 2020).

Therefore, the integration of big data and sustainable infrastructure design fosters the development of more intelligent, resilient urban environments that prioritize ecological health while accommodating growth and urban development.

Data Collection Methods for Urban Environments

The ongoing process of urbanization compels the exploration of novel techniques for data gathering that can sufficiently underpin effective management of cities. The amalgamation of AI and machine learning within urban landscapes permits the accumulation of substantial data quantities through sophisticated sensors and Internet of Things (IoT) apparatuses. Such technologies make possible the real-time oversight of environmental factors, traffic flow, and public health markers, consequently furnishing city planners with actionable intelligence to bolster green and blue infrastructures. Per (Singar et al., 2024), the installation of proprietary IoT frameworks adheres to a methodical structure comprising the Endpoint, the Edge, and the Cloud. This layered methodology enhances the precision of data acquisition and processing, which is vital for the formation of responsive urban ecosystems capable of adapting to citizen needs.

A thorough inquiry into the methodologies utilized for urban data collection elucidates both prospects and obstacles associated with their application. While the growing dependence on digital platforms augments the richness of the data environment, it concurrently engenders apprehensions pertaining to privacy and data protection, often complicating the governance of urban infrastructures. The systematic review discussed in (Singar et al., 2024) delineates critical facets for prospective exploration, underscoring that grasping the complexities of urban data is essential to surmounting these challenges. Additionally, the incorporation of AI not solely aids in refining resource distributions but also bolsters predictive modeling, which is paramount for sustainable urban development. Without addressing obstacles cited, such as the disparate capacities of urban structures alongside the difficulties of data interoperability, the complete potential of smart city ventures might persist unactualized.

In conclusion, the disparities in smart city frameworks across various geographical contexts underscore the necessity for localized strategies regarding data collection techniques. The hurdles involved in transitioning from established smart city paradigms —chiefly devised in the developed nations— to their relevance in developing areas, notably in Southern Africa, illustrate the critical need for adaptable solutions.

As indicated, promoting economic competitiveness while ensuring sustainability demands a sophisticated comprehension of local dynamics. This requirement advocates for the integration of community input in conjunction with quantitative data to formulate a comprehensive perspective on urban dilemmas. Ultimately, efficient data collection practices should embrace varied sources, including participatory strategies that actively involve citizens, thus enriching the governance mechanisms fundamental to constructing resilient, sustainable smart cities.

Big Data Analysis for Green and Blue Infrastructure Enhancing

The incorporation of big data analytics into the formulation and management of green and blue infrastructure notably bolsters the planning and management of urban ecosystems. New technologies extract useful knowledge from extensive datasets, allowing urban planners to refine land utilization, manage resources effectively, and evaluate environmental repercussions. As an example, machine learning frameworks can scrutinize rainfall and runoff patterns, which enables cities to create more efficient stormwater management frameworks. With urban areas confronting the difficulties brought about by climate change, such data-centric strategies are crucial for ensuring that both green areas and water bodies are thoughtfully developed and preserved to bolster ecosystem services and fortify urban resilience. This scenario coincides with the strategic planning model suggested for smart sustainable cities, underscoring the interconnectedness of urban development and technological implementation in achieving sustainable solutions (Bibri and Krogstie, 2021).

In addition, the application of big data in the observation and improvement of green infrastructure can result in considerable advancements in both sustainability and public health outcomes. Sophisticated data analytics can monitor the performance of urban green areas in air purification and carbon capture, providing instantaneous responses for policy modifications. For instance, employing satellite imagery along with urban heat mapping allows planners to identify regions that would most benefit from expanded tree canopies or green roofing solutions. This anticipatory methodology not only amplifies ecological advantages but also fosters social equity by ensuring that marginalized communities secure access to healthier surroundings. As the urban populace expands, the utilization of data-informed insights to enhance blue infrastructure -such as wetlands and urban watercourses- will increasingly become crucial for sustaining biodiversity and improving livability in urban contexts (Ramakrishna, 2021).

The application of big data analytics within the realm of green and blue infrastructure brings forth both prospects and hurdles in the pursuit of sustainable smart cities. This strategy necessitates solid data governing frameworks that assure privacy and ethical management of information while encouraging cooperation among various stakeholders. As elaborated in dialogues on the fourth industrial revolution, the assimilation of technology needs to be inclusive, making sure that all communities, and particularly those susceptible to marginalization, are enabled to partake in the decision-making processes (Ramakrishna, 2021). By concentrating on transparency and stakeholder involvement, urban areas can effectively leverage the complete potential of big data to propel equitable sustainable development. Ultimately, a unified plan that integrates AI and big data with green and blue infrastructures has the capacity to significantly transform urban settings, crafting resilient environments that support the health and welfare of both current inhabitants and forthcoming generations.

INTEGRATING AI WITH GREEN INFRASTRUCTURE

The capability of AI to enhance green infrastructure presents a noteworthy chance for urban areas contending with issues like pollution and resource depletion. With the application of AI technologies, municipalities can utilize extensive datasets gathered from various sensors and IoT devices to better manage their resources. For instance, AI algorithms are capable of forecasting patterns in water usage, which facilitates the optimization of irrigation systems used for urban green spaces, thereby encouraging sustainability while maintaining visual appeal. In the end, merging AI with green infrastructure has the potential to result in noteworthy advancements in urban planning, fostering a comprehensive approach that not only minimizes negative environmental effects but also elevates the living conditions for urban residents. Additionally, the acceleration of the incorporation of sustainable methodologies within urban settings can be notably enhanced by the collaboration of AI and intelligent technologies. This interplay is particularly apparent in the rise of smartscapes that integrate renewable energy sources alongside effective waste management solutions. By interpreting environmental data in real-time, AI can guide decision-making processes and set priorities for interventions, leading to the successful execution of green infrastructure initiatives. As indicated by research, embedding advanced technologies, including AI and machine learning, can tackle urban issues like overcrowding and pollution, creating a platform for the development of resilient habitats that underscore environmental accountability (Khlie and Benmamoun, 2024).

In a time characterized by increasing vulnerability in urban ecosystems, the deliberate use of machine learning and big data analytics can spawn innovative methods for sustainable urban development. Such technologies enable a more profound comprehension of local contexts, thereby permitting customized approaches that account for cultural and economic differences among urban communities. The research illuminates that the relationship between AI and green infrastructure is vital for optimizing resources and promoting community inclusiveness, thus forging a pathway toward a more sustainable tomorrow (Judijanto et al., 2024). Through fostering collaborative efforts among various stakeholders, cities have the potential to establish integrated blue and green infrastructure systems that are technologically advanced while remaining ecologically sound, adeptly addressing the necessities of modern urbanization while safeguarding long-term environmental integrity.

AI-Driven Solutions for Urban Green Spaces

The potential of AI to transform urban green spaces is becoming more and more apparent, as new technologies present chances for managing cities in ways that are more sustainable and effective. In examining large sets of data, AI has the capability to enhance horticultural methods by forecasting the health of plants, their growth behaviors, and how biodiversity reacts within various environmental settings.

Machine learning methods can amalgamate information from weather trends, soil quality, and citizen input, which allows urban planners to make educated choices concerning plant choices and landscape design. This kind of data-based method not only leverages the ecological advantages of green spaces but also encourages community participation, as residents can share their thoughts through mobile applications and platforms that are set up for immediate feedback. Thus, the use of AI-centered solutions leads to more robust urban plant life that can adjust to changing climate conditions and improve the livability of cities. Beyond enhancing plant selection, AI technologies also improve the management of urban green areas using predictive analytics and maintenance planning.

Smart sensors that are part of these environments can keep track of variables like soil moisture, temperature, and pollution levels, making it feasible for on-the-spot irrigation and quick reactions to poor conditions. By applying AI models to assess past maintenance information alongside current environmental feedback, city officials can focus on regions that require urgent attention, leading to better allocation of resources and lower operational expenses. This forward-thinking strategy not only reduces the chance of plants suffering but also aids the aim of sustainability in urban development. Improving maintenance timelines helps urban green spaces flourish while also cutting down the ecological impact typically associated with established maintenance methods.

Moreover, the inclusion of AI-based solutions in urban green spaces can bolster biodiversity by aiding in the creation of ecological pathways that support wildlife. By using geographical information systems (GIS) together with machine learning models, planners can pinpoint essential habitats and migration routes, shedding light on how urban expansion affects local plant and animal life. City planners may utilize this knowledge to formulate interconnected green areas that not only benefit urban dwellers for recreational activities but also act as crucial ecosystems. By ensuring that these pathways are woven into the fabric of urban layouts, aided by AI, cities can move toward a broader understanding of biodiversity while supporting both ecological stability and social health. Therefore, the integration of innovative AI technologies into the oversight of urban green spaces is vital for nurturing sustainable smart cities that emphasize the importance of environmental well-being and urban quality of life.

Predictive Analytics for Environmental Management

The emergence of sophisticated methodologies in data analytics has become vital in configuring effective strategies for environmental management, especially through the utilization of predictive analytics. Environmental management grapples with a multitude of complex issues, which range from resource distribution to urban development and climate resilience. Employing predictive analytical methods enables stakeholders to foresee environmental transformations, pinpoint potential hazards, and fine-tune resource application. By utilizing extensive datasets amassed from varied origins, such as IoT devices and satellite imaging, predictive models can recognize trends and deliver practical insights, thus promoting timely responses. This form of analytics plays a significant role in boosting energy efficiency, as evidenced by different implementations, including independent greenhouse operations where AI's ability to scrutinize historical data culminates in diminished energy consumption (Hoseinzadeh, 2024).

As a result, the application of predictive analytics presents an inventive route towards sustainable methodologies in urban locales, endowing decision-makers with the foresight essential for navigating the intricate and fluid ecological frameworks. The amalgamation of AI and machine learning with predictive analytics marks a significant shift in how both green and blue infrastructures are managed.

This collaboration enhances the forecasting capabilities concerning environmental dynamics, such as fluctuations in water quality within urban waterways or the capacity of green roofs to alleviate urban heat islands. For example, machine learning algorithms can simulate the relationships between climate factors and ecological reactions, giving rise to more nuanced strategies that weave sustainability into urban planning. Furthermore, the fusion of these technological advancements promotes continual oversight and resource management, thereby encouraging climate adaptability within smart city paradigms. Recent scholarly discourse regarding intelligent housing solutions indicates that the convergence of various technological strategies can result in considerable improvements in environmental efficiency (Alotaibi et al., 2024). Thus, this all-encompassing approach to predictive analytics sets the foundation for integrated and responsive environmental management, which is critical for nurturing ecosystems in urban areas. In summation, the effective implementation of predictive analytics in environmental management critically depends on tackling common challenges that include data integrity, regulatory conditions, and workforce skill enhancement.

The swift evolution of digital technologies prompts an informed methodology for assimilating these innovations into pre-existing infrastructural systems. Regulatory impediments frequently obstruct the embrace of state-of-the-art technologies in environmental management; hence, stakeholders must champion modernized policies that align with the capabilities of contemporary analytical tools. Additionally, addressing the skills deficit through educational initiatives and training is crucial for maximizing the effectiveness of predictive analytics. By arming professionals with the means to leverage these opportunities, we can more efficiently deploy AI-driven solutions that bolster sustainability in urban contexts. Ultimately, a thorough strategy that integrates technological advancements with educational components will guarantee the successful incorporation of predictive analytics into robust environmental management within sustainable smart cities.

INTEGRATING AI WITH BLUE INFRASTRUCTURE

The utilization of advanced technologies, especially AI, is significant for the enhancement of the functionality and sustainability of blue infrastructures. The incorporation of AI within the realm of blue infrastructure- which includes water systems, rain gardens, and waterfronts -facilitates the capability for real-time monitoring and predictive analytics, resulting in more effective management of water resources. For example, AI algorithms can scrutinize data gathered from hydrological sensors that are able to predict flood occurrences, thus allowing urban areas to take proactive measures that can lessen damage and safeguard public safety. These anticipatory strategies not only improve urban resilience but also bolster ecological health through maintaining the fragile equilibrium of aquatic ecosystems.

Additionally, machine learning techniques could be utilized to refine the design of green-blue corridors, enhancing biodiversity while achieving urban drainage and aesthetic objectives. This reciprocal relationship between AI and blue infrastructure has the potential not only to uplift urban living environments but also to substantially contribute to strategies for climate adaptation. Evaluating the socio-economic ramifications of AI-powered blue infrastructure unveils its capability to encourage equity and community involvement. The application of participatory design methods alongside the integration of local knowledge into AI frameworks allows planners to devise blue infrastructure that reflects the needs and values of the community. This synchrony guarantees that marginalized groups gain equitable access to essential resources, such as clean water and recreational facilities, tackling systemic inequities frequently present in urban locales. Furthermore, the deployment of AI in scheduling maintenance and resource allocation can result in fairer service delivery. It effectively reduces bureaucratic redundancies, thus enabling faster response times in scenarios of water-related emergencies and maintenance operations.

As urban areas increasingly embrace smart technologies, the involvement of stakeholders becomes crucial, fostering transparency and trust. Integrating AI within participatory governance structures permits communities to cultivate ownership regarding their public spaces, which in turn enhances social cohesion and collective responsibility. Given the persistent challenges posed by urbanization and climate change, the significance of AI in improving blue infrastructure is projected to escalate. Future investigative endeavors should delve into the convergence of big data analytics and AI in shaping more resilient urban water systems that can adapt to varying climate situations. For instance, AI models could be educated using historical data on precipitation and water quality to reveal subtle correlations and trends, fostering better decision-making processes. Such integrated frameworks might also facilitate sensitive modifications in policy and planning as dictated by emerging environmental trends. Furthermore, collaborations among technologists, urban planners, and environmental scientists will be imperative to construct robust AI-driven frameworks that contend with the intricacies of urban water systems. By emphasizing sustainability and resilience, urban areas stand to benefit from AI not merely as an efficiency tool but as a transformative driver for creating more adaptive and inclusive urban environments.

Smart Water Management Systems

The convergence of sophisticated technologies within urban water management frameworks is essential for the progress towards sustainable smart cities. Smart Water Management Systems, which utilize AI and extensive data analysis, augment decision-making processes pertinent to the effective allocation and utilization of water resources. Such systems enable contemporaneous oversight and predictive analysis capabilities, thereby allowing urban areas to react preemptively to water deficiencies and to manage infrastructure competently. Through the examination of data derived from diverse sources including meteorological trends, consumption metrics, and the condition of infrastructure, local governments can enhance water distribution and curtail wastage. This anticipatory methodology not only addresses pressing issues but also cultivates prolonged resilience concerning the repercussions of climate alteration, as it empowers municipalities to adapt with agility to environmental variations and urban expansion obstacles.

Moreover, the imperative for proficient water management becomes exacerbated considering urbanization and demographic increases, particularly in underdeveloped locales. The difficulties articulated in the context of Southern Africa illuminate the confluence of these circumstances, wherein the necessity for sustainable methodologies is pressing. Acknowledging that most African administrations are deprived of dependable data (Blanco-Montero, 2021), smart systems play a vital role in fostering more knowledgeable management strategies. They possess the capacity to transmute conventional frameworks into intelligent solutions by integrating localized water scarcity information with overarching global data trends, thus tackling socio-economic inequalities.

As urban centers globally confront constrained water supplies, the incorporation of AI in water management engenders prospects for improving habitability and accessibility within cities, ensuring that the most disadvantaged communities receive priority in terms of water distribution and conservation initiatives. In addition, the deployment of Smart Water Management Systems can stimulate innovation in urban ecological practices. By aligning with eco-friendly infrastructure projects, these systems can optimize resource use and minimize environmental impacts. The holistic insights regarding urban interactions, as underscored in (Wang and Fu, 2021), indicate that municipalities require comprehensive strategies to confront ecological challenges while promoting economic advancement. By integrating technological innovations with sustainable methodologies, urban planners can more effectively oversee stormwater, mitigate flooding, and enhance water quality. This unified strategy not only addresses the immediate issues resulting from contamination and habitat loss but also motivates community participation in conservation activities. Ultimately, the utilization of AI and sophisticated analytics has the potential to foster more adaptable and self-sustaining urban ecosystems that can resiliently counter both present and forthcoming water-oriented challenges.

AI Applications in Flood Risk Assessment

Progressions within the field of AI have decidedly altered the terrain concerning the evaluation of flood risks, thereby boosting both the precision and the efficiency linked to the modeling of probable flood scenarios. By harnessing the potentials offered by machine learning algorithms alongside big data analytics, scholars and field practitioners now possess the capability to handle extensive quantities of hydrological, meteorological, and geographical data, which ultimately paves the way for more exact forecasting of flood incidents. Conventional methodologies frequently depended on constrained datasets and hand-calculated methods, which were prone to yield substantial inaccuracies. Conversely, methodologies driven by AI amalgamate real-time information sourced from diverse outlets, such as satellite imagery and meteorological forecast models, which enables assessments that are fluid and responsive to fluctuating environmental circumstances. These enhanced capabilities augment stakeholders' comprehension of flood risks and encourage improved decision-making in contexts of urban development and infrastructural funding (Ehwi, 2020).

Furthermore, the capacity to emulate numerous flood scenarios via AI frameworks enhances the formulation of anticipatory flood management techniques. Such hypothetical modeling not only imparts knowledge regarding potential flood effects but also assists in pinpointing areas deemed particularly susceptible to flood occurrences. This geographic analysis serves to direct the allocation of green and blue infrastructural interventions, like rain gardens or urban wetlands, which can alleviate the repercussions of flooding whilst simultaneously bolstering urban robustness (Cao, 2023). By weaving such infrastructural elements into the fabric of city planning, decision-makers can adeptly reconcile ecological advantages with the immediate needs of urban living. As metropolitan areas increasingly battle the adversities engendered by climate change, garnering insights into flood mechanisms through AI applications is rendered essential in reforming urban environments geared towards sustainability. Lastly, the amalgamation of AI within the scope of flood risk evaluation denotes a transformative shift towards increasingly intelligent and sustainable urban settings. This transition not merely alters immediate strategies regarding flood management, but also adds to the overarching urban resilience by advocating for adaptive approaches that adjust in concert with novel data. The cyclical nature of AI-based models permits ongoing learning and refinement, thereby guaranteeing that flood risk assessments uphold their relevance amid evolving climatic patterns and urban progressions. As urban locales persist in expanding, the urgency for coherent flood management tactics that incorporate insights from AI becomes critical, nurturing a comprehensive methodology towards urban sustainability. By engaging such advancements, cities can lead the charge for inclusive flood risk management that aligns human settlements with the natural world, ultimately contributing to healthier and more sustainable urban ecosystems (Lucatello and Alcántara, 2024)

POLICY AND GOVERNANCE FOR AI INTEGRATION IN GREEN AND BLUE INFRASTRUCTURE PLANNING

Effective governance is crucial in relation to the successful incorporation of AI within the planning frameworks of green and blue infrastructure, particularly as urban areas encounter growing stresses stemming from environmental deterioration and urban expansion. It is necessary for policymakers to set up structures that stimulate the uptake of AI technologies while maintaining sustainability and fairness. This necessitates a shift away from conventional bureaucratic models towards a governance strategy that is more adaptable, which facilitates the cooperation of various stakeholders, including governmental bodies, private sector actors, and local communities. Such collaboration holds significant importance for enhancing transparency, as it empowers citizens to partake in decision-making related to urban development. For example, the amalgamation of real-time information derived from AI-centric systems can amplify community awareness and involvement, ultimately encouraging a sense of ownership over projects concerning green and blue infrastructure that resonate with local values and priorities, thus aligning with the aims of smart city frameworks as delineated in (Singar et al., 2024).

The merging of AI into green and blue infrastructure also brings forth considerable regulatory obstacles that call for unique policy responses. Policymakers must confront the ethical ramifications and data privacy issues that encompass the collection and analysis of extensive volumes of data produced by smart city technologies. The establishment of distinct regulations to manage data utilization, protective measures, and ownership is critical for alleviating risks, while concurrently nurturing a setting where AI can flourish. In addition, as pointed out in the insights from (Samrat and Manjunath, 2024), predictive analytics driven by AI can yield pivotal information that aids urban planners and decision-makers in the effective distribution of resources, confronting challenges such as traffic congestion and pollution. By instituting comprehensive governance frameworks that guarantee accountability, uphold ethical norms, and adhere to privacy regulations, urban areas are positioned to harness AI to enhance the design and operational effectiveness of their green and blue initiatives whilst nurturing public confidence in these technologies. An aligned strategy toward policy and governance is paramount to optimization of AI benefits in the realm of green and blue infrastructure formulation. Policymakers need to place emphasis on educational and capability-enhancing endeavors aimed at equipping local governance and stakeholders with the requisite understanding for effective use of AI and machine learning. By channeling investments into training initiatives and cultivating collaborations with educational institutions as well as industry specialists, urban areas can develop a workforce skilled in emerging technologies. Furthermore, as articulated in (Singar et al., 2024), the urgency for confronting environmental issues through AI-utilizing strategies cannot be overstated, thus demanding prompt and knowledgeable policy interventions. The integration of AI into planning activities not merely bolsters operational efficacy but also uplifts urban resilience, climate adaptation, and the overall welfare of their residents. Hence, a comprehensive policy framework that embeds the tenets of sustainability, equity, and technological innovation is crucial for the creation of smart cities adept at navigating the intricacies of the 21st century.

Regulatory Frameworks Supporting AI in Urban Planning

Integrating AI into the realm of urban planning demands the establishment of comprehensive regulatory frameworks capable of managing the multifaceted challenges inherent to these technologies. Such frameworks ought to find a balance between innovative pursuits and governance structures, to ensure that urban planners are able to utilize AI in a manner that is congruent with sustainable development objectives and ethical considerations.

Regulations that are effective should advocate for the incorporation of AI solutions within green and blue infrastructure, intended to transform urban locales into zones resilient to climate impacts. For example, initiatives like the smart blue-green roof system, which aims to enhance urban water management alongside boosting biodiversity, serve as illustrations of the potential advantages linked to the integration of AI. By formulating clear stipulations concerning data utilization, privacy protections, and environmental evaluations, regulatory bodies have the capacity to cultivate an environment that supports technological adoption, while simultaneously defending public interests. In addition, regulatory frameworks hold significant importance in promoting interdisciplinary collaboration among various stakeholders engaged in AI-infused urban planning endeavors. The convergence of disciplinary fields such as technology, municipal governance, and environmental considerations necessitates a systematic methodology to unite engineers, urban planners, ecologists, and policymakers. By nurturing partnerships that highlight the necessity of data sharing paired with cooperative decision-making, these frameworks can facilitate the infusion of diverse perspectives into AI applications. A noteworthy instance of successful collaboration can be observed in water-wise cities, where the integration of numerous disciplines is pivotal to addressing challenges related to water resource management and infrastructural resiliency. These initiatives may be influenced by inventive regulations that delineate performance criteria, thereby establishing benchmarks for sustainable practices in the development of urban areas and prompting a transition towards more adaptive management strategies. In the end, the transformation of regulatory frameworks is crucial to keep step with the rapid advancements in AI technologies. It is essential that emerging technologies are subjected to pilot schemes within regulatory sandboxes, which permit experimentation and iterative development without imposing excessively burdensome compliance tasks. This kind of flexibility allows urban planners to conduct trials with AI-powered simulations or predictive models aimed at flood management, thus improving the efficacy of projects like smart blue-green roofs in the context of mitigating the impacts of extreme weather events.

Expanding the understanding of how regulatory measures impact the field can aid in the formulation of extensive guidelines governing the merger of AI with urban planning tasks, aligning these processes with sustainable development aspirations. These frameworks ought to embody key principles derived from prevailing environmental regulations - reflecting ideas found in efforts aimed at enhancing urban water infrastructure and the overall resilience of cities, as elaborated in (Plegt, 2022) and (Wang and Fu, 2021).

Ethical Considerations in AI and Urban Development with green and blue infrastructure

In the process of integrating AI into urban development, there are substantial ethical considerations that need to be tackled for the sake of responsible incorporation. A primary issue pertains to the biases that might be present within AI algorithms, especially regarding decisions that influence marginalized groups. If machine learning systems utilize historical datasets that are indicative of systemic disparities, there is a risk that these models could unintentionally reinforce or even exacerbate these inequities in the distribution of green and blue infrastructure resources. As municipalities implement AI for the purpose of enhancing environmental services, it is crucial that they critically assess the sources of their data and apply methods that emphasize inclusivity and equity. Such strategies not only satisfy ethical obligations but also bolster community acceptance and involvement in urban planning efforts, rendering them essential for the judicious application of AI in sustainable development settings (Tan and Cugurullo, 2020).

Another significant ethical factor hinges on the need for transparency within AI methodologies, particularly in relation to urban development endeavors that employ green and blue infrastructure. It is vital for stakeholders -including residents, city planners, and environmental advocacy groups- to gain a clear understanding of the processes underlying AI decision-making. A deficiency in transparency could result in distrust and skepticism towards these technological developments, which could hinder wider community engagement and cooperation. The establishment of comprehensive frameworks for elucidating AI decision protocols can aid in clarifying these technologies and fostering accountability amongst urban planners and city officials. Furthermore, the participation of varied community perspectives in the formulation and execution phases can further enhance this transparency, ensuring that the values and necessities of all urban inhabitants are integrated into AI-directed initiatives. These strategies are imperative not only for adhering to ethical standards but also for bolstering the legitimacy of urban policies that incorporate AI within environmental approaches (Adanma and Ogunbiyi, 2024).

Ultimately, it is essential to develop ethical governance structures that will oversee the utilization of AI in green and blue infrastructure projects. This necessitates the creation of guidelines addressing privacy issues, data protection, and the risk of surveillance within urban environments. As AI technologies increasingly depend on extensive data pools, preserving personal privacy becomes a paramount ethical obligation in urban planning and administration. Additionally, regulatory systems ought to be designed to alleviate the dangers associated with an excessive reliance on AI, whereby urban planners might neglect crucial human and ecological factors in preference for efficiency driven by algorithms. Achieving a balance between the innovative capabilities of AI and ethical governance lays the groundwork for the development of smart cities that are both resilient and fair. The formulation of such frameworks will not only defend individual rights but will also advocate for a comprehensive perspective on urban development that emphasizes environmental sustainability and community well-being.

CASE STUDIES OF SUCCESSFUL AI INTEGRATION WITH GREEN AND BLUE INFRASTRUCTURE

The potentiality of AI in augmenting urban sustainability has been evidenced via a diversity of case studies that amalgamate green and blue infrastructure initiatives. For example, in Amsterdam, AI algorithms are utilized for the optimization of water management frameworks, taking advantage of real-time data to surveil water levels as well as predict flooding events. This melding not only assists in prompt responses to hydrological alterations but also boosts the ecological vitality of the canals and adjacent green areas. Through the application of AI, urban planners can anticipate the ramifications of climate change on these systems, guaranteeing the useful implementation of green roofs and permeable pavements for flood reduction, thereby improving urban resilience. Such forward-thinking interventions exemplify the capacity of AI to nurture a synergistic connection between urban progression and environmental sustainability. Furthermore, the incorporation of AI into landscape management strategies provides another noteworthy instance of efficacious application of intelligent technologies within green and blue infrastructure. In Singapore, for instance, the Smart Nation initiative employs machine learning to scrutinize and decode data sourced from vegetation health sensors alongside rainfall patterns. This facility allows for meticulous management of parks and green spaces, ensuring the most favorable conditions for growth and judicious allocation of resources. By harmonizing ecological requisites with urban necessities, AI-derived insights not only preserve biodiversity but also uplift urban visual appeal. These advancements not only engender lively public spaces but also diminish the ecological footprint of urban locales, thereby manifesting that sophisticated technology can align cohesively with sustainable urban methodologies. Finally, inter-sector collaboration is paramount for the efficacious integration of AI within frameworks of green and blue infrastructure. As elucidated in the scrutiny of urban dynamics in Southern Africa, the necessity for bespoke technological solutions accentuates the significance of local stakeholder involvement (Blanco-Montero, 2021).

Through synergistic partnerships among academic entities, governmental bodies, and private sectors, municipalities can concoct holistic strategies that harness AI to tackle specific environmental dilemmas. These cooperative ventures can promote information dissemination and stimulate innovation, thus crafting a robust ecosystem for sustainable advancement. As countries contend with the repercussions of swift urbanization, the assimilation of intelligent technologies lays out a roadmap to boost the habitability of urban realms while fostering environmental guardianship, hence responding to the urgent demand for more sustainable methodologies in the epoch of globalization (Patai and Horvath, 2021).

Future Trends in AI and Green-Blue Urban Infrastructure

Urban localities are progressively incorporating AI ostensibly to improve the operational capacity and efficiency of green and blue infrastructure, which is congruent with sustainability objectives in urban planning. These technological frameworks can serve to optimize the allocation of resources, monitor shifts in environmental conditions, and amplify urban resilience against the repercussions of climate change. For example, machine learning algorithms possess the capability to scrutinize data in real-time from urban ecosystems, like aquatic systems and plant life, to project pollution metrics and biodiversity patterns. Such deployments are of paramount importance in densely populated metropolises, such as Dhaka, where swift urbanization is jeopardizing green spaces and exacerbating heat levels and pollution (Yamasaki et al., 2024).

By utilizing AI in the stewardship of green and blue assets, urban centers are not merely able to alleviate the ramifications of climate change but also enhance public health and living standards, thereby creating urban environments that are more conducive to livability while emphasizing ecological equilibrium. The import of AI in the sphere of urban infrastructure is further illuminated by its use in microclimate modeling, which aids in evaluating prospective climate circumstances and informs adaptive frameworks for urban design. Methodologies such as computational fluid dynamics (CFD) simulations empower planners to elucidate how various urban setups may affect temperature, airflow, and human comfort (Yamasaki et al., 2024). Through the convergence of AI with environmental modeling, urban planners are equipped to forecast how specific modifications in building inventories, green spaces, and aquatic features will influence localized climatic conditions. This can facilitate the formulation of targeted strategies that address urban heat anomalies and promote the feasibility of green roofs along with urban agriculture ventures. As urban areas endeavor to improve their sustainability constructs, the interaction of AI with green-blue infrastructure is anticipated to be crucial in nurturing resilient urban ecosystems capable of accommodating climatic fluctuations. In projecting into the future, the amalgamation of AI with green-blue urban infrastructure is set to fundamentally alter urban landscapes, rendering them more adaptable and reactive to environmental hurdles.

Predicted trends are likely to underline collaboration between government sectors, private enterprises, and academic institutions to enable the advancement of more complex AI applications for urban planning. In increasingly urbanized regions, robust initiatives such as rooftop agriculture optimize the utilization of rooftop areas while advancing sustainability (Sultana et al., 2024). This confluence of technological and ecological paradigms accentuates the potential for substantial progress in urban sustainability methodologies. The persistent endeavor to merge AI into the governance of green and blue infrastructure not only promises enhanced ecological dividends but also safeguards that growing urban populace can thrive within resilient and sustainable settings, as illustrated by the inventive strategies delineated in recent research.

CONCLUSIONS AND KEY FINDINGS

The incorporation of AI into green and blue infrastructure has revealed the significant potential for the advancement of sustainable smart cities. With the ongoing urbanization of cities, challenges brought about by swift population growth and environmental degradation underscore the necessity for inventive strategies in urban planning. By utilizing AI, stakeholders can extract insights regarding urban ecosystems, which supports the creation of infrastructure that underpins environmental sustainability as well as community resilience. The recent increase in academic discourse, indicates a rising acknowledgment of AI's ability to optimize the utilization of resources and improve decision-making processes within urban contexts. This suggests a crucial transition towards adopting technology as an essential element of efficient urban management, pointing to the need for future research to further spotlight practical applications of AI that maximize the advantages of green and blue infrastructure. Navigating the intricacies inherent in urban environments mandates a contextual grasp that accounts for both local and global viewpoints.

The case studies from developing areas, such as those presented in (Blanco-Montero, 2021), illustrate that the adaptation of smart city frameworks must be customized to address specific local requirements and obstacles, especially in regions grappling with socio-economic inequalities. These observations highlight the importance of interdisciplinary cooperation amongst urban planners, environmental scientists, and policymakers in generating infrastructure solutions that resonate with local populations. Future studies should concentrate on formulating frameworks that direct the ethical application of AI in urban contexts, ensuring that progress in green and blue infrastructure not only alleviates environmental issues but also enhances inclusivity and community well-being. Hence, the culmination of this research highlights the critical need for a comprehensive approach to AI integration, nurturing a sustainable and equitable future for urban residents.

The inclusion of AI into both green and blue infrastructures is progressively viewed as essential for the evolution of sustainable smart cities. A primary observation is that AI technologies play a crucial role in enhancing resource management and energy efficiency within urban settings. For example, deploying AI can significantly decrease carbon emissions through systems that combine and analyze data from diverse sources to adjust energy usage in real-time. In the sphere of manufacturing, AI boosts not solely production efficiency but also aids environmental objectives by innovating technology and management practices. Such developments are vital for organizations aiming to fulfill environmental, social, and governance standards, as indicated in research correlating AI usage with enhanced ESG outcomes concerning environmental and governance matters, especially among lower-polluting enterprises (Wang et al., 2024).

An additional notable observation underscores the significance of nascent technologies, such as 3D printing and the Internet of Things (IoT), alongside AI to create eco-friendlier intelligent structures. This tripartite integration allows the construction industry to strive for sustainability without sacrificing efficiency or cost. Recent literature points out that 3D printing enables swift prototyping and customization of building components, while IoT technologies facilitate the live monitoring of energy usage and safety protocols. Moreover, the collaboration among these technologies not only elevates the living standards in urban areas but also aligns with the goals of smart city initiatives that emphasize environmental sustainability and resource preservation (Alotaibi et al., 2024).

Final Thoughts on AI's Role in Sustainable Urban Development

The convergence of AI and sustainable urban development has sparked novel methodologies for tackling the predicaments confronted by contemporary urban areas. As cities persist in struggling with dilemmas like excessive population density, environmental pollution, and the depletion of resources, AI's instruments present potent solutions for optimizing the design and application of both green and blue infrastructures.

Utilizing algorithms capable of analyzing extensive data collections permits city planners to gain a more profound comprehension of environmental dynamics, anticipate patterns of growth, and evaluate the effectiveness of green spaces and water features in alleviating urban heat island effects or enhancing water quality. The capability of AI to devise predictive models empowers cities to proactively distribute resources and adjust to fluctuating circumstances, thereby not only ensuring the sustainability of urban territories but also elevating the quality of life experienced by inhabitants. Furthermore, the collaboration between AI and community stakeholders remains vital in molding a successful narrative for sustainable urban advancement. Involving residents in the decision-making continuum can yield significant local perspectives that algorithms might fail to recognize. Such an inclusive strategy synchronizes technological innovations with the principles, necessities, and ambitions of the community, culminating in more efficacious and accepted measures.

Additionally, perpetual feedback mechanisms facilitated by AI systems enable continuous observation and refinement of urban initiatives, underscoring the critical nature of flexibility in addressing shifting environmental challenges. As technology evolves, promoting a culture of teamwork among AI creators, urban designers, and community members could function as a paradigm for engendering a more inclusive and sustainable urban future. In the end, the triumphant integration of AI in sustainable urban development is contingent upon ethical considerations and policy frameworks that guide its deployment. The surge of AI technologies presents fundamental inquiries regarding data privacy, equity, and environmental justice, which legislators must tackle to ensure that intelligent urban solutions do not intensify pre-existing inequalities. By instituting explicit protocols that emphasize inclusive practices and ecological stewardship, municipalities are positioned to leverage the potential of AI while protecting the rights of underrepresented communities. In summation, as we contemplate the prospective evolution of urban locales, it becomes essential that AI is perceived not solely as a technological instrument but as a catalyst for nurturing collaborative, ethical, and sustainable urban settings that prioritize both human and ecological welfare.

REFERENCES

Abdulaziz, A., Al-Khalifa, K. N., & Hamouda, A. M. (2023). Re-Thinking Data Strategy and Integration for Artificial Intelligence: Concepts, Opportunities, and Challenges. *Applied Sciences (Basel, Switzerland)*, 13(12), 7082. DOI: 10.3390/app13127082

Alotaibi, B. S., Shema, A. I., & Ibrahim, A. U.. (2024). Assimilation of 3D Printing, Artificial Intelligence (AI), and Internet of Things (IoT) for the Construction of Eco-Friendly Intelligent Homes: An Explorative Review. *Heliyon*, 10(17), e36846. DOI: 10.1016/j.heliyon.2024.e36846 PMID: 39286162

Bårnes, V. J., & Pordoğan, B. (2022). *Sustainable Development and Citizen Participation*. Molde University College.

Bibri, S. E., & Krogstie, J. (2021). A Novel Model for Data-Driven Smart Sustainable Cities of the Future: A Strategic Roadmap to Transformational Change in the Era of Big Data. *Future Cities and Environment*, 7(1), 1–25. DOI: 10.5334/fce.116

Biscaya, S. (2020). Impact of Big Data from BIM and GIS on Smart Urban Futures. *Applied Science University Journal*, 4(1), 11–18. DOI: 10.18576/jasu/040102

Blanco-Montero, A. (2021). *Development of an Integrated Model for Urban Sustainable Resilience through Smart City Projects in the Southern African Context*. University of KwaZulu-Natal.

Bressane, A., Pinto, K. P., Garcia, A. P., & Castro, L. C. (2024). Which Dimensions of Nature Contact in Urban Green Spaces Most Significantly Contribute to Mental Wellbeing? *Health & Place*, 89, 103335. DOI: 10.1016/j.healthplace.2024.103335 PMID: 39154413

Cao, F., Xu, X., Zhang, C., & Kong, W. (2023). Evaluation of Urban Flood Resilience and its Space-Time Evolution: A Case Study of Zhejiang Province, China. *Ecological Indicators*, 110643, 110643. Advance online publication. DOI: 10.1016/j.ecolind.2023.110643

Ehwi, R. (2020). Stakeholder Engagement in Smart City Initiatives: Insights from a Review of Smart City Initiatives in Four UK Cities. *University of Cambridge*. https://www.cchpr.landecon.cam.ac.uk/system/files/documents/stakeholder_engagement_review_0.pdf

Hodson, E., Vainio, T., Sayún, M. N., Tomitsch, M., Jones, A., Jalonen, M., Börütecene, A., Hasan, M. T., Paraschivoiu, I., Wolff, A., Yavo-Ayalon, S., Yli-Kauhaluoma, S., & Young, G. W. (2023). Evaluating Social Impact of Smart City Technologies and Services: Methods, Challenges, Future Directions. *Multimodal Technologies and Interaction*, 7(3), 33. DOI: 10.3390/mti7030033

Hoseinzadeh, S., & Garcia, D. A. (2024). AI-Driven Innovations in Greenhouse Agriculture. *Reanalysis of Sustainability and Energy Efficiency Impacts. Energy Conversion and Management: X*, 24, 100701. DOI: 10.1016/j.ecmx.2024.100701

Ige, A. B., Kupa, E., & Ilori, O. (2024). Developing Comprehensive Cybersecurity Frameworks for Protecting Green Infrastructure: Conceptual Models and Practical Applications. *GSC Advanced Research and Reviews*, 20(1), 25–41. DOI: 10.30574/gscarr.2024.20.1.0237

Judijanto, L., Suparwata, D., Cahya, D., & Widodo, I. (2024). The Effect of Green Technology Implementation, Agricultural Product Innovation, and Green Technology Infrastructure Investment on Local Economic Growth in West Java. *West Science Nature and Technology*, 2(3), 142–152. DOI: 10.58812/wsnt.v2i03.1280

Khlie, K., & Benmamoun, Z. (2024). Towards Smarter and Greener Cities: Harnessing AI and Green Technology for Urban Sustainability. *Journal of Infrastructure, Policy and Development.* DOI: 10.58812/wsnt.v2i03.1280

Lacroix, S., Kuller, M., Gougeon, G., Petrucci, J., Lemieux-Chalifour, F., Rioux, A., Dagenais, D., & Bichai, F. (2024). Can We Stop Reinventing the Wheel in Blue-Green Infrastructure Planning? Using Value-Focused Thinking to Enable Transferability of a Multicriteria Planning Support System. *Landscape and Urban Planning*, 252, 105188. DOI: 10.1016/j.landurbplan.2024.105188

Lucatello, S., & Alcántara-Ayala, I. (2024). Sustainable Synergy: Strengthening Disaster Risk Reduction in Latin America and the Caribbean through Nature-Based Solutions. *International Journal of Disaster Risk Reduction*, 113, 104860. DOI: 10.1016/j.ijdrr.2024.104860

Mohammadian, H.D., Wittberg, V., Castro, M., & Bolandian, G. (2020). The 5th Wave and i-Sustainability Plus Theories as Solutions for SocioEdu Consequences of COVID-19. *2020 IEEE Learning With MOOCs (LWMOOCS)*, 118–123. DOI: 10.1016/j.ijdrr.2024.104860

Nirupama, N. (2023). Integrated Aquaculture as an Element of Blue-Green Infrastructure to Enhance Sustainability. *Opus 4.*https://opus4.kobv.de/opus4-hof/files/184/Nair_Nirupama_MA.pdf

Soleimani-Chamkhorami, K., Karbalaie, A., Kasraei, A., Haghighi, E., Famurewa, S. M., & Garmabaki, A. H. S. (2024). Identifying Climate-Related Failures in Railway Infrastructure Using Machine Learning. *Transportation Research Part D, Transport and Environment*, 135, 104371. DOI: 10.1016/j.trd.2024.104371

Song, G., Ifaei, P., Ha, J., Kang, D., Won, W., Liu, J. J., & Na, J. (2024). The AI Circular Hydrogen Economist: Hydrogen Supply Chain Design via Hierarchical Deep Multi-Agent Reinforcement Learning. *Chemical Engineering Journal*, 497, 154464. DOI: 10.1016/j.cej.2024.154464

Suresh, S., Renukappa, S., & Abdul-Aziz, A. (2020). Developments in the United Kingdom Road Transport from a Smart Cities Perspective. DOI: 10.1016/j.ijdrr.2024.104860

Tan, Y., & Cugurullo, F. (2020). The Sustainability of Artificial Intelligence: An Urbanistic Viewpoint from the Lens of Smart and Sustainable Cities. *Sustainability (Basel)*, 12(20), 8548. DOI: 10.3390/su12208548

Wang, X. C., & Fu, G. (2021). *Water-Wise Cities and Sustainable Water Systems: Concepts, Technologies, and Applications*. IWA Publishing.

Zaidi, A., Ajibade, S. S., Musa, M., & Bekun, R. V. (2023). New Insights into the Research Landscape on the Application of Artificial Intelligence in Sustainable Smart Cities. *International Journal of Energy Economics and Policy*, 13(4), 287–299. DOI: 10.32479/ijeep.14683

Compilation of References

Aani, S., Bonny, T., Hasan, S., & Hilal, N. (2019). Can machine language and artificial intelligence revolutionize process automation for water treatment and desalination? *Desalination*, 458, 84–96. Advance online publication. DOI: 10.1016/j.desal.2019.02.005

Abbas, A., Zhang, Z., Zheng, H., Alami, M. M., Alrefaei, A. F., Abbas, Q., Naqvi, S. A. H., Rao, M. J., Mosa, W. F., Abbas, Q., Hussain, A., Hassan, M. Z., & Zhou, L. (2023). Drones in plant disease assessment, efficient monitoring, and detection: A way forward to smart agriculture. *Agronomy (Basel)*, 13(6), 1524. DOI: 10.3390/agronomy13061524

Abdelmageed, S., Tariq, S., Boadu, V., & Zayed, T. (2022). Criteria-based critical review of artificial intelligence applications in water-leak management. *Environmental Reviews*, 30(2), 280–297. Advance online publication. DOI: 10.1139/er-2021-0046

Abdulaziz, A., Al-Khalifa, K. N., & Hamouda, A. M. (2023). Re-Thinking Data Strategy and Integration for Artificial Intelligence: Concepts, Opportunities, and Challenges. *Applied Sciences (Basel, Switzerland)*, 13(12), 7082. DOI: 10.3390/app13127082

Abhilash, R. A., Kumari, A., Singh, R. N., & Kumari, K. (2021). Climate-Smart Agriculture: An Integrated Approach for Attaining Agricultural Sustainability. In V. K. Hebsale Mallappa & M. Shirur (Eds.), *Climate Change and Resilient Food Systems* (pp. 141–189). Springer Singapore. DOI: 10.1007/978-981-33-4538-6_5

Abhilasha, A., Sreenivasulu, A., & Manimozhi, T. Satheesh kumar P., Selvakumar P. and Singh P.,(2022). "The Model of Smart Sensing Device For Sensitive Nano-clusters Modification in Sensing Properties," 2022 2nd International Conference on Advance Computing and Innovative Technologies in Engineering (ICACITE), Greater Noida, India, 2022, pp. 1043-1047, DOI: 10.1109/ICACITE53722.2022.9823846

Abrahart, R. J., & See, L. M. (2000). Neural network modeling of nonlinear hydrological relationships. *Hydrology and Earth System Sciences*, 4(3), 551–560. DOI: 10.5194/hess-4-551-2000

Adamowski, J., Chan, H. F., & Prasher, S. O. (2012). Comparison of multivariate regression and artificial neural networks for peak urban water demand forecasting: Evaluation of different ANN learning algorithms. *Journal of Hydrologic Engineering*, 17(5), 498–511. DOI: 10.1061/(ASCE)HE.1943-5584.0000492

Adamowski, J., & Karapataki, C. (2010). Water demand forecasting in Mediterranean regions using socio-economic and climatic variables. *Water Resources Management*, 24(11), 2417–2432. DOI: 10.1007/s11269-009-9566-4

Adedeji, K., Hamam, Y., Abe, B., & Abu-Mahfouz, A. (2017). Towards Achieving a Reliable Leakage Detection and Localization Algorithm for Application in Water Piping Networks: An Overview. *IEEE Access : Practical Innovations, Open Solutions*, 5, 20272–20285. DOI: 10.1109/ACCESS.2017.2752802

Adewusi, A. O., Asuzu, O. F., Olorunsogo, T., Olorunsogo, T., Adaga, E., & Daraojimba, D. O. (2024). Ai in precision agriculture: A review of technologies for sustainable farming practices. *World Journal of Advanced Research and Reviews*, (1), 2276–2285. DOI: 10.30574/wjarr.2024.21.1.0314

Adhikary, S., Biswas, B., Naskar, M. K., Mukherjee, B., Singh, A. P., & Atta, K. (2022). Remote sensing for agricultural applications. In *Arid Environment-Perspectives, Challenges and Management.* IntechOpen. https://www.intechopen.com/chapters/83417

Agarwal, M., Singh, K., Dohare, R., & Upadhyaya, S. (2016). Process control and optimization of wastewater treatment plants using simulation softwares: A review. *International Journal of Advanced Technology in Engineering and Exploration*, 3(2), 145–149. DOI: 10.19101/IJATEE.2016.322006

Ahmad, A. A., Kasron, N., Noh, N. H. M., Azmon, H., & Chuang, T. C. (2023). *The acceptance of decision support system (DSS) in vegetable cultivation among farmers in Malaysia.* http://etmr.mardi.gov.my/Content/ETMR%20Vol.21(2023)/Aimi%20Athirah_Vol21.pdf

Ahmed, A., Othman, F., Afan, H., Ibrahim, R., Fai, C., Hossain, S., Ehteram, M., & Elshafie, A. (2019). Machine learning methods for better water quality prediction. *Journal of Hydrology (Amsterdam)*, 578, 124084. Advance online publication. DOI: 10.1016/j.jhydrol.2019.124084

Ahmed, A., Salih, A., & Zhang, Z. (2018). Reinforcement learning for optimizing valve operations in water distribution networks. *Journal of Hydroinformatics*, 20(6), 1235–1248. DOI: 10.2166/hydro.2018.098

Aigües de Barcelona. (2024). *Aigues Chatbot.* Retrieved July 12, 2024 from https://www.aiguesdebarcelona.cat/ca/aigues-de-barcelona

Aivazidou, E., Banias, G., Lampridi, M., Vasileiadis, G., Anagnostis, A., Papageorgiou, E. I., & Bochtis, D. (2021). Smart technologies for sustainable water management: An urban analysis. *Sustainability (Basel)*, 13(24), 13940. DOI: 10.3390/su132413940

Akhai, S., Singh, V. P., & John, S. (2016). Human performance in industrial design centers with small unit air conditioning systems. *Journal of Advanced Research in Production Industrial Engineering*, 3(2), 5–11.

Akhai, S., Singh, V. P., & John, S. (2016). Investigating Indoor Air Quality for the Split-Type Air Conditioners in an Office Environment and Its Effect on Human Performance. *Journal of Mechanical Civil Engineering*, 13(6), 113–118.

Akkem, Y., Biswas, S. K., & Varanasi, A. (2023). Smart farming using artificial intelligence: A review. *Engineering Applications of Artificial Intelligence*, 120, 105899. DOI: 10.1016/j.engappai.2023.105899

Al-Adhaileh, M., & Alsaade, F. (2021). Modelling and Prediction of Water Quality by Using Artificial Intelligence. *Sustainability (Basel)*, 13(8), 4259. DOI: 10.3390/su13084259

Aldegheishem, A., Alrajeh, N., García, L. R., & Lloret, J. (2022). Swap: Smart water protocol for the irrigation of urban gardens in smart cities. *IEEE Access : Practical Innovations, Open Solutions*, 10, 39239–39247. DOI: 10.1109/ACCESS.2022.3165579

Aldhyani, T., Al-Yaari, M., Alkahtani, H., & Maashi, M. (2020). Water Quality Prediction Using Artificial Intelligence Algorithms. *Applied Bionics and Biomechanics*, 2020, 1–12. Advance online publication. DOI: 10.1155/2020/6659314 PMID: 33456498

Alexander Obaigbena, Michael Ayorinde DadaMichael Tega MajemiteOnyeka Henry DaraojimbaJohnson Sunday OlihaZamathula Queen Sikhakhane Nwokediegwu. (2024). Review of Smart Water Management: Iot and ai in water and wastewater treatment. *World Journal of Advanced Research and Reviews*, 21(1), 1373–1382. DOI: 10.30574/wjarr.2024.21.1.0171

Alghadeer, S. G. (2023). The use of artificial intelligence in water management projects in the kingdom of saudi arabia. *American Journal of Technology*, 2(1), 54–65. DOI: 10.58425/ajt.v2i1.208

Ali, S., Amir, S., Ali, S., Rehman, M. U., Majid, S., &Yatoo, A. M. (2021). Water pollution: Diseases and health impacts. *Freshwater Pollution and Aquatic Ecosystems: Environmental Impact and Sustainable Management*, 1.

Ali, G., Mijwil, M. M., Buruga, B. A., Abotaleb, M., & Adamopoulos, I. (2024). A Survey on Artificial Intelligence in Cybersecurity for Smart Agriculture: State-of-the-Art, Cyber Threats, Artificial Intelligence Applications, and Ethical Concerns. *Mesopotamian Journal of Computer Science*, 2024, 71–121. DOI: 10.58496/MJCSC/2024/007

Al-Juaidi, A., Kharabsheh, S., & Al-Taani, A. (2017). Real-time water usage management in smart cities using clustering algorithms. *Journal of Water Resources Planning and Management*, 143(10), 04017054. DOI: 10.1061/(ASCE)WR.1943-5452.0000814

Allawi, M., Salih, S., Kassim, M., Ramal, M., Mohammed, A., & Yaseen, Z. (2022). Application of Computational Model Based Probabilistic Neural Network for Surface Water Quality Prediction. *Mathematics*, 10(21), 3960. Advance online publication. DOI: 10.3390/math10213960

Alotaibi, B. S., Shema, A. I., & Ibrahim, A. U.. (2024). Assimilation of 3D Printing, Artificial Intelligence (AI), and Internet of Things (IoT) for the Construction of Eco-Friendly Intelligent Homes: An Explorative Review. *Heliyon*, 10(17), e36846. DOI: 10.1016/j.heliyon.2024.e36846 PMID: 39286162

Alprol, A. E., Mansour, A. T., Ibrahim, M. E., & Ashour, M. (2024). Artificial Intelligence Technologies Revolutionizing Wastewater Treatment: Current Trends and Future Prospective. *Water (Basel)*, 16(2), 314. DOI: 10.3390/w16020314

Alshehri, F., & Rahman, A. (2023). Coupling machine and deep learning with explainable artificial intelligence for improving prediction of groundwater quality and decision-making in arid region, saudi arabia. *Water (Basel)*, 15(12), 2298. DOI: 10.3390/w15122298

Alvarez-Cano, J., Cavero, V., & Diez-Canseco, F. (2022). Idas y venidas del diseño de la política de alimentación saludable en el perú: Análisis comparativo de sus documentos regulatorios. *Revista Peruana de Medicina Experimental y Salud Pública*, •••, 480–488. DOI: 10.17843/rpmesp.2022.394.11896 PMID: 36888812

Amorós, A. M. R., Cantos, J. O., & Castiñeira, C. J. B. (2014). Competencias por el uso del agua en la provincia de alicante: experiencias de gestión en la armonización de usos urbano-turísticos y agrícolas - (competències per l'ús de l'aigua a la província d'alacant: experiències de gestió en l'harmonització...). *Documents d'Analisi Geografica*, 60(3), 523–548. Advance online publication. DOI: 10.5565/rev/dag.136

Anglian Water Services. (2023). *Anglian Water chatbot*. Retrieved July 10, 2024 from https://www.anglianwater.co.uk/

Angon, P. B., Mondal, S., Jahan, I., Datto, M., Antu, U. B., Ayshi, F. J., & Islam, M. (2023). S. (2023). Integrated Pest Management (IPM) in Agriculture and Its Role in Maintaining Ecological Balance and Biodiversity. *Advances in Agriculture*, 2023, 1–19. DOI: 10.1155/2023/5546373

Antunes, A., Andrade-Campos, A., Sardinha-Lourenço, A., & Oliveira, M. S. (2018). Short-term water demand forecasting using machine learning techniques. *Journal of Hydroinformatics*, 20(6), 1343–1366. DOI: 10.2166/hydro.2018.163

Apaza, A. A. Z., Gárate, S. S. O., Cuadros, A. E. J. C., & Ccasa, P. M. C. (2022). Modelo predictivo de la potabilidad del agua mediante un árbol de decisión en Inteligencia Artificial. *Innovación y Software*, 3(2), 121–131. DOI: 10.48168/innosoft.s9.a72

Asadi, A., Verma, A., Yang, K., & Mejabi, B. (2017). Wastewater treatment aeration process optimization: A data mining approach. *Journal of Environmental Management*, 203(Pt 2), 630–639. DOI: 10.1016/j.jenvman.2016.07.047 PMID: 27460213

Ashwini, R. M., Pallavi, C. V., Jayashree, S., Geetha, L. S., Jalaja, G., & Raghavendra, C. K & Santosh Reddy (2022). Artificial Intelligence-Based Techniques to Monitor and Maintain the Quality of Drinking Water Supplied to Households [Patent]. https://patentscope.wipo.int/search/es/detail.jsf?docId=IN379761934&_cid=P20-LZCWA8-96371-1

Asociación de Tecnología, A. (2007). Smart water solutions. *Sanctuary: Modern Green Homes*, 1, 34–35.

Australian Government. Water sensitive urban desing. Available online: https://cdn.environment.sa.gov.au/environment/docs/water-sensitive-urban-design-policy-gen.pdf (accessed on 21 March 2023).

Awuchi, C. G., Igwe, V. S., & Amagwula, I. O. (2020). Nutritional diseases and nutrient toxicities: A systematic review of the diets and nutrition for prevention and treatment. *International Journal of Advanced Academic Research*, 6(1), 1–46. DOI: 10.46654/ij.24889849.e61112

BABLE. (2024). *Smart Water Management.* Retrieved July 2, 2024 from https://www.bable-smartcities.eu/explore/solutions/solution/smart-water-management.html

Badar, S., Labghough, S., Al-Abdulghani, A., Mohammed, E., Bouhali, O., & Qaraqe, K. (2023). Machine Learning Assisted Approach for Water Leaks Detection. *2023 International Conference on Information Networking (ICOIN)*, 433-437. https://doi.org/DOI: 10.1109/ICOIN56518.2023.10048954

Bae, K., Kwon, S., & Hwang, M. (2020). CNN-based urban water demand forecasting. *Sustainable Cities and Society*, 60, 102275. DOI: 10.1016/j.scs.2020.102275

Bahad, P., & Saxena, P. (2019). Study of AdaBoost and Gradient Boosting Algorithms for Predictive Analytics., 235-244. https://doi.org/. DOI: 10.1109/ACCESS.2021.3052032

Bakker, M., Vreeburg, J., Schagen, K., & Rietveld, L. (2013). A fully adaptive forecasting model for short-term drinking water demand. *Environmental Modelling & Software*, 48, 141–151. DOI: 10.1016/j.envsoft.2013.06.012

Balafoutis, A., Beck, B., Fountas, S., Vangeyte, J., Wal, T., Soto, I., Gómez-Barbero, M., Barnes, A., & Eory, V. (2017a). Precision Agriculture Technologies Positively Contributing to GHG Emissions Mitigation, Farm Productivity and Economics. *Sustainability (Basel)*, 9(8), 1339. DOI: 10.3390/su9081339

Balali-Mood, M., Naseri, K., Tahergorabi, Z., Khazdair, M. R., & Sadeghi, M. (2021). Toxic mechanisms of five heavy metals: Mercury, lead, chromium, cadmium, and arsenic. *Frontiers in Pharmacology*, 12, 227. DOI: 10.3389/fphar.2021.643972 PMID: 33927623

Balduzzi, D., Frean, M., & Leary, L.. (2017). The shattered gradients problem: if resnets are the answer, then what is the question? In: Proceedings of the 34th international conference on machine learning. *Proceedings of Machine Learning Research*, 70, 342–350.

Balkrishna, A., Pathak, R., Kumar, S., Arya, V., & Singh, S. K. (2023). A comprehensive analysis of the advances in Indian Digital Agricultural architecture. *Smart Agricultural Technology*, 5, 100318. DOI: 10.1016/j.atech.2023.100318

Balta, D., Balta Kaç, S., Balta, M., & Eken, S. (2023). Water Quality Estimation and Anomaly Detection: A Review. *EAI Endorsed Transactions on Internet of Things*, 9(4), e2. DOI: 10.4108/eetiot.v9i4.3660

Banerjee, K., Bali, V., Nawaz, N., Bali, S., Mathur, S., Mishra, R. K., & Rani, S. (2022, February). A machine-learning approach for prediction of water contamination using latitude, longitude, and elevation. *Water (Basel)*, 14(5), 728. DOI: 10.3390/w14050728

Bao-lei, Y., De'an, Z., & Jun, Z. (2011). Prediction system of sewage outflow COD based on LS-SVM. *2011 2nd International Conference on Intelligent Control and Information Processing,* 1, 399–402. https://doi.org/DOI: 10.1109/ICICIP.2011.6008273

Barbedo, J. G. A. (2019). A review on the use of unmanned aerial vehicles and imaging sensors for monitoring and assessing plant stresses. *Drones (Basel)*, 3(2), 40. DOI: 10.3390/drones3020040

Barbosa Júnior, M. R., Moreira, B. R. D. A., Carreira, V. D. S., Brito Filho, A. L. D., Trentin, C., Souza, F. L. P. D., Tedesco, D., Setiyono, T., Flores, J. P., Ampatzidis, Y., Silva, R. P. D., & Shiratsuchi, L. S. (2024). Precision agriculture in the United States: A comprehensive meta-review inspiring further research, innovation, and adoption. *Computers and Electronics in Agriculture*, 221, 108993. DOI: 10.1016/j.compag.2024.108993

Bårnes, V. J., & Pordoğan, B. (2022). *Sustainable Development and Citizen Participation.* Molde University College.

Barrile, V., Simonetti, S., Citroni, R., Fotia, A., & Bilotta, G. (2022). Experimenting agriculture 4.0 with sensors: A data fusion approach between remote sensing, UAVs and self-driving tractors. *Sensors (Basel)*, 22(20), 7910. DOI: 10.3390/s22207910 PMID: 36298261

Baruch, I., Georgieva, P., Barrera-Cortés, J., & Azevedo, S. (2005). Adaptive recurrent neural network control of biological wastewater treatment. *International Journal of Intelligent Systems*, 20(2), 173–193. Advance online publication. DOI: 10.1002/int.20061

Barzilay, J. I., Weinberg, W. G., & Eley, J. W. (1999). *The water we drink: Water quality and its effects on health.* Rutgers University Press.

Begna, T. (2020). Major challenging constraints to crop production farming system and possible breeding to overcome the constraints. [IJRSAS]. *International Journal of Research Studies in Agricultural Sciences*, 6(7), 27–46.

Beiranvand, B., & Rajaee, T. (2022). Application of artificial intelligence-based single and hybrid models in predicting seepage and pore water pressure of dams: A state-of-the-art review. *Advances in Engineering Software*, 173, 103268. DOI: 10.1016/j.advengsoft.2022.103268

Bennehalli, B. (2024). Machine learning approach to battery management and balancing techniques for enhancing electric vehicle battery performance. *Journal of Electrical Systems*, 20(2s), 885–892. DOI: 10.52783/jes.1685

Bertel, M. E. C., Morales, J. D. C. J., Barrios, R. L. A., & Sierra, C. A. S. (2014). Precisión y exactitud para las determinaciones de la demanda bioquímica de oxigeno en aguas por luminiscencia. *Ingenium*, 8(19), 55. DOI: 10.21774/ing.v8i19.386

Bhunia, S., Karmakar, S., Bhattacharjee, S., Roy, K., Kanthal, S., Pramanick, M., Baishya, A., & Mandal, B. (2021). Optimization of energy consumption using data envelopment analysis (DEA) in rice-wheat-green gram cropping system under conservation tillage practices. *Energy*, 236, 121499. DOI: 10.1016/j.energy.2021.121499

Bibri, S. E., & Krogstie, J. (2021). A Novel Model for Data-Driven Smart Sustainable Cities of the Future: A Strategic Roadmap to Transformational Change in the Era of Big Data. *Future Cities and Environment*, 7(1), 1–25. DOI: 10.5334/fce.116

Bid, I. (2023, noviembre 28). *Cómo las nuevas tecnologías están transformando el abastecimiento de agua y saneamiento en Latinoamérica y el Caribe*. https://idbinvest.org/es/publicaciones/como-las-nuevas-tecnologias-estan-transformando-el-abastecimiento-de-agua-y

Bino, ovo. (2023, abril 13). *Sensores de flujo para medir la velocidad de flujo, volumen...* Binoovo. https://binoovo.com/blog/sensores-de-flujo-medir-flujo-volumen-masa-liquidos-gases

Biscaya, S. (2020). Impact of Big Data from BIM and GIS on Smart Urban Futures. *Applied Science University Journal*, 4(1), 11–18. DOI: 10.18576/jasu/040102

Blanco-Montero, A. (2021). *Development of an Integrated Model for Urban Sustainable Resilience through Smart City Projects in the Southern African Context*. University of KwaZulu-Natal.

Bob, D. (2022, junio 8). *What is a magnetic sensor? What are the categories?* https://www.linkedin.com/pulse/what-magnetic-sensor-categories-bob-dong

Bongiovanni, R., & Lowenberg-Deboer, J. (2004). Precision Agriculture and Sustainability. *Precision Agriculture*, 5(4), 359–387. DOI: 10.1023/B:PRAG.0000040806.39604.aa

Boomgaard, M., Langeveld, J., & Clemens, F. (2001). Wastewater system optimization using genetic algorithms. *Proceedings of the World Environmental and Water Resources Congress,* 1–9. https://doi.org/DOI: 10.1061/40569(2001)498

Bougadis, J., Adamowski, K., & Diduch, R. (2005). Short-term municipal water demand forecasting. *Hydrological Processes*, 19(1), 137–148. DOI: 10.1002/hyp.5763

Boujelben, M., Benmessaoud, Z., Abid, M., & Elleuchi, M. (2023). An efficient system for water leak detection and localization based on IoT and lightweight deep learning. *Internet of Things : Engineering Cyber Physical Human Systems*, 24, 100995. DOI: 10.1016/j.iot.2023.100995

Boutin Mesa, J. M. (2018). Combatiendo la sequía con inteligencia artificial. *Tecnoaqua*, (30), 98–101.

Bozorg-Haddad, O., Delpasand, M., & Loáiciga, H. A. (2021). Water quality, hygiene, and health. In *Economical, Political, and Social Issues in Water Resources* (pp. 217-257). Elsevier. DOI: 10.1016/B978-0-323-90567-1.00008-5

Brentan, B., Meirelles, G., Manzi, D., & Izquierdo, J. (2017). Hybrid regression model for near real-time urban water demand forecasting. *Journal of Hydroinformatics*, 19(2), 293–306. DOI: 10.2166/hydro.2016.059

Bressane, A., Pinto, K. P., Garcia, A. P., & Castro, L. C. (2024). Which Dimensions of Nature Contact in Urban Green Spaces Most Significantly Contribute to Mental Wellbeing? *Health & Place*, 89, 103335. DOI: 10.1016/j.healthplace.2024.103335 PMID: 39154413

Brooks, D. B. (2006). An Operational Definition of Water Demand Management. *International Journal of Water Resources Development*, 22(4), 521–528. DOI: 10.1080/07900620600779699

Brown, E. C., & Horton, A. A. (Eds.). (2018). *Advances in water quality monitoring and assessment*. Academic Press.

Brown, R. R., Keath, N., & Wong, T. H. (2009). Urban water management in cities: Historical, current and future regimes. *Water Science and Technology*, 59(5), 847–855. DOI: 10.2166/wst.2009.029 PMID: 19273883

Cabarcas Álvarez, A., Puello Marrugo, P., & Lechuga Sierra, J. E. (2023). *Sistema de monitoreo para cuerpos de agua en la ciudad de cartagena basado en internet de las cosas*. Encuentro Internacional De Educación en Ingeniería., DOI: 10.26507/paper.3383

Cairone, S., Hasan, S., Zarra, T., Belgiorno, V., & Naddeo, V. (2023). Optimization of wastewater treatment by integration of artificial intelligence techniques: Recent progress and future perspectives. *Global NEST International Conference on Environmental Science & Technology.*https://doi.org/DOI: 10.30955/gnc2023.00353

Campisi-Pinto, S., Adamowski, J., & Oron, G. (2012). Forecasting Urban Water Demand Via Wavelet-Denoising and Neural Network Models. Case Study: City of Syracuse, Italy. *Water Resources Management*, 26, 3539–3558. DOI: 10.1007/s11269-012-0089-y

Canteli, S., García, J., Sanz, D., & Hernández, F. (2019). Deep learning for residential water consumption forecasting. *Procedia CIRP*, 80, 259–264. DOI: 10.1016/j.procir.2019.01.104

Cao, F., Xu, X., Zhang, C., & Kong, W. (2023). Evaluation of Urban Flood Resilience and its Space-Time Evolution: A Case Study of Zhejiang Province, China. *Ecological Indicators*, 110643, 110643. Advance online publication. DOI: 10.1016/j.ecolind.2023.110643

Cartegraph for Asset Management Software for Water Treatment Plants. Retrieved Available online: https://www.cartegraph.com/software-for/water-treatment-plants (accessed on 20 Jan, 2023).

Carvajal, L. J. O. and Mazo, W. H. A. (2019). Oportunidades y desafíos: una mirada holística a las iniciativas y proyectos en curso para ciudades inteligentes en colombia y medellín. *Unaciencia Revista De Estudios E Investigaciones*, 12(22), 53-65. .DOI: 10.35997/runacv12n22a6

Catacora, A., & César, C. (2021). Implementación de un sistema de internet de las cosas para optimizar la gestión del agua en la agricultura de la región tacna. *Ingenieria e Investigacion*, 3(1), 519–533. DOI: 10.47796/ing.v3i1.478

CEPAL. (2024). *Water resources*. Retrieved July 17, 2024 from https://www.cepal.org/en/subtopics/water-resources

Chang, F., Chang, L., & Chen, J. (2023). Artificial intelligence techniques in hydrology and water resources management. *Water (Basel)*, 15(10), 1846. DOI: 10.3390/w15101846

Chang, F., & Guo, S. (2020). Advances in hydrologic forecasts and water resources management. *Water (Basel)*, 12(6), 1819. DOI: 10.3390/w12061819

Chang, M., & Liu, J. (2009). Water Demand Prediction Model Based on Radial Basis Function Neural Network. *2009 First International Conference on Information Science and Engineering*, 5295-5298. https://doi.org/DOI: 10.1109/ICISE.2009.1343

Chang, T.-Y., Cho, W.-T., Tseng, S.-Y., Ouyang, Y., & Lai, C.-F. (2020). *Predictive Maintenance of Water Purification Unit for Smart Factories*. Cognitive Cities.

Chan, T., Chin, C., & Zhong, X. (2018). Review of Current Technologies and Proposed Intelligent Methodologies for Water Distributed Network Leakage Detection. *IEEE Access : Practical Innovations, Open Solutions*, 6, 78846–78867. DOI: 10.1109/ACCESS.2018.2885444

Chen, J., Wu, Q., & Ahmadi, C. (2022). AIoT@Water - An Intelligent Water Resources Management System. 2022 5th International Conference on Information and Communications Technology (ICOIACT), 199-204. https://doi.org/DOI: 10.1109/ICOIACT55506.2022.9971964

Chen, K., Chen, H., Zhou, C., Huang, Y., Qi, X., Shen, R., Liu, F., Zuo, M., Zou, X., Wang, J., Zhang, Y., Chen, D., Chen, X., Deng, Y., & Ren, H. (2019). Comparative analysis of surface water quality prediction performance and identification of key water parameters using different machine learning models based on big data. *Water Research*, 171, 115454. DOI: 10.1016/j.watres.2019.115454 PMID: 31918388

Chen, L., Yao, Y., & Zhang, W. (2016). Anomaly detection in urban water distribution systems using machine learning algorithms. *Procedia Engineering*, 154, 627–632. DOI: 10.1016/j.proeng.2016.07.530

Chen, Y., & Liu, C. (2021). Deep learning applications in water quality monitoring: A review. *The Science of the Total Environment*, 761, 144306. DOI: 10.1016/j.scitotenv.2020.144306

Chere-Quiñónez, B. F., Souza, R. C. U., Reyna-Tenorio, L. J., Canchingre-Bone, M. E., & Mosquera-Quintero, G. A. (2022). Energía eléctrica a partir de aguas residuales industriales en guayaquil, ecuador. *Sapienza: International Journal of Interdisciplinary Studies*, 3(6), 202–210. DOI: 10.51798/sijis.v3i6.509

Choudhary, M., Peter, C., Shukla, S., Govender, P., Joshi, G., & Wang, R. (2019). Environmental issues: A challenge for wastewater treatment. *Green Materials for Wastewater Treatment*, 1–13. https://doi.org/DOI: 10.1016/j.scitotenv.2020.138415 PMID: 32348944

Chowdhury, S., & Karanfil, T. (2024). Applications of artificial intelligence (AI) in drinking water treatment processes: Possibilities. *Chemosphere*, 356, 141958. DOI: 10.1016/j.chemosphere.2024.141958 PMID: 38608775

Cirelli, A. F. (2012). El agua: Un recurso esencial. *Química Viva*, 11(3), 147–170.

Codd, E. F. (1970). A relational model of data for large shared data banks. *Communications of the ACM*, 13(6), 377–387. DOI: 10.1145/362384.362685

Cohen, L. Quality function deployment: how to make QFD work for you. Prentice Hall, Hoboken, New, Jersey, U.S.A, pp 211-231, 1995.

Collado, I. (2024), IA y Machine Learning en las disciplinas del agua y el medioambiente: del hidrólogo al "hidroinformático". Sitio Sener, https://www.group.sener/insights/ia-machine-learning-disciplinas-del-agua-medioambiente-del-hidrologo-al-hidroinformatico/

Comisión Nacional del Agua (Conagua). (2020) Usos del agua en México 2020. https://sinav30.conagua.gob.mx:8080/UsosAgua/

Contreras Clavijo, S. A., & Rojano Mueses, X. G. (2023). Sistema Inteligente de Monitoreo y Control para la Planta de Tratamiento de Agua Potable El Carrizal Salcedo basado en IOT e Inteligencia Artificial [Tesis]. Universidad Técnica de Ambato. https://repositorio.uta.edu.ec/handle/123456789/38419

Council, E. P. a. o. t. (2024). *Directive 2000/60/EC* Retrieved July 12, 2024 from https://eur-lex.europa.eu/legal-content/EN/TXT/PDF/?uri=CELEX:02000L0060-20141120

Crini, G., & Lichtfouse, E. (2018). Wastewater treatment: An overview. *Environmental Chemistry for a Sustainable World*, 18, 1–21. DOI: 10.1007/978-3-319-92111-2_1

Cubero, R. (1991). Neural Networks for Water Demand Time Series Forecasting. 453-460. https://doi.org/.DOI: 10.1007/BFb0035927

Cui, T., & Zhang, H. (2020). Artificial intelligence in water quality monitoring and control: Applications, challenges, and perspectives. *Water Supply*, 20(5), 1837–1852. DOI: 10.2166/ws.2020.224

Cutore, P., Campisano, A., Kapelan, Z., Modica, C., & Savic, D. (2008). Probabilistic prediction of urban water consumption using the SCEM-UA algorithm. *Urban Water Journal*, 5(2), 125–132. DOI: 10.1080/15730620701754434

Danu, R. (2018). Optimization of water distribution algorithms using AI. *International Journal of Civil Engineering and Technology*, 9(13), 2130–2146. https://doi.org/https://doi.org/10.17605/OSF.IO/FU2PD

Dasallas, L., Lee, J., Jang, S., & Jang, S. (2024). Development and application of technical key performance indicators (kpis) for Smart Water Cities (swcs) global standards and certification schemes. *Water (Basel)*, 16(5), 741. DOI: 10.3390/w16050741

Das, S., Bai, H., Wu, C., Kao, J., Barney, B., Kidd, M., & Kuettel, M. (2016). Improving the performance of industrial clarifiers using three-dimensional computational fluid dynamics. *Engineering Applications of Computational Fluid Mechanics*, 10(1), 130–144. DOI: 10.1080/19942060.2015.1121518

Dedja, K., Nakano, F., Pliakos, K., & Vens, C. (2022). Explaining random forest prediction through diverse rulesets. ArXiv, abs/2203.15511. https://doi.org//arXiv.2203.15511.DOI: 10.48550

Dehghani, R., & Poudeh, H. (2021). Application of novel hybrid artificial intelligence algorithms to groundwater simulation. *International Journal of Environmental Science and Technology*. Advance online publication. DOI: 10.1007/s13762-021-03596-5

Del-Coco, M., Leo, M., & Carcagnì, P. (2024). Machine Learning for Smart Irrigation in Agriculture: How Far along Are We? *Information (Basel)*, 15(6), 306. DOI: 10.3390/info15060306

Devi, R. R., Dharshini, P., Hemala, R., & Swetha, D. "Heart Disease Prediction using Random Forest Classifier," *2023 International Conference on Data Science, Agents & Artificial Intelligence (ICDSAAI)*, Chennai, India, 2023, pp. 1-8, DOI: 10.1109/ICDSAAI59313.2023.10452459

DEWA. (2024). *RAMMAS Chatbot*. Retrieved July 14, 2024 from https://www.dewa.gov.ae/en/rammas

Dhananjayan, V., Jayanthi, P., Jayakumar, S., & Ravichandran, B. (2020). Agrochemicals Impact on Ecosystem and Bio-monitoring. In Kumar, S., Meena, R. S., & Jhariya, M. K. (Eds.), *Resources Use Efficiency in Agriculture* (pp. 349–388). Springer Singapore., DOI: 10.1007/978-981-15-6953-1_11

Dhillon, R., & Moncur, Q. (2023). Small-Scale Farming: A Review of Challenges and Potential Opportunities Offered by Technological Advancements. *Sustainability (Basel)*, 15(21), 15478. DOI: 10.3390/su152115478

Di Nardo, A., Boccelli, D. L., Herrera, M., Creaco, E., Cominola, A., Sitzenfrei, R., & Taormina, R. (2021). Smart urban water networks: Solutions, trends and challenges. *Water (Basel)*, 13(4), 501. DOI: 10.3390/w13040501

Dou, J., Yunus, A., Bui, D., Merghadi, A., Sahana, M., Zhu, Z., Chen, C., Han, Z., & Pham, B. (2019). Improved landslide assessment using support vector machine with bagging, boosting, and stacking ensemble machine learning framework in a mountainous watershed, Japan. *Landslides*, 17(3), 641–658. DOI: 10.1007/s10346-019-01286-5

Drogkoula, M., Kokkinos, K., & Samaras, N. (2023). A comprehensive survey of machine learning methodologies with emphasis in water resources management. *Applied Sciences (Basel, Switzerland)*, 13(22), 12147. DOI: 10.3390/app132212147

Dubey, A. (2020, agosto 4). Working of an Ultrasonic Sensor. *NerdyElectronics*. https://nerdyelectronics.com/working-of-an-ultrasonic-sensor/

Duerr, I., Merrill, H., Wang, C., Bai, R., Boyer, M., Dukes, M., & Bliznyuk, N. (2018). Forecasting urban household water demand with statistical and machine learning methods using large space-time data: A Comparative study. *Environmental Modelling & Software*, 102, 29–38. DOI: 10.1016/j.envsoft.2018.01.002

Duff, H., Hegedus, P., Loewen, S., Bass, T., & Maxwell, B. (2021). Precision Agroecology. *Sustainability (Basel)*, 14(1), 106. DOI: 10.3390/su14010106

Dwivedi, Y., Hughes, L., Ismagilova, E., Aarts, G., Coombs, C., & Crick, T.. (2019). Artificial intelligence (AI): Multidisciplinary perspectives on emerging challenges, opportunities, and agenda for research, practice, and policy. *International Journal of Information Management*. Advance online publication. DOI: 10.1016/j.ijinfomgt.2019.08.002

Ebisi, F., Nikolakakos, I., Karunamurthi, J., Alnuaimi, A., Buraimi, E., & Alblooshi, S. (2023). Machine Learning Schemes for Leak Detection in IoT-enabled Water Transmission System. 2023 International Conference on IT Innovation and Knowledge Discovery (ITIKD), 1-7. https://doi.org/DOI: 10.1109/ITIKD56332.2023.10100175

Edwards, P. C. (2020). The Beginnings of Agriculture. In Hollander, D., & Howe, T. (Eds.), *A Companion to Ancient Agriculture* (1st ed., pp. 119–148). Wiley., DOI: 10.1002/9781118970959.ch7

Egbuna, C., Amadi, C. N., Patrick-Iwuanyanwu, K. C., Ezzat, S. M., Awuchi, C. G., Ugonwa, P. O., & Orisakwe, O. E. (2021). Emerging pollutants in Nigeria: A systematic review. *Environmental Toxicology and Pharmacology*, 85, 103638. DOI: 10.1016/j.etap.2021.103638 PMID: 33757839

Ehret, D. L., Hill, B. D., Helmer, T., & Edwards, D. R. (2011). Neural network modeling of greenhouse tomato yield, growth and water use from automated crop monitoring data. *Computers and Electronics in Agriculture*, 79(1), 82–89. DOI: 10.1016/j.compag.2011.07.013

Ehwi, R. (2020). Stakeholder Engagement in Smart City Initiatives: Insights from a Review of Smart City Initiatives in Four UK Cities. *University of Cambridge*. https://www.cchpr.landecon.cam.ac.uk/system/files/documents/stakeholder_engagement_review_0.pdf

Elenes, J. R. F., Escobar, R. R., & Urbano, P. M. (2023). La gestión de los recursos hídricos en el municipio de culiacán, sinaloa, méxico. Revistas Cuadernos De Trabajo De Estudios Regionales en Economía, Población Y Desarrollo, 13(73), 3-42. .DOI: 10.20983/epd.2023.73.1

Engineering, O. (2023). *What is a Coriolis Flow Meter and How Does it Work*

Engineering, O. (2024). *Lessons about Positive Displacement Flow Meters*. https://www.omega.com/en-us/resources/positive-displacement-flow-meter

EPA. EPANET: Application for Modeling Drinking Water Distribution Systems. Retrieved. Available online: https://www.epa.gov/water-research/epanet (accessed on 22 Jan, 2023).

Erdal, H., & Karakurt, O. (2013). Advancing monthly streamflow prediction accuracy of CART models using ensemble learning paradigms. *Journal of Hydrology (Amsterdam)*, 477, 119–128. DOI: 10.1016/j.jhydrol.2012.11.015

Erdal, H., Karakurt, O., & Namli, E. (2013). High performance concrete compressive strength forecasting using ensemble models based on discrete wavelet transform. *Engineering Applications of Artificial Intelligence*, 26(4), 1246–1254. DOI: 10.1016/j.engappai.2012.10.014

Eryiğit, M. (2014). Cost optimization of water distribution networks by using artificial immune systems. *Journal of Water Supply: Research & Technology - Aqua*, 64(1), 47–63. DOI: 10.2166/aqua.2014.031

Escolero, O., Kralisch, S., Martínez, S. E., & Perevochtchikova, M. (2016). Diagnóstico y análisis de los factores que influyen en la vulnerabilidad de las fuentes de abastecimiento de agua potable a la Ciudad de México, México. *Boletín de la Sociedad Geológica Mexicana*, 68(3), 409–427. DOI: 10.18268/BSGM2016v68n3a3

Esteves Francisco, A., Cardoso, J., Leitão, S., & Pires, E. (2023). Impact of artificial intelligence in the reduction of electrical consumption in wastewater treatment plants: A review. *Journal of Information Systems Engineering & Management*, 8(3). https://doi.org/https://doi.org/10.55267/iadt.07.13623

EWS. (2011). *European Water Stewardship (EWS) Standard*. European Water Partnership. Retrieved July 3, 2024 from https://ews.info/media/1007/document1.pdf

Facultad de Medicina Veterinaria y Zootecnia UNAM. (2024) Estructura e importancia del agua en el funcionamiento molecular. https://fmvz.unam.mx/fmvz/p_estudios/apuntes_bioquimica/Unidad_2.pdf

Fältström, E., & Anderberg, S. (2020). Towards control strategies for microplastics in urban water. *Environmental Science and Pollution Research International*, 27(32), 40421–40433. DOI: 10.1007/s11356-020-10064-z PMID: 32666462

Fang, X., Zhai, Z., Zang, J., & Zhu, Y. (2022). An intelligent dosing algorithm model for wastewater treatment plants. *Journal of Physics: Conference Series*, 2224(1), 1–6. DOI: 10.1088/1742-6596/2224/1/012027

Farooqui, N. A. Haleem, Mohd., Khan, W., & Ishrat, M. (2024). Precision Agriculture and Predictive Analytics: Enhancing Agricultural Efficiency and Yield. In N. Singh, S. Birla, M. D. Ansari, & N. K. Shukla (Eds.), *Intelligent Techniques for Predictive Data Analytics* (1st ed., pp. 171–188). Wiley. DOI: 10.1002/9781394227990.ch9

Faster, C. (2024). *Monitoreo De Calidad Del Agua*. FasterCapital. https://fastercapital.com/keyword/monitoreo-de-calidad-del-agua.html

Fathi, M., Haghi Kashani, M., Jameii, S. M., & Mahdipour, E. (2022). Big data analytics in weather forecasting: A systematic review. *Archives of Computational Methods in Engineering*, 29(2), 1247–1275. DOI: 10.1007/s11831-021-09616-4

Feiy, L. (2021). Intelligent water drinking equipment based on artificial intelligence [Patent]. https://patentscope.wipo.int/search/es/detail.jsf?docId=CN345721710&_cid=P20-LZCWA8-96371-1

Fels, A., Mandi, L., Kammoun, A., Ouazzani, N., Monga, O., & Hbid, M. (2023). Artificial intelligence and wastewater treatment: A global scientific perspective through text mining. *Water (Basel)*, 15(19), 3487. DOI: 10.3390/w15193487

Felsche, E., & Ludwig, R. (2021). Applying machine learning for drought prediction in a perfect model framework using data from a large ensemble of climate simulations. *Natural Hazards and Earth System Sciences*, 21(12), 3679–3691. DOI: 10.5194/nhess-21-3679-2021

Fereidooni, Z., Tahayori, H., & Bahadori-Jahromi, A. (2020). A hybrid model-based method for leak detection in large scale water distribution networks. *Journal of Ambient Intelligence and Humanized Computing*, 12(2), 1613–1629. DOI: 10.1007/s12652-020-02233-2

Fernández, A. S., Sutil, M. F., Aparicio, M. J. d. l. F., & Cruz, P. V. (2022). Control predictivo no lineal tolerante a fallos en una planta de tratamiento de aguas residuales. Actas De Las XXXVII Jornadas De Automática 7, 8 Y 9 De Septiembre De 2016, Madrid. .DOI: 10.17979/spudc.9788497498081.0623

Fernández-Jáuregui, C. A. (1999). El agua como fuente de conflictos: Repaso de los focos de conflictos en el mundo. *Revista CIDOB d'Afers Internacionals*, 45/46, 179–194.

Ferrari, E., Verda, D., Pinna, N., & Muselli, M. (2023). Optimizing Water Distribution through Explainable AI and Rule-Based Control. *Computers*, 12(6), 123. DOI: 10.3390/computers12060123

Filho, J. V., Scortegagna, A., Vieira, A. P. S. D., & Jaskowiak, P. A. (2024). Machine learning for water demand forecasting: Case study in a Brazilian coastal city. *Water Practice & Technology*, 19(5), 1586–1602. DOI: 10.2166/wpt.2024.096

Firat, M., Turan, M. E., & Yurdusev, M. A. (2010). Comparative analysis of neural network techniques for predicting water consumption time series. *Journal of Hydrology (Amsterdam)*, 384(1-2), 46–51. DOI: 10.1016/j.jhydrol.2010.01.005

Fleming, S., Garen, D., Goodbody, A., McCarthy, C., & Landers, L. (2021). Assessing the new Natural Resources Conservation Service water supply forecast model for the American West: A challenging test of explainable, automated, ensemble artificial intelligence. *Journal of Hydrology (Amsterdam)*, 602, 126782. DOI: 10.1016/j.jhydrol.2021.126782

Foster, L., Szilagyi, K., Wairegi, A., Oguamanam, C., & de Beer, J. (2023). Smart farming and artificial intelligence in East Africa: Addressing indigeneity, plants, and gender. *Smart Agricultural Technology*, 3, 100132. DOI: 10.1016/j.atech.2022.100132

Fragoso Sandoval, L., Roberto Ruiz, J., Flores, Z., & Juárez León, A. B. (2013). Sistema para control y gestión de redes de agua potable de dos localidades de México. Ingeniería Hidráulica y Ambiental, 34(1), 112-126. Recuperado en 31 de julio de 2024, de http://scielo.sld.cu/scielo.php?script=sci_arttext&pid=S1680-03382013000100009&lng=es&tlng=es

Franco, E. F., Ramos, R., Ovando-Javier, A., Montero-Espaillat, E., Bonilla, S., & Veda, A. (2023). Sensores de calidad de agua para el control de la contaminación fisicoquímica en los acuíferos de latinoamérica: una revisión. *Ciencia, Ambiente Y Clima*, 6(1), 45-70. .DOI: 10.22206/cac.2023.v6i1.pp45-70

Freier, A., Mazzalay, V. H., & Rolando, A. (2020). Calentamiento global y política energética. incidencias de la gobernanza multinivel en el caso argentino. *Td. Temas y Debates (Rosario)*, (40), 33–58. DOI: 10.35305/tyd.v0i40.471

Fuentes-Peñailillo, F., Gutter, K., Vega, R., & Silva, G. C. (2024). Transformative Technologies in Digital Agriculture: Leveraging Internet of Things, Remote Sensing, and Artificial Intelligence for Smart Crop Management. *Journal of Sensor and Actuator Networks*, 13(4), 39. DOI: 10.3390/jsan13040039

Fu, G., Sun, S., Hoang, L., Yuan, Z., & Butler, D. (2023). Artificial intelligence underpins urban water infrastructure of the future: A holistic perspective. Cambridge Prisms. *Water (Basel)*, 1, e14. Advance online publication. DOI: 10.1017/wat.2023.15

Galelli, S., & Castelletti, A. (2013). Assessing the predictive capability of randomized tree-based ensembles in streamflow modelling. *Hydrology and Earth System Sciences*, 17(7), 2669–2684. DOI: 10.5194/hess-17-2669-2013

Ganthavee, V., & Trzcinski, A. P. (2024). Artificial intelligence and machine learning for the optimization of pharmaceutical wastewater treatment systems: A review. *Environmental Chemistry Letters*, 22(5), 2293–2318. Advance online publication. DOI: 10.1007/s10311-024-01748-w

Gao, F., & Zhang, X. (2021). Mapping Crop Phenology in Near Real-Time Using Satellite Remote Sensing: Challenges and Opportunities. *Yaogan Xuebao*, 2021/8379391, 2021/8379391. Advance online publication. DOI: 10.34133/2021/8379391

Gao, X., Zeng, W., Shen, Y., Guo, Z., Yang, J., Cheng, X., Hua, Q., & Yu, K. (2020). Integrated deep neural networks-based complex system for urban water management. *Complexity*, 2020, 1–12. DOI: 10.1155/2020/8848324

Gao, Y., Liu, X., Zhang, Y., & Wang, L. (2020). Deep learning-based real-time water demand prediction model. *Journal of Hydrology (Amsterdam)*, 588, 124892. DOI: 10.1016/j.jhydrol.2020.124892

García et ai. (2010). *Transporte de Contaminantes en Aguas Subterráneas mediante Redes Neuronales Artificiales*. https://www.scielo.cl/scielo.php?script=sci_arttext&pid=S0718-07642010000500011

Garcia-Cuerva, L., & Minguez, R. (2023). Artificial intelligence techniques for smart water quality monitoring. *Sensors (Basel)*, 23(1), 118. DOI: 10.3390/s23010118 PMID: 38202979

Garillos-Manliguez, C. A., & Chiang, J. Y. (2021). Multimodal deep learning and visible-light and hyperspectral imaging for fruit maturity estimation. *Sensors (Basel)*, 21(4), 1288. DOI: 10.3390/s21041288 PMID: 33670232

Garzón, A., Kapelan, Z., Langeveld, J., & Taormina, R. (2022). Machine learning-based surrogate modeling for urban water networks: Review and future research directions. *Water Resources Research*, 58(5), e2021WR031808. Advance online publication. DOI: 10.1029/2021WR031808

Gato, S., Jayasuriya, N., & Roberts, P. (2007). Forecasting residential water demand: Case study of Melbourne, Australia. *Journal of Water Resources Planning and Management*, 133(4), 309–319. DOI: 10.1061/(ASCE)0733-9496(2007)133:4(309)

Gautam, J., Chakrabarti, A., Agarwal, S., Singh, A., Gupta, S., & Singh, J. (2020). Monitoring and forecasting water consumption and detecting leakage using an IoT system. *Water Science and Technology: Water Supply*, 20(3), 1103–1113. DOI: 10.2166/ws.2020.035

Geetha, S., & Gouthami, S. (2016). Internet of things enabled real time water quality monitoring system. *Smart Water*, 2(1), 1. Advance online publication. DOI: 10.1186/s40713-017-0005-y

Ghiassi, M., & Zimbra, D. K. (2008). Modeling and forecasting short-term water demand using neural networks and dynamic regression. *Journal of Water Resources Planning and Management*, 134(2), 119–128. DOI: 10.1061/(ASCE)0733-9496(2008)134:2(138)

Godo Pla, L., Emiliano, P., Valero, F., Poch, M., & Monclús Sales, H. (2018). Aplicación de la Inteligencia Artificial en la potabilización: Proyecto DrinkIA. © RETEMA: revista técnica de medio ambiente, 2020, núm. 210, p. 96-102.

Gonçalves, R., Soares, J. J. M., & Lima, R. M. F. (2020). An IoT-Based Framework for Smart Water Supply Systems Management. *Future Internet*, 12(7), 114. DOI: 10.3390/fi12070114

González, L. F. D., Casierra, G. A. T., & Moreno, E. del R. (2017). Confinated horizontal vortex generator. *Ingeniería y Desarrollo*, 35(1), 38–57.

González, T., Domínguez, J. R., Beltrán-Heredia, J., García, H. M., & Sanchez-Lavado, F. (2007). Aluminium sulfate as coagulant for highly polluted cork processing wastewater: Evaluation of settleability parameters and design of a clarifier-thickener unit. *Journal of Hazardous Materials*, 148(1–2), 6–14. DOI: 10.1016/j.jhazmat.2006.09.049 PMID: 17049726

Goodarzi, D., Abolfathi, S., & Borzooei, S. (2020). Modelling solute transport in water disinfection systems: Effects of temperature gradient on the hydraulic and disinfection efficiency of serpentine chlorine contact tanks. *Journal of Water Process Engineering*, 37(Oct), 101411. DOI: 10.1016/j.jwpe.2020.101411

Gozzer, E., Canchihuamán, F., Chalco, V., Huari, G., Cisneros-Dionisio, A., Quispe, W., & Espinoza, R. (2021). Duración de la gestión ministerial y logros en salud en el perú. *Revista Peruana de Medicina Experimental y Salud Pública*, 38(2), 326–336. DOI: 10.17843/rpmesp.2021.382.6721 PMID: 34468584

Granata, F., Papirio, S., Esposito, G., Gargano, R., & De Marinis, G. (2017, February). Machine learning algorithms for the forecasting of wastewater quality indicators. *Water (Basel)*, 9(2), 105. DOI: 10.3390/w9020105

Gregory, D. (30-Dec-19). *Differential Pressure Transmitter Explained—RealPars*. https://www.realpars.com/blog/differential-pressure-transmitter

Guan, Y. (2023). *Development of UAS and Geophysical Techniques for Agriculture and Environmental Site Characterization* [PhD Thesis, Missouri University of Science and Technology]. https://search.proquest.com/openview/311861e9f4884b5d27fb61399e5ff507/1?pq-origsite=gscholar&cbl=18750&diss=y

Guo, Y., Wang, J., & Cheng, J. (2015). Prediction of water demand in China's urban areas based on nonlinear regression and meteorological data. *Journal of Water Resources Planning and Management*, 141(7), 04014083. DOI: 10.1061/(ASCE)WR.1943-5452.0000493

Gupta, S. K., & Deshmukh, R. R. (Eds.). (2019). *Emerging trends in water quality monitoring and assessment*. Springer.

Gurung, T., Stewart, R., Beal, C., & Sharma, A. (2015). Smart meter enabled water end-use demand data: Platform for the enhanced infrastructure planning of contemporary urban water supply networks. *Journal of Cleaner Production*, 87, 642–654. DOI: 10.1016/j.jclepro.2014.09.054

Gutiérrez, J. D., Riss, W., & Ospina, R. (2006). Bioindicación De La Calidad Del Agua En La Sabana De Bogota - Colombia, Mediante La Utilización De La Lógica Difusa Neuro-Adaptativa Como Herramienta. *Caldasia*, 28(1), 45–56.

Guzmán-Avalos, P., Molinero-Hernández, D., Galván-González, S., Herrera-Sandoval, N., Solorio-Díaz, G., & Rubio-Maya, C. (2023, January). Numerical design and optimization of a hydraulic micro-turbine adapted to a wastewater treatment plant. *Alexandria Engineering Journal*, 62, 555–565. DOI: 10.1016/j.aej.2022.07.004

Gyarmati, G., & Mizik, T. (2020). The present and future of the precision agriculture. *2020 IEEE 15th International Conference of System of Systems Engineering (SoSE)*, 593–596. DOI: 10.1109/SoSE50414.2020.9130481

Hameed, M., Sharqi, S., Yaseen, Z., Afan, H., Hussain, A., & El-Shafie, A. (2017). Application of artificial intelligence (AI) techniques in water quality index prediction: A case study in tropical region, Malaysia. *Neural Computing & Applications*, 28(S1), 893–905. DOI: 10.1007/s00521-016-2404-7

Hamilton, R., Braun, B., Dare, R., Koopman, B., & Svoronos, S. (2006). Control issues and challenges in wastewater treatment plants. *IEEE Control Systems*, 26(4), 63–69. DOI: 10.1109/MCS.2006.1657877

Hananda, N., Kamul, A., Harito, C., Djuana, E., Elwirehardja, G. N., Pardamean, B., … & Pasang, T. (2023). Battery optimization by machine learning algorithms: research gap via bibliometric analysis. E3S Web of Conferences, 388, 01020. DOI: 10.1051/e3sconf/202338801020

Harrou, F., Dairi, A., Dorbane, A., & Sun, Y. (2023). Energy consumption prediction in water treatment plants using deep learning with data augmentation. *Results in Engineering*, 20, 101428. DOI: 10.1016/j.rineng.2023.101428

Hasan, A., & Twala, B. (2015). Improving single classifiers prediction accuracy for underground water pump station in a gold mine using ensemble techniques. IEEE EUROCON 2015 - International Conference on Computer as a Tool (EUROCON), 1-7. https://doi.org/DOI: 10.1109/EUROCON.2015.7313694

Hasan, M., Khatun, R., & Ahmed, F. (2020). Regression analysis for predicting rural water quality in developing countries. *Water Research*, 170, 115373. DOI: 10.1016/j.watres.2019.115373

Hashim, H. Q., & Sayl, K. N. (2021). Detection of suitable sites for rainwater harvesting planning in an arid region using geographic information system. *Applied Geomatics*, 13(2), 235–248. DOI: 10.1007/s12518-020-00342-3

Hassan, M., Kowalska, A., & Ashraf, H. (2023). Advances in deep learning algorithms for agricultural monitoring and management. *Applied Research in Artificial Intelligence and Cloud Computing*, 6(1), 68–88.

Hassan, T. A., & Suhail, Z. A. (2023). The Efficacy of Ozonized Water Versus Ringer Lactate Arthrocentesis for the Treatment of Temporomandibular Joint Internal Derangement. *The Journal of Craniofacial Surgery*, 34(3), e238–e241. DOI: 10.1097/SCS.0000000000009085 PMID: 36730935

Hastie, T., Tibshirani, R., & Friedman, J. (2009). *The elements of statistical learning: Data mining, inference, and prediction* (2nd ed.). Springer. DOI: 10.1007/978-0-387-84858-7

Haupt, S. (2021). The history and practice of AI in the environmental sciences. *Bulletin of the American Meteorological Society*. Advance online publication. DOI: 10.1175/BAMS-D-20-0234.1

Hauser, J. R., & Clausing, D. (1988). *The house of quality*. Harvard Business Review.

Hernández, E. V., Lalaleo, L., & Cáceres, W. C. (2022). ¿se pueden prevenir las pandemias? epidemiología basada en aguas residuales. *Investigación Y Desarrollo*, 12(1), 88-90. .DOI: 10.31243/id.v12.2020.991

Herrera, M., Torgo, L., Izquierdo, J., & Pérez-García, R. (2010). Predictive models for forecasting hourly urban water demand. *Journal of Hydrology (Amsterdam)*, 387(1-2), 141–150. DOI: 10.1016/j.jhydrol.2010.04.005

Herrick, C., Dymond, R. L., & Anderson, M. R. (2015). Climate variability and water demand: Evidence from a southwestern United States urban setting. *Urban Climate*, 12, 1–15. DOI: 10.1016/j.uclim.2015.01.001

HidraSoftware. Available online: https://www.hidrasoftware.com/ (accessed on 22 Jan, 2023)

Hoban, A. Chapter 2 Water Sensitive Urban Design Approaches and Their Description. In Book Approaches to Water Sensitive Urban Design A. K. Sharma, T. Gardner, & D. Begbie., Woodhead Publishing, Sawston, Cambridge, England, pp. 25-47, 2019

Hodson, E., Vainio, T., Sayún, M. N., Tomitsch, M., Jones, A., Jalonen, M., Börütecene, A., Hasan, M. T., Paraschivoiu, I., Wolff, A., Yavo-Ayalon, S., Yli-Kauhaluoma, S., & Young, G. W. (2023). Evaluating Social Impact of Smart City Technologies and Services: Methods, Challenges, Future Directions. *Multimodal Technologies and Interaction*, 7(3), 33. DOI: 10.3390/mti7030033

Holguín, H. M., & Dutrénit, G. (2021). La regulación como instrumento de fomento a la innovación para la atención de la diabetes en méxico. *Brazilian Journal of Business*, 3(3). Advance online publication. DOI: 10.34140/bjbv3n3-032

Hoque, A., Padhiary, M., Prasad, G., & Kumar, K. (2025). Real-Time Data Processing in Agricultural Robotics. In Bora, D. J., & Bania, R. K. (Eds.), *Computer Vision Techniques for Agricultural Advancements* (pp. 431–468). IGI Global Scientific Publishing.

Hoseinzadeh, S., & Garcia, D. A. (2024). AI-Driven Innovations in Greenhouse Agriculture. *Reanalysis of Sustainability and Energy Efficiency Impacts. Energy Conversion and Management: X*, 24, 100701. DOI: 10.1016/j.ecmx.2024.100701

House-Peters, L. A., & Chang, H. (2011). Urban water demand modeling: Review of concepts, methods, and organizing principles. *Water Resources Research*, 47(5), W05401. Advance online publication. DOI: 10.1029/2010WR009624

Hu, Y., Zhang, Y., Jian, D., & Li, J. (2023). Prediction of Coagulant Dosage in Waterworks Based on Ensemble Learning. 2023 12th International Conference of Information and Communication Technology (ICTech), 184-189. https://doi.org/ DOI: 10.1109/ICTech58362.2023.00046

Huang, G., & Wang, J. (2022). Intelligent water quality monitoring based on machine learning and big data analysis. *Environmental Science and Pollution Research International*, 29(10), 15742–15755. DOI: 10.1007/s11356-022-18789-6

Huang, H., Zhang, Z., & Song, F. (2021). An ensemble-learning-based method for short-term water demand forecasting. *Water Resources Management*, 35(6), 1757–1773. DOI: 10.1007/s11269-021-02808-4

Huang, Y., Yang, J., & Li, M. (2018). Leak detection in urban water pipelines using support vector machines. *Journal of Water Resources Planning and Management*, 144(9), 04018062. DOI: 10.1061/(ASCE)WR.1943-5452.0000984

Hu, H., Li, S., & Wen, Y. (2020). Review of artificial intelligence technologies in water quality management. *Water Environment Research*, 92(12), 2041–2052. DOI: 10.1002/wer.1361 PMID: 32449943

Hu, P., Tong, J., Wang, J., Yang, Y., & Turci, L. (2019). A hybrid model based on CNN and Bi-LSTM for urban water demand prediction. *2019 IEEE Congress on Evolutionary Computation (CEC)*, 1088-1094. https://doi.org/DOI: 10.1109/CEC.2019.8790060

Hutton, C. J., Veldhuis, M.-C., & Allsop, W. (2014). Real-time demand forecasting for urban water systems. *Water Resources Research*, 50(5), 3723–3743. DOI: 10.1002/2014WR015563

Iddio, E., Wang, L., Thomas, Y., McMorrow, G., & Denzer, A. (2020). Energy efficient operation and modeling for greenhouses: A literature review. *Renewable & Sustainable Energy Reviews*, 117, 109480. DOI: 10.1016/j.rser.2019.109480

Ige, A. B., Kupa, E., & Ilori, O. (2024). Developing Comprehensive Cybersecurity Frameworks for Protecting Green Infrastructure: Conceptual Models and Practical Applications. *GSC Advanced Research and Reviews*, 20(1), 25–41. DOI: 10.30574/gscarr.2024.20.1.0237

Ihuhwa Marta Tau. (2014). *Efecto Coriolis | PDF | Física | Física Aplicada e Interdisciplinaria*. Scribd. https://es.scribd.com/presentation/199377222/Efecto-Coriolis

Inalegwu, O. C., Roman, M., Mumtaz, F., Zhang, B., Nambisan, A., & Huang, J. (2024). Optical fiber specklegram sensor for water leak detection and localization. *Optical Waveguide and Laser Sensors*, III, 15. Advance online publication. DOI: 10.1117/12.3014056

Inoue, Y. (2020). Satellite- and drone-based remote sensing of crops and soils for smart farming – a review. *Soil Science and Plant Nutrition*, 66(6), 798–810. DOI: 10.1080/00380768.2020.1738899

Instituto Nacional de Salud Pública (INSP). (2022) Importancia del agua para el buen funcionamiento de nuestro organismo. https://www.insp.mx/resources/images/stories/INSP/Docs/cts/210131_cts2.pdf

Intergovernmental Agreement On A National Water Initiative. Available online: https://www.pc.gov.au/inquiries/completed/water-reform/national-water-initiative-agreement-2004.pdf (accessed on 15 April 2023).

Ioffe, S., & Szegedy, C. (2015). Batch normalization: accelerating deep network training by reducing internal covariate shift. In: Proceedings of the 32nd international conference on machine learning. *Proceedings of Machine Learning Research*, 37, 448–456.

ISO. (2016, November). *ISO 24516-1:2016 Guidelines for the management of assets of water supply and wastewater systems Part 1: Drinking water distribution networks*. Retrieved July 3, 2024 from https://www.iso.org/standard/64117.html

ISO. (2019, July). *ISO 46001:2019 Water efficiency management systems — Requirements with guidance for use*. Retrieved July 5, 2024 from https://www.iso.org/standard/68286.html

ISO. (2024a, January). *ISO 24510:2024 Activities relating to drinking water and wastewater services — Guidelines for the assessment and for the improvement of the service to users*. Retrieved July 4, 2024 from https://www.iso.org/standard/81484.html

ISO. (2024b, January). *ISO 24591-1:2024 Smart water management Part 1: General guidelines and governance*. Internacional Organization for Standardization. Retrieved July 4, 2024 from https://www.iso.org/standard/79033.html

ISO. (2024c, Apri). *ISO 24591-2:2024 Smart water management Part 2: Data management guidelines*. Internacional Organization for Standardization. Retrieved July 4, 2024 from https://www.iso.org/standard/79034.html

ISO. (2024d, February). *ISO 46001:2019/Amd 1:2024 Water efficiency management systems — Requirements with guidance for use*. Retrieved July 5, 2024 from https://www.iso.org/standard/68286.html

Iyer, S., Kaushik, S., & Nandal, P. (2023). *Water Quality Prediction Using Machine Learning*. MR International Journal of Engineering and Technology., DOI: 10.58864/mrijet.2023.10.1.8

Izazola, H. (2001). Agua y sustentabilidad en la Ciudad de México. *Estudios Demográficos y Urbanos*, 16(2 (47)), 285–320. DOI: 10.24201/edu.v16i2.1121

Jafar, R., Awad, A., Jafar, K., & Shahrour, I. (2022). Predicting effluent quality in full-scale wastewater treatment plants using shallow and deep artificial neural networks. *Sustainability (Basel)*, 14(23), 15598. Advance online publication. DOI: 10.3390/su142315598

Jain, A., & Ormsbee, L. (2002). Short-term water demand forecast modeling techniques—CONVENTIONAL METHODS VERSUS AI. *Journal - American Water Works Association*, 94(7), 64–72. Advance online publication. DOI: 10.1002/j.1551-8833.2002.tb09507.x

Jain, A., Varshney, A. K., & Joshi, U. C. (2001). Short-term water demand forecast modeling at IIT Kanpur using artificial neural networks. *Water Resources Management*, 15(5), 299–321. DOI: 10.1023/A:1014415503476

Jalal and T. Ezzedine. ''Performance analysis of machine learning algorithms for water quality monitoring system,'' in Proc. Int. Conf. Internet Things, Embedded Syst. Commun. (IINTEC), Dec. 2019, pp. 86–89.

Janssen, S. J. C., Porter, C. H., Moore, A. D., Athanasiadis, I. N., Foster, I., Jones, J. W., & Antle, J. M. (2017). Towards a new generation of agricultural system data, models and knowledge products: Information and communication technology. *Agricultural Systems*, 155, 200–212. DOI: 10.1016/j.agsy.2016.09.017 PMID: 28701813

JAPAMA. (2024, January 29). *JAPAMA's chatbot, the AI tool where you report leaks and even corrupt employees*. Retrieved July 1, 2024 from https://www.luznoticias.mx/2024-01-29/sinaloa/chatbot-de-japama-la-herramienta-con-ia-en-donde-reportas-fugas-y-hasta-empleados-corruptos/193696

Jararweh, Y., Fatima, S., Jarrah, M., & AlZu'bi, S. (2023). Smart and sustainable agriculture: Fundamentals, enabling technologies, and future directions. *Computers & Electrical Engineering*, 110, 108799. DOI: 10.1016/j.compeleceng.2023.108799

Jawad, Z. N., & Balázs, V. (2024). Machine learning-driven optimization of enterprise resource planning (ERP) systems: A comprehensive review. *Beni-Suef University Journal of Basic and Applied Sciences*, 13(1), 4. DOI: 10.1186/s43088-023-00460-y

Jenny, H., Wang, Y., Garcia Alonso, E., & Minguez, R. (2020). Using Artificial Intelligence for Smart Water Management Systems. *ADB BRIEFS*(143). http://dx.doi.org/10.22617/BRF200191-2

Ji, Y., Wu, J., Wang, Y., Elumalai, V., & Subramani, T. (2020). Seasonal variation of drinking water quality and human health risk assessment in Hancheng City of Guanzhong Plain, China. *Exposure and Health*, 12(3), 469–485. DOI: 10.1007/s12403-020-00357-6

Johnson, L. E., & Smith, R. D. (2017). Remote sensing technologies for water quality monitoring. *Environmental Monitoring and Assessment*, 189(8), 407. DOI: 10.1007/s10661-017-6099-4 PMID: 28730581

Jothiprakash, V., & Sasikumar, K. (2009). Water demand forecasting for a water supply project using optimization techniques. *Journal of Water Resources Planning and Management*, 135(5), 371–381. DOI: 10.1061/(ASCE)0733-9496(2009)135:5(371)

Judijanto, L., Suparwata, D., Cahya, D., & Widodo, I. (2024). The Effect of Green Technology Implementation, Agricultural Product Innovation, and Green Technology Infrastructure Investment on Local Economic Growth in West Java. *West Science Nature and Technology*, 2(3), 142–152. DOI: 10.58812/wsnt.v2i03.1280

Kalpana, A. V., Chandrasekar, T., Ramalingam, R., Tamizhselvi, S. P., Ramesh, T., & Chitra, M. (2024). Enhancing Healthcare Integration With IoT for Seamless and Responsive Patient Care. In Khade, S., & Mishra, R. (Eds.), *Future of AI in Biomedicine and Biotechnology* (pp. 147–181). IGI Global., DOI: 10.4018/979-8-3693-3629-8.ch008

Kambalimath, S., & Deka, P. C. (2020). A basic review of fuzzy logic applications in hydrology and water resources. *Applied Water Science*, 10(8), 191. DOI: 10.1007/s13201-020-01276-2

Kamyab, H., Khademi, T., & Chelliapan, S. (2023). The latest innovative avenues for the utilization of artificial Intelligence and big data analytics in water resource management. *Results in Engineering, 20*, 101566. https://doi.org/10.1016/j.rineng.2023.101566

Kanamori, T., Takenouchi, T., Eguchi, S., & Murata, N. (2007). Robust Loss Functions for Boosting. *Neural Computation*, 19(8), 2183–2244. DOI: 10.1162/neco.2007.19.8.2183 PMID: 17571942

Kang, M., Kim, K., Yu, J., Bae, H., & Jang, I. (2023). Comparison of Leak Detection Techniques Using Water Pipeline Vibration Sensor Data. 2023 IEEE International Conference on Consumer Electronics-Asia (ICCE-Asia), 1-4. https://doi.org/DOI: 10.1109/ICCE-Asia59966.2023.10326371

Kavya, M., Mathew, A., Shekar, P. R., & P, S. (2023). Short term water demand forecast modelling using artificial intelligence for smart water management. *Sustainable Cities and Society, 95*, 104610. https://doi.org/10.1016/j.scs.2023.104610

Kelley, C. T. (1960). *Iterative methods for optimization*. SIAM.

Khammari, A., Nashashibi, F., Abramson, Y., & Laurgeau, C. (2005). Vehicle detection combining gradient analysis and AdaBoost classification. Proceedings. 2005 IEEE Intelligent Transportation Systems, 2005., 66-71. https://doi.org/.DOI: 10.1109/ITSC.2005.1520202

Khan, A., Vibhute, A. D., Mali, S., & Patil, C. H. (2022). A systematic review on hyperspectral imaging technology with a machine and deep learning methodology for agricultural applications. *Ecological Informatics*, 69, 101678. DOI: 10.1016/j.ecoinf.2022.101678

Kharel, T. P., Ashworth, A. J., Owens, P. R., & Buser, M. (2020). Spatially and temporally disparate data in systems agriculture: Issues and prospective solutions. *Agronomy Journal*, 112(5), 4498–4510. DOI: 10.1002/agj2.20285

Kher, L., Khanna, P., & Agarwal, S. (1979). Nonlinear Optimization of Water Supply Systems. *Journal of the Environmental Engineering Division*, 105(4), 781–784. DOI: 10.1061/JEEGAV.0000939

Kim, D., Choi, S., Kang, S., & Noh, H. (2023). A Study on Developing an AI-Based Water Demand Prediction and Classification Model for Gurye Intake Station. *Water (Basel)*, 15(23), 4160. DOI: 10.3390/w15234160

Kisi, O. (2013). Streamflow forecasting using different artificial neural network algorithms. *Journal of Hydrologic Engineering*, 18(1), 88–100. DOI: 10.1061/(ASCE)HE.1943-5584.0000625

Kleiman, N., Hillard, T., Forman, A., Gold, A., Weng, D., Adibhatla, V., Carillo, S., Coffy, A., Gichinga, R., Kleinfeld, Z., Prakash, M., Riaz, S., & Sokolsky, J. (2016). *Smart Water Management* (Innovation and the City, pp. 26-27). Center for an Urban Future. https://www.jstor.org/stable/resrep14850.15

Kliestik, T., Nica, E., Durana, P., & Popescu, G. H. (2023). Artificial intelligence-based predictive maintenance, time-sensitive networking, and big data-driven algorithmic decision-making in the economics of Industrial Internet of Things. *Oeconomia Copernicana*, 14(4), 1097–1138. DOI: 10.24136/oc.2023.033

Koyama, H., Kamogashira, T., & Yamasoba, T. (2024). Heavy Metal Exposure: Molecular Pathways, Clinical Implications, and Protective Strategies. *Antioxidants*, 13(1), 76. DOI: 10.3390/antiox13010076 PMID: 38247500

Kuğuoğlu, B., Voort, H., & Janssen, M. (2021). The giant leap for smart cities: Scaling up smart city artificial intelligence of things (aiot) initiatives. *Sustainability (Basel)*, 13(21), 12295. DOI: 10.3390/su132112295

Kumar, A., & Kaur, M. (2021). Applications of artificial intelligence in water quality management. *Environmental Technology & Innovation*, 21, 101313. DOI: 10.1016/j.eti.2020.101313

Kumarappa, S., & Manjunatha, H. M. (2024). Machine learning-based prediction of lithium-ion battery life cycle for capacity degradation modelling. *World Journal of Advanced Research and Reviews*, 21(2), 1299–1309. DOI: 10.30574/wjarr.2024.21.2.0553

Kumar, M., & Puri, A. (2012). A review of permissible limits of drinking water. *Indian Journal of Occupational and Environmental Medicine*, 16(1), 40. DOI: 10.4103/0019-5278.99696 PMID: 23112507

Labadie, J. W. (2004). Optimal operation of multireservoir systems: State-of-the-art review. *Journal of Water Resources Planning and Management*, 130(2), 93–111. DOI: 10.1061/(ASCE)0733-9496(2004)130:2(93)

Lacroix, S., Kuller, M., Gougeon, G., Petrucci, J., Lemieux-Chalifour, F., Rioux, A., Dagenais, D., & Bichai, F. (2024). Can We Stop Reinventing the Wheel in Blue-Green Infrastructure Planning? Using Value-Focused Thinking to Enable Transferability of a Multicriteria Planning Support System. *Landscape and Urban Planning*, 252, 105188. DOI: 10.1016/j.landurbplan.2024.105188

Lakhiar, I. A., Yan, H., Zhang, C., Wang, G., He, B., Hao, B., Han, Y., Wang, B., Bao, R., Syed, T. N., Chauhdary, J. N., & Rakibuzzaman, M. (2024). A Review of Precision Irrigation Water-Saving Technology under Changing Climate for Enhancing Water Use Efficiency, Crop Yield, and Environmental Footprints. *Agriculture*, 14(7), 1141. DOI: 10.3390/agriculture14071141

Landers, J. N., De Freitas, P. L., De Oliveira, M. C., Da Silva Neto, S. P., Ralisch, R., & Kueneman, E. A. (2021). Next Steps for Conservation Agriculture. *Agronomy (Basel)*, 11(12), 2496. DOI: 10.3390/agronomy11122496

Lassiter, A., & Leonard, N. D. (2022). A systematic review of municipal smart water for climate adaptation and mitigation. *Environment and Planning. B, Urban Analytics and City Science*, 49(5), 1406–1430. DOI: 10.1177/23998083211072864

Lechgar, H., Mallouk, A., Malaainine, M., Nahhal, T., & Rhinane, H. (2016). Artificial Intelligence (AI) Contribution to GIS in Optimal Positioning of Hydrophone Sensors Using Genetic Algorithm (Case Study: Water Network, Casablanca, Morocco). 11-19. https://doi.org/.DOI: 10.1007/978-3-319-30298-0_2

Lee, D.-S., Fu, S.-K., & Lai, C.-W. (2024). Establishment and analysis of online energy baseline for water purification plants: A case study in Kaohsiung city, Taiwan. *Heliyon*, 10(14), e33981. DOI: 10.1016/j.heliyon.2024.e33981 PMID: 39092258

Lee, J., Kim, H., & Lee, K. (2019). IoT-based water quality monitoring system using machine learning techniques. *Sustainability*, 11(15), 4135. DOI: 10.3390/su11154135

Leitão, J., Simões, N., Sá Marques, J. A., Gil, P., Ribeiro, B., & Cardoso, A. (2019). Detecting urban water consumption patterns: A time-series clustering approach. *Water Science and Technology: Water Supply*, 19(8), 2323–2329. DOI: 10.2166/ws.2019.113

León, L., & Rolando Israel, M. S. Juan, & Gabriel. (2024). *Construcción de un medidor de agua inteligente mediante Arduino para reducir las pérdidas en la distribución de agua*. https://dspace.ucacue.edu.ec/items/94539eb1-9644-437a-9aef-e00acf619889

Liang, B., Verma, S., Xu, J., Liang, S., Liu, Z., Wang, Y., . . . Chen, F. (2020). A data driven approach for leak detection with smart sensors. 2020 16th International Conference on Control, Automation, Robotics and Vision (ICARCV). DOI: 10.1109/ICARCV50220.2020.9305403

Li, J., & Li, Z. (2022). Application of artificial intelligence in water quality monitoring based on multi-source data fusion. *Environmental Monitoring and Assessment*, 194(7), 445. DOI: 10.1007/s10661-022-10190-1

Li, L., & Chen, H. (2023). Artificial Intelligence and Internet of Things-Based Leak Detection Method for the Water Supply Network. *International Transactions on Electrical Energy Systems*, 2023(1), 3443047. DOI: 10.1155/2023/3443047

Li, L., Rong, S., Wang, R., & Yu, S. (2021). Recent advances in artificial intelligence and machine learning for nonlinear relationship analysis and process control in drinking water treatment: A review. *Chemical Engineering Journal*, 405, 126673. Advance online publication. DOI: 10.1016/j.cej.2020.126673

Lima, M. F. d., Freitas, L. L. G., Henning, E., & Kalbusch, A. (2021). *Avaliação da utilização de hidrômetros com aquisição de dados por telemetria para monitoramento do consumo de água de uma agência bancária em joinville/sc*. Simpósio Nacional De Sistemas Prediais., DOI: 10.46421/sispred.v1i.1578

Lim, C. L. (2020). Fundamental concepts of human thermoregulation and adaptation to heat: A review in the context of global warming. *International Journal of Environmental Research and Public Health*, 17(21), 7795. DOI: 10.3390/ijerph17217795 PMID: 33114437

Linaza, M. T., Posada, J., Bund, J., Eisert, P., Quartulli, M., Döllner, J., Pagani, A. G., Olaizola, I., Barriguinha, A., Moysiadis, T., & Lucat, L. (2021). Data-Driven Artificial Intelligence Applications for Sustainable Precision Agriculture. *Agronomy (Basel)*, 11(6), 1227. DOI: 10.3390/agronomy11061227

Lin, M., & Chu, C. (2008). Applying Artificial Immune System to Minimize Construction Cost of Water Distribution Networks. *2008 Fourth International Conference on Natural Computation*, 6, 628-632. https://doi.org/DOI: 10.1109/ICNC.2008.376

Liu, R. (2019). Artificial intelligence system for water dispenser [Patent]. https://patentscope.wipo.int/search/es/detail.jsf?docId=CN276502343&_cid=P20-LZCWA8-96371-1

Liu, C., & Liu, H. (2020). Water quality monitoring and management based on internet of things and artificial intelligence. *IEEE Internet of Things Journal*, 8(3), 1702–1712. DOI: 10.1109/JIOT.2020.2961604

Liu, F., Zhang, Y., & Zhao, X. (2020). Hybrid AI algorithms for sustainable water management in urban environments. *Sustainability*, 12(4), 1452. DOI: 10.3390/su12041452

Liu, Q., Yang, L., & Yang, M. (2021). Digitalisation for water sustainability: Barriers to implementing circular economy in Smart Water Management. *Sustainability (Basel)*, 13(21), 11868. DOI: 10.3390/su132111868

Liu, Y., Ma, X., Li, Y., Tie, Y., Zhang, Y., & Gao, J. (2019). Water pipeline leakage detection based on machine learning and wireless sensor networks. *Sensors (Basel)*, 19(23), 5086. DOI: 10.3390/s19235086 PMID: 31766356

Li, W., Zhang, H., Liu, L., & Ye, J. (2022). A review on AI technologies for water quality monitoring. *Frontiers in Environmental Science*, 10, 780412. DOI: 10.3389/fenvs.2022.780412

Li, X., Zhang, Y., & Wang, L. (2018). Real-time water quality monitoring using IoT-enabled sensors in river ecosystems. *Water Science and Technology*, 78(1), 185–194. DOI: 10.2166/wst.2018.091

Li, Y., Feng, L., & Wang, R. (2019). Deep learning models for drought prediction and water availability assessment in arid regions. *Journal of Hydrology (Amsterdam)*, 575, 370–383. DOI: 10.1016/j.jhydrol.2019.05.027

Li, Z., Ma, J., & Wang, P. (2019). Artificial intelligence for smart water quality monitoring and control. *Journal of Hydroinformatics*, 21(1), 171–188. DOI: 10.2166/hydro.2018.169

Li, Z., Wang, J., & Chen, X. (2017). Water quality classification using k-means clustering algorithm: A case study of the Pearl River, China. *Water Science and Technology*, 76(8), 2193–2203. DOI: 10.2166/wst.2017.379

Llanos-Zavalaga, F., Quiroz, J. A. C., & Silva, C. C. (2022). Análisis de políticas públicas en salud: El caso de recursos humanos en el perú. *Revista Médica Herediana*, 33(3), 178–186. DOI: 10.20453/rmh.v33i3.4338

Lobbrecht, A. H., & Solomatine, D. P. (2002). Machine learning in real-time control of water systems. *Urban Water Journal*, 4(3), 257–263. DOI: 10.1080/15730620500246236

López, I. D., Figueroa, A., & Corrales, J. C. (2016). Un mapeo sistemático sobre predicción de calidad del agua mediante técnicas de inteligencia computacional. *Revista Ingenierías Universidad de Medellín*, 15(28), 35–51. DOI: 10.22395/rium.v15n28a2

Lorente-Leyva, L., Pavón-Valencia, J., Montero-Santos, Y., Herrera-Granda, I., Herrera-Granda, E., & Peluffo-Ordóñez, D. (2019). Artificial Neural Networks for Urban Water Demand Forecasting: A Case Study. *Journal of Physics: Conference Series*, 1284(1), 012004. Advance online publication. DOI: 10.1088/1742-6596/1284/1/012004

Loures, L., Chamizo, A., Ferreira, P., Loures, A., Castanho, R., & Panagopoulos, T. (2020). Assessing the Effectiveness of Precision Agriculture Management Systems in Mediterranean Small Farms. *Sustainability (Basel)*, 12(9), 3765. DOI: 10.3390/su12093765

Lowe, M., Qin, R., & Mao, X. (2022). A Review on Machine Learning, Artificial Intelligence, and Smart Technology in Water Treatment and Monitoring. *Water (Basel)*, 14(9), 1384. Advance online publication. DOI: 10.3390/w14091384

Lowenberg-DeBoer, J., Franklin, K., Behrendt, K., & Godwin, R. (2021). Economics of autonomous equipment for arable farms. *Precision Agriculture*, 22(6), 1992–2006. DOI: 10.1007/s11119-021-09822-x PMID: 34075304

Lowenberg-DeBoer, J., Huang, I. Y., Grigoriadis, V., & Blackmore, S. (2020). Economics of robots and automation in field crop production. *Precision Agriculture*, 21(2), 278–299. DOI: 10.1007/s11119-019-09667-5

Luan, Y., & Liu, J. (2017). A distributed model for urban water demand prediction. *Kexue Tongbao*, 62(24), 2770–2779. DOI: 10.1360/N972016-01154

Lucatello, S., & Alcántara-Ayala, I. (2024). Sustainable Synergy: Strengthening Disaster Risk Reduction in Latin America and the Caribbean through Nature-Based Solutions. *International Journal of Disaster Risk Reduction*, 113, 104860. DOI: 10.1016/j.ijdrr.2024.104860

Lu, H., & Ma, X. (2020). Hybrid decision tree-based machine learning models for short-term water quality prediction. *Chemosphere*, 249, 126169. DOI: 10.1016/j.chemosphere.2020.126169 PMID: 32078849

Luna, T., Ribau, J., Figueiredo, D., & Alves, R. (2019). Improving energy efficiency in water supply systems with pump scheduling optimization. *Journal of Cleaner Production*, 213, 342–356. Advance online publication. DOI: 10.1016/j.jclepro.2018.12.190

Lu, P., Jiang, B., & Weiner, J. (2020). Crop spatial uniformity, yield and weed suppression. *Advances in Agronomy*, 161, 117–178. DOI: 10.1016/bs.agron.2019.12.003

Luvhimbi, N., Tshitangano, T. G., Mabunda, J. T., Olaniyi, F. C., & Edokpayi, J. N. (2022, April). Water quality assessment and evaluation of human health risk of drinking water from source to point of use at thulamela municipality, limpopo province. *Scientific Reports*, 12(1), 6059. DOI: 10.1038/s41598-022-10092-4 PMID: 35411067

Madhusoodhanan, C. G., Sreeja, K. G., & Eldho, T. I. (2016, October). Climate change impact assessments on the water resources of India under extensive human interventions. *Ambio*, 45(6), 725–741. DOI: 10.1007/s13280-016-0784-7 PMID: 27170012

Madrigal, L. P., Meraz, J. G., Wiegand, C. L., Reyes, J. I., & Delgado, J. M. (2010). Monitoreo De La Salinidad Mediante Sensores Remotos. *Terra Latinoamericana*, 28(1), 15–26.

Maestre-Góngora, G. (2015). Revisión de literatura sobre ciudades inteligentes: Una perspectiva centrada en las tic. *Ingeniare*, (19), 137–149. DOI: 10.18041/1909-2458/ingeniare.19.531

Maestrini, B., & Basso, B. (2018). Drivers of within-field spatial and temporal variability of crop yield across the US Midwest. *Scientific Reports*, 8(1), 14833. DOI: 10.1038/s41598-018-32779-3 PMID: 30287846

Mahadev Madgule, N. (2023). Vinayaka, Yeshwant M. Sonkhaskar, Dhiren Ramanbhai Patel, R. Karthikeyan, P. Selvakumar,(2023). Mechanical properties and microstructure of activated TIG welded similar joints of Inconel alloys by desirability approaches. *Materials Today: Proceedings*, 77, 528–533. DOI: 10.1016/j.matpr.2022.12.250

Mahadeva, R., Patel, V., Ghosh, A., Dixit, S., Suthar, B., Gupta, V., Awaar, V., Bharadwaj, Y., Gupta, M., Haranatti, J., & Arora, R. (2023). Artificial Intelligence in Water Desalination: A Novel Approach for Global Sustainability. E3S Web of Conferences. https://doi.org/.DOI: 10.1051/e3sconf/202343001193

Mahardhika, S., & Putriani, O. (2023). Deployment and use of Artificial Intelligence (AI) in water resources and water management. *IOP Conference Series. Earth and Environmental Science*, 1195(1), 012056. Advance online publication. DOI: 10.1088/1755-1315/1195/1/012056

Maitra, A., & Damle, M. (2024). Revolutionizing Plant Health Management with Technological Digital Transformation to Enhance Disease Control & Fortifying Plant Resilience. *2024 3rd International Conference for Innovation in Technology (INOCON)*, 1–8. DOI: 10.1109/INOCON60754.2024.10511728

Makropoulos, C. K., & Butler, D. (2008). A neurofuzzy modeling framework for predicting urban water demand. *Journal of Water Resources Planning and Management*, 134(2), 112–118. DOI: 10.1061/(ASCE)0733-9496(2008)134:2(112)

Malek, N., Yaacob, W., Wah, Y., Nasir, S., Shaadan, N., & Indratno, S. (2022). Comparison of ensemble hybrid sampling with bagging and boosting machine learning approach for imbalanced data. *Indonesian Journal of Electrical Engineering and Computer Science*, 29(1), 598. Advance online publication. DOI: 10.11591/ijeecs.v29.i1.pp598-608

Manasa, S., Anusha, K. N., Sharma, S., Thakur, A., & Sood, A. (2023). An Overview of Precision Farming. *International Journal of Environment and Climate Change*, 13(12), 441–456. DOI: 10.9734/ijecc/2023/v13i123701

Manawadu, L., Gunathilaka, M. D. K. L., Wijerathne, V. P. I. S., & Gunathilake, K. L. W. I. (2021). How and Why the Land Resources Changed: A Chronological Change of Land Use and Land Cover in Sri Lanka. In Hasnat, G. N. T., & Hossain, M. K. (Eds.), *Examining International Land Use Policies, Changes, and Conflicts* (pp. 88–110). IGI Global. DOI: 10.4018/978-1-7998-4372-6.ch005

Maraveas, C. (2022). Incorporating Artificial Intelligence Technology in Smart Greenhouses: Current State of the Art. *Applied Sciences (Basel, Switzerland)*, 13(1), 14. DOI: 10.3390/app13010014

Marimuthu, V., Khobragade, P. R., & Patil, M. B. (2022). Artificial Intelligence-Based Techniques to Monitor and Maintain the Quality of Drinking Water Supplied to Households [Patent]. https://patentscope.wipo.int/search/es/detail.jsf?docId=IN374031194&_cid=P20-LZCWA8-96371-1

Marvin, G., Grbčić, L., Družeta, S., & Kranjčević, L. (2023). Water distribution network leak localization with histogram-based gradient boosting histogram-based gradient boosting water network leak localization. *Journal of Hydroinformatics*, 25(3), 663–684. Advance online publication. DOI: 10.2166/hydro.2023.102

Massaro, A., Panarese, A., & Galiano, A. (2021, June 7-9). Technological Platform for Hydrogeological Risk Computation and Water Leakage Detection based on a Convolutional Neural Network. *Agronomy (Basel)*, 13(4).

McCulloch, W. S., & Pitts, W. (1943). A logical calculus of the ideas immanent in nervous activity. *The Bulletin of Mathematical Biophysics*, 5(4), 115–133. DOI: 10.1007/BF02478259

Mehmood, H., Liao, D., & Mahadeo, K. (2020). A Review of Artificial Intelligence Applications to Achieve Water-related Sustainable Development Goals. 2020 IEEE / ITU International Conference on Artificial Intelligence for Good (AI4G), 135-141. https://doi.org/DOI: 10.1109/AI4G50087.2020.9311018

Melina. (2018, September). *Fig. 2. Principle of operation for thermal mass flow sensor. Extracted...* ResearchGate. https://www.researchgate.net/figure/Principle-of-operation-for-thermal-mass-flow-sensor-Extracted-from-18_fig1_342260098

Ministry of Environment and Natural Resources. (2020, December 30). *National Water Program 2020-2024*. Retrieved July 5, 20224 from https://www.gob.mx/cms/uploads/attachment/file/642632/PNH_2020-2024__ptimo.pdf

Misra, N. N., Dixit, Y., Al-Mallahi, A., Bhullar, M. S., Upadhyay, R., & Martynenko, A. (2020). IoT, big data, and artificial intelligence in agriculture and food industry. *IEEE Internet of Things Journal*, 9(9), 6305–6324. DOI: 10.1109/JIOT.2020.2998584

Mitchell, T. M. (1997). *Machine Learning*. McGraw-Hill.

Mohammadi, A., Azmi, M., & Araghinejad, S. (2020). Long short-term memory neural networks for predicting reservoir water levels. *Journal of Hydrology (Amsterdam)*, 581, 124412. DOI: 10.1016/j.jhydrol.2019.124412

Mohammed, M., K. Alqahtani, N., Munir, M., & A. Eltawil, M. (2024). Applications of ai and iot for advancing date palm cultivation in saudi arabia. *Internet of Things - New Insights*. .DOI: 10.5772/intechopen.113175

Mondal, S., & Palit, D. (2022). Challenges in natural resource management for ecological sustainability. In *Natural Resources Conservation and Advances for Sustainability* (pp. 29–59). Elsevier., https://www.sciencedirect.com/science/article/pii/B9780128229767000041 DOI: 10.1016/B978-0-12-822976-7.00004-1

Monday, C., Zaghloul, M. S., Krishnamurthy, D., & Achari, G. (2024). A Review of AI-Driven Control Strategies in the Activated Sludge Process with Emphasis on Aeration Control. *Water (Basel)*, 16(2), 305. DOI: 10.3390/w16020305

Montalvo, I., Izquierdo, J., Pérez-García, R., & Tung, M. (2008). Particle Swarm Optimization applied to the design of water supply systems. *Computers & Mathematics with Applications (Oxford, England)*, 56(3), 769–776. DOI: 10.1016/j.camwa.2008.02.006

Morejón, M. A. E., Manobanda, M. E. B., Velasco, S. R. M., & Ballesteros, H. F. V. (2023). Implementación del geoportal para la junta administradora de agua potable y saneamiento – jaapys - santa fé centro, año 2023. *ConcienciaDigital*, 6(2.1), 6-23. .DOI: 10.33262/concienciadigital.v6i2.1.2553

Morgan, K. T., & Kadyampakeni, D. M. (2020). Nutrition of Florida citrus trees. *University of Florida Institute of Food and Agricultural Sciences: Gainesville, FL, USA, 27*. https://journals.flvc.org/edis/article/download/117225/115342/170141

Mosavi, A., Hosseini, F., Choubin, B., Goodarzi, M., Dineva, A., & Sardooi, E. (2020). Ensemble Boosting and Bagging Based Machine Learning Models for Groundwater Potential Prediction. *Water Resources Management*, 35(1), 23–37. DOI: 10.1007/s11269-020-02704-3

Mounce, S. R., Boxall, J. B., & Khan, A. (2016). Novel application of genetic algorithms for optimization of water distribution network design. *Water Resources Research*, 52(6), 4379–4394. DOI: 10.1002/2016WR018896

Mowla, M., Mowla, N., Shah, A. F. M. S., Rabie, K. M., & Shongwe, T. (2023). Internet of Things and Wireless Sensor Networks for Smart Agriculture Applications: A Survey. *IEEE Access : Practical Innovations, Open Solutions*, 11, 145813–145852. DOI: 10.1109/ACCESS.2023.3346299

Msiza, I., Nelwamondo, F., & Marwala, T. (2008). Water Demand Prediction using Artificial Neural Networks and Support Vector Regression. *Journal of Computers*, 3(11), 1–8. DOI: 10.4304/jcp.3.11.1-8

Munteanu, C., Teoibas-Serban, D., Iordache, L., Balaurea, M., & Blendea, C. D. (2021). Water intake meets the Water from inside the human body—Physiological, cultural, and health perspectives-Synthetic and Systematic literature review. *Balneo and PRM Research Journal*, 12(3), 196–209. DOI: 10.12680/balneo.2021.439

Mustafa, H., Mustapha, A., Hayder, G., & Salisu, A. (2021). Applications of IoT and Artificial Intelligence in Water Quality Monitoring and Prediction: A Review. 2021 6th International Conference on Inventive Computation Technologies (ICICT), 968-975. https://doi.org/DOI: 10.1109/ICICT50816.2021.9358675

Naim, A., Aaroud, A., Akodadi, K., & El Hachimi, C. (2021). A fully AI-based system to automate water meter data collection in Morocco country. *Array (New York, N.Y.)*, 10, 100056. DOI: 10.1016/j.array.2021.100056

Najah, A., El-shafie, A., Karim, O., Jaafar, O., & El-Shafie, A. (2011). An application of different artificial intelligences techniques for water quality prediction. *International Journal of Physical Sciences*, 6, 5298–5308. DOI: 10.5897/IJPS11.1180

Narendar, S. D., Murugamani, C., Kshirsagar, P. R., Tirth, V., Islam, S., Qaiyum, S., Suneela, B., Al Duhayyim, M., & Waji, Y. A. (2022, February). IoT based smart wastewater treatment model for Industry 4.0 using artificial intelligence. *Scientific Programming*, 2022, 1–11.

Nasir, N., Kansal, A., Alshaltone, O., Barneih, F., Sameer, M., Shanableh, A., & Al-Shamma'a, A. (2022). Water quality classification using machine learning algorithms. *Journal of Water Process Engineering*, 48, 102920. DOI: 10.1016/j.jwpe.2022.102920

Negm, A., Ma, X., & Aggidis, G. (2023). *Water Pressure Optimisation for Leakage Management Using Q Learning*. DOI: 10.1109/CAI54212.2023.00120

Ni, J., & Zhenxiang, X. (2022). Using Artificial Intelligence for Smart Water Management *Water*(Special Issue). https://www.mdpi.com/journal/water/special_issues/Artificial_Intelligence_Water

Nikoloska, R., Bykerk, L., Vitanage, D., Miró, J. V., Chen, F., Wang, Y., & Liang, B. (2020). Enhancing sydney water's leak prevention through acoustic monitoring. *Water E-Journal*, 5(2), 1–15. DOI: 10.21139/wej.2020.014

Nikolov, G., & Nikolova, B. (2008). *Virtual Techniques For Liquid Level Monitoring Using Differential Pressure Sensors. 9*(2).

Nirupama, N. (2023). Integrated Aquaculture as an Element of Blue-Green Infrastructure to Enhance Sustainability. *Opus 4*.https://opus4.kobv.de/opus4-hof/files/184/Nair_Nirupama_MA.pdf

Nishant, R., Kennedy, M., & Corbett, J. (2020). Artificial intelligence for sustainability: Challenges, opportunities, and a research agenda. *International Journal of Information Management*, 53, 102104. DOI: 10.1016/j.ijinfomgt.2020.102104

Noori, R., Farahani, F., Jun, C., Aradpour, S., Bateni, S. M., Ghazban, F., Hosseinzadeh, M., Maghrebi, M., Naseh, M. R. V., & Abolfathi, S. (2022). A nonthreshold model to estimate carcinogenic risk of nitrate-nitrite in drinking water. *Journal of Cleaner Production*, 363(Aug), 132432. DOI: 10.1016/j.jclepro.2022.132432

Nordin, N., Mohd, N., Koting, S., Ismail, Z., Sherif, M., & El-Shafie, A. (2021). Groundwater quality forecasting modelling using artificial intelligence: A review. *Groundwater for Sustainable Development*, 14, 100643. Advance online publication. DOI: 10.1016/j.gsd.2021.100643

Nova, K. (2023). AI-Enabled Water Management Systems: An Analysis of System Components and Interdependencies for Water Conservation. *Eigenpub Review of Science and Technology, 7*(1), 105-124. https://studies.eigenpub.com/index.php/erst/article/view/12

O'Reilly, G., Bezuidenhout, C. C., & Bezuidenhout, J. J. (2018). Artificial neural networks: Applications in the drinking water sector. *Water Science and Technology: Water Supply*, 18(6), 1869–1887. DOI: 10.2166/ws.2018.016

Okoli, N. J. (2023). Smart water metering system (swms) adoption: a systematic literature review. *International Conference on Artificial Intelligence and Its Applications*, 2023, 172-180. DOI: 10.59200/ICARTI.2023.025

Okoli, N. J., & Kabaso, B. (2024). Building a smart water city: Iot smart water technologies, applications, and future directions. *Water (Basel)*, 16(4), 557. DOI: 10.3390/w16040557

Oliveira, L., Gonçalves, O., & Uchida, C. (2009). Performance evaluation of dual-flush WC cistern in a multifamily building in Brazil. *Building Services Engineering Research and Technology*, 30(1), 27–36. DOI: 10.1177/0143624408098445

Oyebode, O., & Ighravwe, D. (2019). Urban Water Demand Forecasting: A Comparative Evaluation of Conventional and Soft Computing Techniques. *Resources*, 8(3), 156. Advance online publication. DOI: 10.3390/resources8030156

Pacchin, E., Gagliardi, F., Alvisi, S., & Franchini, M. (2019). A comparison of short-term water demand forecasting models. *Water Resources Management*, 33(4), 1481–1497. DOI: 10.1007/s11269-019-02213-y

Padhiary, M., Kumar, R., & Sethi, L. N. (2024). Navigating the Future of Agriculture: A Comprehensive Review of Automatic All-Terrain Vehicles in Precision Farming. *Journal of The Institution of Engineers (India): Series A, 105*(2). DOI: 10.1007/s40030-024-00816-2

Padhiary, M. (2024). Status of Farm Automation, Advances, Trends, and Scope in India. [IJSR]. *International Journal of Science and Research (Raipur, India)*, 13(7), 737–745. DOI: 10.21275/SR24713184513

Padhiary, M., Barbhuiya, J. A., Roy, D., & Roy, P. (2024). 3D Printing Applications in Smart Farming and Food Processing. *Smart Agricultural Technology*, 100553, 100553. Advance online publication. DOI: 10.1016/j.atech.2024.100553

Padhiary, M., & Kumar, R. (2024). Enhancing Agriculture Through AI Vision and Machine Learning: The Evolution of Smart Farming. In Thangam, D. (Ed.), *Advancements in Intelligent Process Automation* (pp. 295–324). IGI Global Scientific Publishing.

Padhiary, M., Kyndiah, A. K., Kumar, R., & Saha, D. (2024). Exploration of electrode materials for in-situ soil fertilizer concentration measurement by electrochemical method. *International Journal of Advanced Biochemistry Research*, 8(4), 539–544. DOI: 10.33545/26174693.2024.v8.i4g.1011

Padhiary, M., Rani, N., Saha, D., Barbhuiya, J. A., & Sethi, L. N. (2023). Efficient Precision Agriculture with Python-based Raspberry Pi Image Processing for Real-Time Plant Target Identification. *International Journal of Research and Analytical Reviews*, 10(3), 539–545. http://doi.one/10.1729/Journal.35531

Padhiary, M., & Roy, P. (2024). Advancements in Precision Agriculture: Exploring the Role of 3D Printing in Designing All-Terrain Vehicles for Farming Applications. [IJSR]. *International Journal of Science and Research (Raipur, India)*, 13(5), 861–868. DOI: 10.21275/SR24511105508

Padhiary, M., Saha, D., Kumar, R., Sethi, L. N., & Kumar, A. (2024). Enhancing Precision Agriculture: A Comprehensive Review of Machine Learning and AI Vision Applications in All-Terrain Vehicle for Farm Automation. *Smart Agricultural Technology*, 8, 100483. DOI: 10.1016/j.atech.2024.100483

Padhiary, M., Sethi, L. N., & Kumar, A. (2024). Enhancing Hill Farming Efficiency Using Unmanned Agricultural Vehicles: A Comprehensive Review. *Transactions of the Indian National Academy of Engineering : an International Journal of Engineering and Technology*, 9(2), 253–268. DOI: 10.1007/s41403-024-00458-7

Padhiary, M., Tikute, S. V., Saha, D., Barbhuiya, J. A., & Sethi, L. N. (2024). Development of an IOT-Based Semi-Autonomous Vehicle Sprayer. *Agricultural Research*. Advance online publication. DOI: 10.1007/s40003-024-00760-4

Pagano, C., Navarra, G., Coppola, L., Savarese, B., Avilia, G., Giarra, A., Pagano, G., Marano, A., Trifuoggi, M., Bifulco, M., & Laezza, C. (2023). Impacts of Environmental Pollution on Brain Tumorigenesis. *International Journal of Molecular Sciences*, 24(5), 5045. DOI: 10.3390/ijms24055045 PMID: 36902485

Pahl-Wostl, C., Tàbara, D., Bouwen, R., Craps, M., Dewulf, A., Mostert, E., Ridder, D., & Taillieu, T. (2008). The importance of social learning and culture for sustainable water management. *Ecological Economics*, 64(3), 484–495. DOI: 10.1016/j.ecolecon.2007.08.007

Pakeerathan, K. (2023). Image Processing: A Smart Technology for Early Detection of Crop Pests and Diseases. In Pakeerathan, K. (Ed.), *Smart Agriculture for Developing Nations* (pp. 135–150). Springer Nature Singapore., DOI: 10.1007/978-981-19-8738-0_10

Palaniappan, P., & Choo, K. K. R. (2020). Applications of artificial intelligence in water quality monitoring and control. *Journal of Computational Science*, 41, 101092. DOI: 10.1016/j.jocs.2019.101092

Palansooriya, K. N., Dissanayake, P. D., Igalavithana, A. D., Tang, R., Cai, Y., & Chang, S. X. (2023). Converting food waste into soil amendments for improving soil sustainability and crop productivity: A review. *The Science of the Total Environment*, 881, 163311. DOI: 10.1016/j.scitotenv.2023.163311 PMID: 37044338

Paredes Paucar, H. A. (2022). *Diseño de un prototipo de nodo inteligente integrable a una red inalámbrica para telemedición del consumo de agua potable como propuesta para el GADM de Guano.*http://dspace.espoch.edu.ec/handle/123456789/20774

Park, S., Lee, J., & Kang, D. (2021). Self-organizing neural networks for water usage optimization in smart buildings. *Automation in Construction*, 122, 103458. DOI: 10.1016/j.autcon.2020.103458

Pashaei, R., Dzingelevičienė, R., Abbasi, S., Szultka-Młyńska, M., & Buszewski, B. (2022). Determination of the pharmaceuticals–nano/microplastics in aquatic systems by analytical and instrumental methods. *Environmental Monitoring and Assessment*, 194(2), 1–33. DOI: 10.1007/s10661-022-09751-w

Pathak, S., Shukla, A. K., Ojha, C. S. P., & Garg, R. D. Identifying hot spots for stormwater harvesting and reuse in urban areas applying remote sensing, GIS and system optimization techniques. *IEEE International Geoscience and Remote Sensing Symposium (IGARSS)*, pp 967-970, 2016. DOI: 10.1109/IGARSS.2016.7729245

Paudel, D., Boogaard, H., de Wit, A., Janssen, S., Osinga, S., Pylianidis, C., & Athanasiadis, I. N. (2021). Machine learning for large-scale crop yield forecasting. *Agricultural Systems*, 187, 103016. DOI: 10.1016/j.agsy.2020.103016

Paul, D., Goswami, A., Chetri, R., Roy, R., & Sen, P. (2022). Bayesian Optimization-Based Gradient Boosting Method of Fault Detection in Oil-Immersed Transformer and Reactors. *IEEE Transactions on Industry Applications*, 58(2), 1910–1919. DOI: 10.1109/TIA.2021.3134140

Pelletier, G. J., Chapra, S. C., & Tao, H. (2005). *QUAL2Kw—A framework for modeling water quality in streams and rivers using a genetic algorithm for calibration*. https://apps.ecology.wa.gov/publications/summarypages/0503044.html

Penereiro, J. C., & Araujo, G. de S. (2018). Proposal of using sensor networks for measurements of the efficiencies of a low cost vertical wind turbine. *Acta Scientiarum. Technology*, 40(1), 35490. https://www.redalyc.org/journal/3032/303258327039/. DOI: 10.4025/actascitechnol.v40i1.35490

Pepperl & Fuchs. (2020, octubre 14). *Ultrasonic Sensors Knowledge (Part 6) – Synchronizing Ultrasonic Sensors*. Pepperl+Fuchs. https://www.pepperl-fuchs.com/mexico/es/27967.htm

Perea, R. G., Poyato, E. C., Montesinos, P., & Díaz, J. A. A. (2016). Optimización de la predicción de demanda de agua mediante algoritmos neuro-genéticos para un conjunto de datos reducido. *XXXIV Congreso Nacional De Riegos*. DOI: 10.21151/CNRiegos.2016.C18

Perea, R., Poyato, E., Montesinos, P., & Díaz, J. (2015). Irrigation Demand Forecasting Using Artificial Neuro-Genetic Networks. *Water Resources Management*, 29(15), 5551–5567. DOI: 10.1007/s11269-015-1134-4

Pereira, L. S., Paredes, P., & Jovanovic, N. (2020). Soil water balance models for determining crop water and irrigation requirements and irrigation scheduling focusing on the FAO56 method and the dual Kc approach. *Agricultural Water Management*, 241, 106357. DOI: 10.1016/j.agwat.2020.106357

Pérez-Beltrán, C. H., Robles, A. D., Rodriguez, N. A., Ortega-Gavilán, F., & Jiménez-Carvelo, A. M. (2024). Artificial intelligence and water quality: From drinking water to wastewater. *Trends in Analytical Chemistry*, 172, 117597. DOI: 10.1016/j.trac.2024.117597

Pérez-Padillo, J., García Morillo, J., Ramirez-Faz, J., Torres Roldán, M., & Montesinos, P. (2020). Design and Implementation of a Pressure Monitoring System Based on IoT for Water Supply Networks. *Sensors (Basel)*, 20(15), 4247. DOI: 10.3390/s20154247 PMID: 32751617

Pesantez, J., Berglund, E., & Kaza, N. (2020). Smart meters data for modeling and forecasting water demand at the user-level. *Environmental Modelling & Software*, 125, 104633. DOI: 10.1016/j.envsoft.2020.104633

Pichs, R. (2012). *Recursos naturales, economía mundial y crisis ambiental*. Editorial Científico-Técnica.

Ponte, B., Fuente, D., Parreño, J., & Pino, R. (2016). Intelligent Decision Support System for Real-Time Water Demand Management. *Int. J. Comput. Intell. Syst.*, 9(1), 168–183. DOI: 10.1080/18756891.2016.1146533

Popartan, L. A., Cortés, À., Garrido-Baserba, M., Verdaguer, M., Poch, M., & Gibert, K. (2022). The digital revolution in the urban water cycle and its ethical–political implications: A critical perspective. *Applied Sciences (Basel, Switzerland)*, 12(5), 2511. DOI: 10.3390/app12052511

Pushpalatha, N., Gunapriya, D., Sri Sangeetha, R., Anbarasu, P., Rajan Babu, W., Premkumar, R., & Hemananth, B. (2022). Artificial Intelligence-Based Smart Water Bottle with Drinking Water Consumption Remainder and Water Quality Monitor [Patent]. https://patentscope.wipo.int/search/es/detail.jsf?docId=IN379762001&_cid=P20-LZCWA8-96371-1

Qi, J., Chang, N.-B., & Yang, Y. J. (2016). Development of a web-based decision support system for improved urban water demand forecasting and management. *Environmental Modelling & Software*, 76, 3–14. DOI: 10.1016/j.envsoft.2015.11.005

Qiu, Y., Ekström, S., Valverde-Pérez, B., Smets, B. F., Climent, J., Domingo-Félez, C., Cuenca, R. M., & Plósz, B. G. (2024). Numerical modelling of surface aeration and N2O emission in biological water resource recovery. *Water Research*, 255(May), 121398. DOI: 10.1016/j.watres.2024.121398 PMID: 38503179

Quorum Business Solutions. (2023). Quorum EnergySuite. Available online: https://www.quorumsoftware.com/solutions/ (accessed on 23 Jan, 2023)

Rahbaralam, M., Modesto, D., Card'us, J., Abdollahi, A., & Cucchietti, F. M. (2020). Predictive Analytics for Water Asset Management: Machine Learning and Survival Analysis. *ArXiv, abs/2007.03744*.

Rajak, P., Ganguly, A., Adhikary, S., & Bhattacharya, S. (2023). Internet of Things and smart sensors in agriculture: Scopes and challenges. *Journal of Agriculture and Food Research*, 14, 100776. DOI: 10.1016/j.jafr.2023.100776

Rajakumar, A., Anthony, A., & Kumar, V. (2021). Application of deep learning methods for urban water demand forecast modelling. https://doi.org/DOI: 10.5194/egusphere-egu21-9572

Raj, E. F. I., Appadurai, M., & Athiappan, K. (2021). Precision Farming in Modern Agriculture. In Choudhury, A., Biswas, A., Singh, T. P., & Ghosh, S. K. (Eds.), *Smart Agriculture Automation Using Advanced Technologies* (pp. 61–87). Springer Singapore., DOI: 10.1007/978-981-16-6124-2_4

Ramakrishnan, T., Mohan Gift, M. D., Chitradevi, S., Jegan, R., & Subha Hency Jose, P. Nagaraja, Rajneesh Sharma H.N., Selvakumar P., Sintayehu Mekuria Hailegiorgis, (2022). "Study of Numerous Resins Used in Polymer Matrix Composite Materials", Advances in Materials Science and Engineering, vol. 2022, Article ID 1088926, 8 pages, 2022. DOI: 10.1155/2022/1088926

Ramalingam, R. (2024). An Innovative Investigation on Predicting Forest Fire Using Machine Learning Approach. In Ouaissa, M., Ouaissa, M., Boulouard, Z., Iwendi, C., & Krichen, M. (Eds.), *AI and IoT for Proactive Disaster Management* (pp. 61–77). IGI Global., DOI: 10.4018/979-8-3693-3896-4.ch004

Rana, R., Kalia, A., Boora, A., Alfaisal, F. M., Alharbi, R. S., Berwal, P., Alam, S., Khan, M. A., & Qamar, O. (2023). Artificial Intelligence for Surface Water Quality Evaluation, Monitoring and Assessment. *Water (Basel)*, 15(22), 3919. DOI: 10.3390/w15223919

Rapp Alyson, H., Capener Annelise, M., & Sowby Robert, B. (2023). Adoption of Artificial Intelligence in Drinking Water Operations: A Survey of Progress in the United States. *Journal of Water Resources Planning and Management*, 149(7), 06023002. DOI: 10.1061/JWRMD5.WRENG-5870

Rathi, B. S., Kumar, P. S., & Vo, D.-V.-N. (2021). Critical review on hazardous pollutants in water environment: Occurrence, monitoring, fate, removal technologies and risk assessment. *The Science of the Total Environment*, 797(Nov), 149134. DOI: 10.1016/j.scitotenv.2021.149134 PMID: 34346357

Ravichandran, T., Gavahi, K., Ponnambalam, K., Burtea, V., & Mousavi, S. (2021). Ensemble-based machine learning approach for improved leak detection in water mains. *Journal of Hydroinformatics*, 23(2), 307–323. Advance online publication. DOI: 10.2166/hydro.2021.093

RealPars. (2023). *Turbine Flow Meter Explained | Operation and Calibration—RealPars*. https://www.realpars.com/blog/turbine-flow-meter

Rebahi, Y., Hilliger, B., Lowin, P., Zheng, B., Da Bormida, G., & Ladjeri, K. (2023). Towards AI-Based Condition Monitoring and Predictive Maintenance for Water Smart Pipes: The SANDMAN Approach. *Artificial Intelligence and Applications (Commerce, Calif.)*, 2(2), 100–110. DOI: 10.47852/bonviewAIA32021513

Renugadevi, R. (2024). Disease Monitoring and Management: Healthcare Information Systems (HIS) in Smart Cities. In Baral, S., & Goel, R. (Eds.), *Analyzing Current Digital Healthcare Trends Using Social Networks* (pp. 178–192). IGI Global., DOI: 10.4018/979-8-3693-1934-5.ch010

Renugadevi, R., & Rajeswari, M. A. R, R. Rajalakshmi and G. Ramkumar, "A Distinctive Explainable Deep Learning Framework to Predict Pulmonary Function at Chest CT Scan," *2023 First International Conference on Advances in Electrical, Electronics and Computational Intelligence (ICAEECI)*, Tiruchengode, India, 2023, pp. 1-7, DOI: 10.1109/ICAEECI58247.2023.10370935

Riahi, S., Nasrabadi, T., & Shahsavari, P. (2020). Multi-objective optimization of wastewater treatment plants using evolutionary algorithms. *Journal of Cleaner Production*, 246, 118964. DOI: 10.1016/j.jclepro.2019.118964

Richards, C. E., Tzachor, A., Avin, S., & Fenner, R. (2023). Rewards, risks and responsible deployment of artificial intelligence in water systems. *Nature Water*, 1(5), 422–432. DOI: 10.1038/s44221-023-00069-6

Richter, B. D., Abell, D., Bacha, E., Brauman, K., Calos, S., Cohn, A., Disla, C., O'Brien, S. F., Hodges, D., Kaiser, S., Loughran, M., Mestre, C., Reardon, M., & Siegfried, E. (2013). Tapped out: How can cities secure their water future? *Water Policy*, 15(3), 335–363. DOI: 10.2166/wp.2013.105

Rivero, R. M., Mittler, R., Blumwald, E., & Zandalinas, S. I. (2022). Developing climate-resilient crops: Improving plant tolerance to stress combination. *The Plant Journal*, 109(2), 373–389. DOI: 10.1111/tpj.15483 PMID: 34482588

Rodriguez, D. J., & Singh, J. (2020). Machine learning and AI for water quality monitoring and control. *Journal of Water Process Engineering*, 36, 101309. DOI: 10.1016/j.jwpe.2020.101309

Rogers, B. C., Dunn, G., Hammer, K., & Novalia, W. DE Haan FJ., Brown L., Brown R.R., Lloyd S., Urich C., Wong T.H.F., Chesterfield, C. Water Sensitive Cities Index: A diagnostic tool to assess water sensitivity and guide management actions. In Book Water research, Eberhard Morgenroth, Guang-Hao Chen, Ana Deletic, David Dixon, Xiaodi Hao, Peter Jarvis, Gregory Korshin, Hui Lu, Sun Yat-Sen, Yoshihiko Matsui, Eberhard Morgenroth, Ameet J. Pinto, Maria Reis, Hang-Sik Shin, Thomas Ternes, Stefan Wuertz, Zhiguao Yuan, Yan Zhou, Julie Zilles., Publisher: ELSEVIER Publisher location, Country, Amsterdam, Netherlands, Volume **186**, pp. 116411, **2020**.

Rojek, I. (2014). Models for Better Environmental Intelligent Management within Water Supply Systems. *Water Resources Management*, 28(12), 3875–3890. DOI: 10.1007/s11269-014-0654-7

Romano, M., & Kapelan, Z. (2014). Adaptive water demand forecasting for near real-time management of smart water distribution systems. *Environmental Modelling & Software*, 60, 265–276. DOI: 10.1016/j.envsoft.2014.06.016

Romero, G. (2024) Binomio IA - Agua: la revolución digital de la gestión hídrica, en Retema, Revista Técnica de Medio Ambiente https://drive.google.com/file/d/10dcNgc5YSGdDodxlslMjRLJ1Su8sqM4G/view

Romero, C., & Shan, J. (2005). Development of an artificial neural network-based software for prediction of power plant canal water discharge temperature. *Expert Systems with Applications*, 29(4), 831–838. DOI: 10.1016/j.eswa.2005.06.009

Ronzoni, M., Accorsi, R., Botti, L., & Manzini, R. (2021). A support-design framework for Cooperative Robots systems in labor-intensive manufacturing processes. *Journal of Manufacturing Systems*, 61, 646–657. DOI: 10.1016/j.jmsy.2021.10.008

Ruiz-Vanoye, J. A., Díaz-Parra, O., Ramos-Fernández, J. C., Trejo-Macotela, F. R., & Xicoténcatl-Pérez, J. M. (2023). Editorial: A Brief Panorama of Water Smart Cities. *International Journal of Combinatorial Optimization Problems and Informatics*, 14(1), 1. https://www.ijcopi.org/ojs/article/view/33. DOI: 10.61467/2007.1558.2023.v14i1.332

Sabamehr, M., & Ekin, C. Ç. (2023). AI-Driven Drought Management System: A Turkish Case Study. 2023 4th International Informatics and Software Engineering Conference (IISEC), DOI: 10.46908/rict.v3i1.67

Sabir, R. M., Sarwar, A., Shoaib, M., Saleem, A., Alhousain, M. H., Wajid, S. A., Rasul, F., Adnan Shahid, M., Anjum, L., Safdar, M., Muhammad, N. E., Waqas, R. M., Zafar, U., & Raza, A. (2024). Managing Water Resources for Sustainable Agricultural Production. In Kanga, S., Singh, S. K., Shevkani, K., Pathak, V., & Sajan, B. (Eds.), *Transforming Agricultural Management for a Sustainable Future* (pp. 47–74). Springer Nature Switzerland., DOI: 10.1007/978-3-031-63430-7_3

Saghiv, M. S., Sagiv, M. S., Saghiv, M. S., & Sagiv, M. S. (2020). Thermoregulation. *Basic Exercise Physiology: Clinical and Laboratory Perspectives*, 437-463.

Saha, D., Padhiary, M., Barbhuiya, J. A., Chakraborty, T., & Sethi, L. N. (2023). Development of an IOT based Solenoid Controlled Pressure Regulation System for Precision Sprayer. *International Journal for Research in Applied Science and Engineering Technology*, 11(7), 2210–2216. DOI: 10.22214/ijraset.2023.55103

Şahín, E., & Yüce, H. (2023). Prediction of water leakage in pipeline networks using graph convolutional network method. *Applied Sciences (Basel, Switzerland)*, 13(13), 7427. DOI: 10.3390/app13137427

Saiz-Rubio, V., & Rovira-Más, F. (2020). From smart farming towards agriculture 5.0: A review on crop data management. *Agronomy (Basel)*, 10(2), 207. DOI: 10.3390/agronomy10020207

Saleem, S. R., Zaman, Q. U., Schumann, A. W., & Abbas Naqvi, S. M. Z. (2023). Variable rate technologies. In *Precision Agriculture* (pp. 103–122). Elsevier., DOI: 10.1016/B978-0-443-18953-1.00010-6

Salloom, T., Kaynak, O., & He, W. (2021). A novel deep neural network architecture for real-time water demand forecasting. *Journal of Hydrology (Amsterdam)*, 126353, 126353. Advance online publication. DOI: 10.1016/j.jhydrol.2021.126353

Salvaris, M., Dean, D., & Tok, W. (2018). Training AI Models., 211-241. https://doi.org/.DOI: 10.1007/978-1-4842-3679-6_9

Samanta, P., Desai, A. V., Let, S., & Ghosh, S. K. (2019). Advanced porous materials for sensing, capture and detoxification of organic pollutants toward water remediation. *ACS Sustainable Chemistry & Engineering*, 7(8), 7456–7478. DOI: 10.1021/acssuschemeng.9b00155

Sanchez, E., Sanchez, E., Acosta, A., & González, E. (2021). Sistema de riego automatizado iot aplicado en el cultivo del pepino (cucumis sativus). Congreso Nacional De Ciencia Y Tecnología – APANAC, 523-529. .DOI: 10.33412/apanac.2021.3234

Saravanan, A., Kumar, P. S., Jeevanantham, S., Karishma, S., Tajsabreen, B., Yaashikaa, P. R., & Reshma, B. (2021). Effective water/wastewater treatment methodologies for toxic pollutants removal: Processes and applications towards sustainable development. *Chemosphere*, 280, 130595. DOI: 10.1016/j.chemosphere.2021.130595 PMID: 33940449

Sarma, H. (2022). Understanding emerging contaminants in soil and water: current perspectives on integrated remediation approaches. *Emerging Contaminants in the Environment*, 1-38.

Schmidt, M. (2022). El cañito de la red llega, lo que no llega es el agua. Hacia una ecología política del agua en el chaco salteño, argentina. *Territorios*, (46). Advance online publication. . DOI: 10.12804/revistas.urosario.edu.co/territorios/a.9250

Scornet, E., Biau, G., & Vert, J. (2014). Consistency of Random Forests. *Annals of Statistics*, 43(4), 1716–1741. DOI: 10.1214/15-AOS1321

Secretaría de Salud. (2016) https://www.gob.mx/salud/articulos/sabes-por-que-es -importante-consumir-agua

Selvakumar, P. (2024). Emerging Engineering Technologies: Design, Development, and Industrial Applications of IoT-Based Aquaculture Monitoring System. In El Kacimi, Y., & Alaoui, K. (Eds.), *Emerging Engineering Technologies and Industrial Applications* (pp. 332–341). IGI Global., DOI: 10.4018/979-8-3693-1335-0.ch016

Selvakumar, P.. (2024). AI-Based Cameras to Detect a Leak of Chemicals or Gas in Real Time. In Satishkumar, D., Sivaraja, M., & Global, I. G. I. (Eds.), *Industry Applications of Thrust Manufacturing: Convergence with Real-Time Data and AI* (pp. 221–236)., DOI: 10.4018/979-8-3693-4276-3.ch009

Selvakumar, P.. (2024). Prediction of Earthquakes, Volcanic Eruptions, Tornadoes, Wildfires, and Droughts Through Artificial Intelligence. In Satishkumar, D., Sivaraja, M., & Global, I. G. I. (Eds.), *Utilizing AI and Machine Learning for Natural Disaster Management* (pp. 133–145)., DOI: 10.4018/979-8-3693-3362-4.ch008

Selvakumar, P., Mariappan, P., Shunmugam, S., Paul, R., & Jabez, S. (2024). Removal of chemicals from effluent using photobioreactor technology to improve environmental and health impacts. *Zeitschrift für Physikalische Chemie*, 2024(7), 1295–1302. Advance online publication. DOI: 10.1515/zpch-2023-0450

Selvakumar, P., Muthusamy, S., Satishkumar, D., Vigneshkumar, P., Selvamurugan, C., & Satheesh Kumar, P. (2024). AI-Powered Tools. In Satishkumar, D., & Sivaraja, M. (Eds.), *Using Real-Time Data and AI for Thrust Manufacturing* (pp. 20–42). IGI Global., DOI: 10.4018/979-8-3693-2615-2.ch002

Selvakumar, P., Palanisamy, S. K., Cinthaikinian, S., Palanisamy, V., Mariappan, R., & Selvakumar, P. K. (2024). Biosensors and its diverse applications in healthcare systems. *Zeitschrift für Physikalische Chemie*, 2024. Advance online publication. DOI: 10.1515/zpch-2023-0406

Seo, Y., Choi, K., Lim, Y., Lee, B., & Choi, Y. (2023). Application of Machine Learning Models for Water Pipeline Leakage Detection. Crisis and Emergency Management: Theory and Praxis. https://doi.org/.DOI: 10.14251/crisisonomy.2023.19.4.45

Shafi, U., Mumtaz, R., García-Nieto, J., Hassan, S. A., Zaidi, S. A. R., & Iqbal, N. (2019). Precision agriculture techniques and practices: From considerations to applications. *Sensors (Basel)*, 19(17), 3796. DOI: 10.3390/s19173796 PMID: 31480709

Shahhosseini, M., Hu, G., & Pham, H. (2019). Optimizing Ensemble Weights and Hyperparameters of Machine Learning Models for Regression Problems. ArXiv, abs/1908.05287. https://doi.org/.DOI: 10.1016/j.mlwa.2022.100251

Shah, M. I., Alaloul, W. S., Alqahtani, A., Aldrees, A., Musarat, M. A., & Javed, M. F. (2021). Predictive Modeling Approach for Surface Water Quality: Development and Comparison of Machine Learning Models. *Sustainability (Basel)*, 13(14), 7515. DOI: 10.3390/su13147515

Shaik, A., & Srinivasan, S. (2018). A Brief Survey on Random Forest Ensembles in Classification Model. *International Conference on Innovative Computing and Communications*. https://doi.org/DOI: 10.1007/978-981-13-2354-6_27

Shaikh, M., & Birajdar, F. (2024). Artificial intelligence in groundwater management: Innovations, challenges, and future prospects. *International Journal of Science and Research Archive*, 11(1), 502–512. DOI: 10.30574/ijsra.2024.11.1.0105

Shaikh, M., & Birajdar, F. (2024). Harmony in hydroinformatics: Integrating ai and iec for sustainable groundwater conservation in solapur. *International Journal of Science and Research Archive*, 11(1), 2163–2175. DOI: 10.30574/ijsra.2024.11.1.0294

Shaikh, T. A., Rasool, T., & Lone, F. R. (2022). Towards leveraging the role of machine learning and artificial intelligence in precision agriculture and smart farming. *Computers and Electronics in Agriculture*, 198, 107119. DOI: 10.1016/j.compag.2022.107119

Sharma, A., Sharma, A., Tselykh, A., Bozhenyuk, A., Choudhury, T., Alomar, M. A., & Sánchez-Chero, M. (2023). Artificial intelligence and internet of things oriented sustainable precision farming: Towards modern agriculture. *Open Life Sciences*, 18(1), 20220713. DOI: 10.1515/biol-2022-0713 PMID: 37854322

Sharma, A., & Sivapragasam, M. (2021). Application of machine learning techniques in water quality assessment: A review. *Environmental Progress & Sustainable Energy*, 40(5), e13313. DOI: 10.1002/ep.13313

Sharma, R., Kamble, S. S., & Gunasekaran, A. (2018). Big GIS analytics framework for agriculture supply chains: A literature review identifying the current trends and future perspectives. *Computers and Electronics in Agriculture*, 155, 103–120. DOI: 10.1016/j.compag.2018.10.001

Shenbaga Lakshmi, G., Revathi T., & Kavi Priya Shanthosam. (2019, diciembre). *Fig. 6. System Architecture.* ResearchGate. https://www.researchgate.net/figure/System-Architecture_fig6_337733237

Shen, Y., & Cheng, W. (2022). A Tree-Based Machine Learning Method for Pipeline Leakage Detection. *Water (Basel)*, 14(18), 2833. Advance online publication. DOI: 10.3390/w14182833

Shravani, D., R, P., B, P., S, G., & G, S. (2019). Cloud-Based Water Leakage Detection and Localization. 2019 IEEE International Conference on Cloud Computing in Emerging Markets (CCEM), 86-91. https://doi.org/DOI: 10.1109/CCEM48484.2019.00018

Sidhu, R. K., Kumar, R., Rana, P. S., & Jat, M. L. (2021). Automation in drip irrigation for enhancing water use efficiency in cereal systems of South Asia: Status and prospects. *Advances in Agronomy*, 167, 247–300. DOI: 10.1016/bs.agron.2021.01.002

Simidjievski, N., Todorovski, L., & Džeroski, S. (2015). Predicting long-term population dynamics with bagging and boosting of process-based models. *Expert Systems with Applications*, 42(22), 8484–8496. DOI: 10.1016/j.eswa.2015.07.004

Singh, S. Amarjeet kaur & Ashmeet kaur (2022). Automatic Tank Cleaning Device [Patent]. https://patentscope.wipo.int/search/es/detail.jsf?docId=IN389950475&_cid=P20-LZCW81-94862-1

Singh, S., & Jain, P. (2022). Applications of Artificial Intelligence for the Development of Sustainable Agriculture. In Kumar, P., Tomar, R. S., Bhat, J. A., Dobriyal, M., & Rani, M. (Eds.), *Agro-biodiversity and Agri-ecosystem Management* (pp. 303–322). Springer Nature Singapore., DOI: 10.1007/978-981-19-0928-3_16

Singh, V. P., Khedikar, S., Mathew, J., & Garg, T. (2022). Harnessing Artificial Intelligence for Drought Management. In Gupta, R. K., Jain, A., Wang, J., Singh, V. P., & Bharti, S. (Eds.), *Artificial Intelligence of Things for Weather Forecasting and Climatic Behavioral Analysis* (pp. 130–143). IGI Global. DOI: 10.4018/978-1-6684-3981-4.ch009

Sishodia, R. P., Ray, R. L., & Singh, S. K. (2020). Applications of Remote Sensing in Precision Agriculture: A Review. *Remote Sensing (Basel)*, 12(19), 3136. DOI: 10.3390/rs12193136

Sitzenfrei, R., Wang, Q., Kapelan, Z., & Savić, D. (2020). Using Complex Network Analysis for Optimization of Water Distribution Networks. *Water Resources Research*, 56(8), e2020WR027929. Advance online publication. DOI: 10.1029/2020WR027929 PMID: 32999510

Smart Sensor, C. (2024). *Comate meter | Flow measurement for compressed air, gases, steam*. Comate Flow Meter. https://www.comateflowmeter.com/

Smith, A. B., & Johnson, C. D. (2020). Advances in water quality monitoring systems. *Environmental Science and Pollution Research International*, 27(4), 4193–4206. DOI: 10.1007/s11356-019-07059-5

Smith, A., Johnson, M., & Turner, C. (2019). Real-time water quality monitoring using distributed sensor networks and artificial neural networks. *Journal of Water Resources Planning and Management*, 145(6), 04019030. DOI: 10.1061/(ASCE) WR.1943-5452.0001060

Sokolov, A. (2022). On expressive abilities of ensembles of decision trees. Proceedings of Academician O.B. Lupanov 14th International Scientific Seminar "Discrete Mathematics and Its Applications". https://doi.org/DOI: 10.20948/dms-2022-72

Soleimani-Chamkhorami, K., Karbalaie, A., Kasraei, A., Haghighi, E., Famurewa, S. M., & Garmabaki, A. H. S. (2024). Identifying Climate-Related Failures in Railway Infrastructure Using Machine Learning. *Transportation Research Part D, Transport and Environment*, 135, 104371. DOI: 10.1016/j.trd.2024.104371

Somayaji, S. R. K., Nallakaruppan, M. K., Chengoden, R., Koppu, S., Iyapparaja, M., Sadhasivam, J., & Sethuraman, S. (2022). Smart water resource management using artificial intelligence—A review. *Sustainability (Basel)*, 14(20), 13384. DOI: 10.3390/su142013384

Song, G., Ifaei, P., Ha, J., Kang, D., Won, W., Liu, J. J., & Na, J. (2024). The AI Circular Hydrogen Economist: Hydrogen Supply Chain Design via Hierarchical Deep Multi-Agent Reinforcement Learning. *Chemical Engineering Journal*, 497, 154464. DOI: 10.1016/j.cej.2024.154464

Spellman, F. R. (2013). *Water & wastewater infrastructure: Energy efficiency and sustainability*. Crc Press. DOI: 10.1201/b13968

Srinivas, R., Singh, A. P., & Shankar, D. (2020). Understanding the threats and challenges concerning Ganges River basin for effective policy recommendations towards sustainable development. *Environment, Development and Sustainability*, 22(4), 3655–3690. DOI: 10.1007/s10668-019-00361-0

Suárez. Patricia (2024), "¿Existe vida en otros planetas?" Gaceta UNAM (5489) https://www.gaceta.unam.mx/g20240725

Subeesh, A., & Mehta, C. R. (2021a). Automation and digitization of agriculture using artificial intelligence and internet of things. *Artificial Intelligence in Agriculture*, 5, 278–291. DOI: 10.1016/j.aiia.2021.11.004

Suchetana, B., Srivastava, B., Gupta, H., & Saharia, M. (2023). Promoting Sustainable Water Usage and Management With Water Data, AI and Policy. Proceedings of the 6th Joint International Conference on Data Science & Management of Data (10th ACM IKDD CODS and 28th COMAD). https://doi.org/DOI: 10.1145/3570991.3571021

Sugumar, S. J., Sahana, R., Phadke, S., Prasad, S., & Srilakshmi, G. R. ''Real time water treatment plant monitoring system using IoT and machine learning approach,'' in Proc. Int. Conf. Design Innov. 3Cs Compute Communicate Control (ICDI3C), Jun. 2021, pp. 286–289. DOI: 10.1109/ICDI3C53598.2021.00064

Sullivan, L. P. (1986). *Quality Function Deployment. Quality Progress*. Springer.

Sumana, B., & Santhanam, T. (2015). Optimizing the Prediction of Bagging and Boosting. *Indian Journal of Science and Technology*, 8(35), 1–13. DOI: 10.17485/ijst/2015/v8i35/78449

Sun, A., & Scanlon, B. (2019). How can Big Data and machine learning benefit environment and water management: A survey of methods, applications, and future directions. *Environmental Research Letters*, 14(7), 073001. Advance online publication. DOI: 10.1088/1748-9326/ab1b7d

Sun, Y., Yeh, W., Hsu, N., & Louie, P. (1995). Generalized Network Algorithm for Water-Supply-System Optimization. *Journal of Water Resources Planning and Management*, 121, 392–398. DOI: 10.1061/(ASCE)0733-9496(1995)121:5(392)

Sun, Z., Zhang, Q., & Wang, X. (2019). River flow prediction using recurrent neural networks for enhanced water management. *Water Resources Research*, 55(3), 2547–2563. DOI: 10.1029/2018WR024567

Surendran, U., Raja, P., Jayakumar, M., & Subramoniam, S. R. (2021). Use of efficient water saving techniques for production of rice in India under climate change scenario: A critical review. *Journal of Cleaner Production*, 309, 127272. DOI: 10.1016/j.jclepro.2021.127272

Su, W.-H. (2020). Crop plant signaling for real-time plant identification in smart farm: A systematic review and new concept in artificial intelligence for automated weed control. *Artificial Intelligence in Agriculture*, 4, 262–271. DOI: 10.1016/j.aiia.2020.11.001

Talaviya, T., Shah, D., Patel, N., Yagnik, H., & Shah, M. (2020). Implementation of artificial intelligence in agriculture for optimisation of irrigation and application of pesticides and herbicides. *Artificial Intelligence in Agriculture*, 4, 58–73. DOI: 10.1016/j.aiia.2020.04.002

Talekar, B. (2020). A Detailed Review on Decision Tree and Random Forest. *Bioscience Biotechnology Research Communications*, 13(14), 245–248. Advance online publication. DOI: 10.21786/bbrc/13.14/57

Tang, Z. Q., Liu, G., Xu, W. N., Xia, Z. Y., & Xiao, H. (2012). Water demand forecasting in Hubei province with BP neural network model. *Advanced Materials Research*, 599, 701–704. . DOI: 10.4028/www.scientific.net/AMR.599.701

Tan, Y., & Cugurullo, F. (2020). The Sustainability of Artificial Intelligence: An Urbanistic Viewpoint from the Lens of Smart and Sustainable Cities. *Sustainability (Basel)*, 12(20), 8548. DOI: 10.3390/su12208548

Tao, M., Zhang, T., Xie, X., & Liang, X. (2023). Water footprint modeling and forecasting of cassava based on different artificial intelligence algorithms in Guangxi, China. *Journal of Cleaner Production*, 382, 135238. DOI: 10.1016/j.jclepro.2022.135238

Tecnologías para el uso sostenible del agua: una contribución a la seguridad alimentaria y la adaptación al cambio climático. Available online: https://www.fao.org/3/i3442s/i3442s.pdf (accessed on 21 Jan 2023).

Thompson, T. M., & Kim, J. (2021). Artificial intelligence applications in water quality monitoring: A review. *Water Research*, 196, 116979. DOI: 10.1016/j.watres.2021.116979

Tinelli, S., & Juran, I. (2019). Artificial intelligence-based monitoring system of water quality parameters for early detection of non-specific bio-contamination in water distribution systems. *Water Science and Technology: Water Supply*, 19(6), 1785–1792. DOI: 10.2166/ws.2019.057

Tinoco, C., Julio, N., Meirelles, B., Pineda, R., Figueroa, R., Urrutia, R., & Parra, Ó. (2022). Water Resources Management in Mexico, Chile and Brazil: Comparative Analysis of Their Progress on SDG 6.5. 1 and the Role of Governance. *Sustainability (Basel)*, 14(10), 5814. DOI: 10.3390/su14105814

Tiwari, M., Adamowski, J., & Adamowski, K. (2016). Water demand forecasting using extreme learning machines. *Journal of Water and Land Development*, 28(1), 37–52. DOI: 10.1515/jwld-2016-0004

Tomlinson, J., Arnott, J., & Harou, J. (2020). A water resource simulator in Python. *Environmental Modelling & Software*, 126, 104635. DOI: 10.1016/j.envsoft.2020.104635

Tosin Michael, O., Fatai Adeshina, A., & Zamathula Queen, S. (2024). A review of smart water management systems from Africa and the United States. *Engineering Science &Technology Journal*, 5(4), 1231–1242. DOI: 10.51594/estj.v5i4.1014

Tsai, Y., Chang, H., Lin, S., Chiou, A., & Lee, T. (2022). Using convolutional neural networks in the development of a water pipe leakage and location identification system. *Applied Sciences (Basel, Switzerland)*, 12(16), 8034. DOI: 10.3390/app12168034

Türkoğlu, M., & Hanbay, D. (2019). Plant disease and pest detection using deep learning-based features. *Turkish Journal of Electrical Engineering and Computer Sciences*, 27(3), 1636–1651. DOI: 10.3906/elk-1809-181

Tyagi, V., Pandey, P., Jain, S., & Ramachandran, P. (2023). A two-stage model for data-driven leakage detection and localization in water distribution networks. *Water (Basel)*, 15(15), 2710. DOI: 10.3390/w15152710

Tylman, W., Kolczynski, J., & Anders, G. (2010). Fully automatic AI-based leak detection system. *Energy*, 35(9), 3838–3848. DOI: 10.1016/j.energy.2010.05.038

Ugran, V., Desai, N. N., Chakraborti, D., Masali, K. A., Mantur, P., Kulkarni, S., Deshmukh, N., Chadchan, K. S., Das, S. N., Tanksali, A. S., Arwikar, A. S., Guggarigoudar, S. P., Vallabha, T., Patil, S. S., & Das, K. K. (2017). Groundwater fluoride contamination and its possible health implications in Indi taluk of Vijayapura District (Karnataka State), India. *Environmental Geochemistry and Health*, 39(5), 1017–1029. DOI: 10.1007/s10653-016-9869-2 PMID: 27591763

Uhlenbrook, S., Yu, W., Schmitter, P., & Smith, D. M. (2022). Optimising the water we eat—Rethinking policy to enhance productive and sustainable use of water in agri-food systems across scales. *The Lancet. Planetary Health*, 6(1), e59–e65. DOI: 10.1016/S2542-5196(21)00264-3 PMID: 34998461

UN. (2024). Goal 6: Ensure access to water and sanitation for all. https://www.un.org/sustainabledevelopment/water-and-sanitation/

United Nations. (2023). *UN 2023 Water Conference* UN 2023 Water Conference, New York. https://sdgs.un.org/conferences/water2023

Valdez, J. E., Pandolfi, D., & Villagra, A. (2018). Red de sensores inteligentes para adquisición de datos de una planta de desalinización de agua. Informes Científicos Técnicos - UNPA, 10(2), 83-95. .DOI: 10.22305/ict-unpa.v10i2.272

Vanijjirattikhan, R., Khomsay, S., Kitbutrawat, N., Khomsay, K., Supakchukul, U., Udomsuk, S., Suwatthikul, J., Oumtrakul, N., & Anusart, K. (2022). AI-based acoustic leak detection in water distribution systems. *Results in Engineering*, 15, 100557. DOI: 10.1016/j.rineng.2022.100557

Varona, J., Tecpoyotl-Torres, M., & Velázquez, R. (2013). Micro sensor-actuador térmico sin baterías para aplicaciones en microelectrónica de ultra-bajo consumo de potencia. *Revista Mexicana de Física*, 59(1), 26–38.

Vijayakumar, G., Rajkumar, M., Rajiv Chandar, N., Selvakumar, P., & Duraisamy, R. (2024). Muniyandi Rajkumar, Rajiv Chandar N, Selvakumar P and Ramesh Duraisamy, (2024). Environmentally friendly TDS removal from waste water by electrochemical ion exchange batch-type recirculation (EIR) technique. *Environmental Science. Water Research & Technology*, 10(4), 826–835. DOI: 10.1039/D3EW00793F

Vijayaragavan, D., Praveen, N., Sudharsan, M., & Vijayan, P. (2022). Machine Learning Model for Water Quality Prediction using Python and AI framework. International Journal of Advanced Research in Science. *Tongxin Jishu*. Advance online publication. DOI: 10.48175/IJARSCT-3749

Vijila, J., & Sahaya Vasanthi, S. (2022). Artificial intelligence-based drinking water purification system with Graphene Nano-material using Machine Learning and Wireless Sensor Networks [Patent]. https://patentscope.wipo.int/search/es/detail.jsf?docId=IN355391255&_cid=P20-LZCWA8-96371-1

Villajulca, J. C. (2009, septiembre 27). Caudal - Flujo: Medidores por desplazamiento positivo. *Instrumentacion y Automatizacion Industrial*. https://instrumentacionycontrol.net/caudal-flujo-medidores-por-desplazamiento-positivo/

Villarin, M. C., & Rodriguez-Galiano, V. F. (2019). Machine learning for modeling water demand. *Journal of Water Resources Planning and Management*, 145(5), 04019017. DOI: 10.1061/(ASCE)WR.1943-5452.0001067

Wan, X., Yang, X., Yu, L., Yang, Y., Hao, F., & Ting, T. (2023). Rural safe drinking water monitoring system [Patent]. https://patentscope.wipo.int/search/es/detail.jsf?docId=CN422248651&_cid=P20-LZCWA8-96371-1

Wang, C., Li, Z., Ni, X., Shi, W., Zhang, J., Bian, J., & Liu, Y. (2023). Residential water and energy consumption prediction at hourly resolution based on a hybrid machine learning approach. *Water Research*, 246, 120733. DOI: 10.1016/j.watres.2023.120733 PMID: 37879283

Wang, F., Zhu, X., & Jiang, Q. (2015). IoT-based agricultural water management using evolutionary optimization algorithms. *Computers and Electronics in Agriculture*, 118, 56–63. DOI: 10.1016/j.compag.2015.08.007

Wang, J., & Azam, W. (2024). Natural resource scarcity, fossil fuel energy consumption, and total greenhouse gas emissions in top emitting countries. *Geoscience Frontiers*, 15(2), 101757. DOI: 10.1016/j.gsf.2023.101757

Wang, J., & Liu, Y. (2016). RVM-BASED urban hourly water consumption forecasting model and its application. *Journal of Simulation*, 4, 37.

Wang, L., & Chen, C. (2021). Advances in artificial intelligence techniques for water quality prediction and control. *Journal of Hydrology and Hydromechanics*, 69(1), 29–40. DOI: 10.2478/johh-2021-0003

Wang, L., Wu, J., Zhang, W., Wang, L., & Cui, W. (2021). Efficient Seismic Stability Analysis of Embankment Slopes Subjected to Water Level Changes Using Gradient Boosting Algorithms. *Frontiers in Earth Science (Lausanne)*, 9, 807317. Advance online publication. DOI: 10.3389/feart.2021.807317

Wang, X. C., & Fu, G. (2021). *Water-Wise Cities and Sustainable Water Systems: Concepts, Technologies, and Applications*. IWA Publishing.

Wang, Y., Cheng, Y., Liu, H., Guo, Q., Dai, C., Zhao, M., & Liu, D. (2023). A review on applications of artificial intelligence in wastewater treatment. *Sustainability (Basel)*, 15(18), 13557. Advance online publication. DOI: 10.3390/su151813557

Wang, Z., Zhang, Y., & Ma, J. (2021). Application of machine learning techniques in water quality management: A review. *The Science of the Total Environment*, 758, 143607. DOI: 10.1016/j.scitotenv.2020.143607

Wang, Z., Zhou, X., Wang, H., Huang, Z., Ji, J., Peng, Z., & Jiang, K. (2024). XGB-SEGA coupled energy saving method for wastewater treatment plants. *Applied Water Science*, 14(2), 29. DOI: 10.1007/s13201-023-02081-3

Water Sensitive Cities Australia. Available online: http://URL https://wscaustralia.org.au/2022/08/09/water-sensitive-cities-australia-our-first-12-months/ (accessed on 11 May 2023).

Wei, S., & Huang, G. (2021). Review on machine learning applications in water quality monitoring and control. *Water Science and Technology*, 84(7), 1522–1537. DOI: 10.2166/wst.2021.358 PMID: 33843740

West, D., & Mangiameli, P. (2000). Identifying process conditions in an urban wastewater treatment plant. *International Journal of Operations & Production Management*, 20(5), 573–590. DOI: 10.1108/01443570010318931

Whang, S., Roh, Y., Song, H., & Lee, J. (2021). Data collection and quality challenges in deep learning: A data-centric AI perspective. *The VLDB Journal*, •••, 1–23. DOI: 10.1007/s00778-022-00775-9

William, P., Oyebode, O., Ramu, G., Gupta, M., Alzubaidi, L., & Srivastava, A. (2023). Architecture for AI-Based Validation of Wastewater Management Using Open Data Exchange Technique. *2023 International Conference on Circuit Power and Computing Technologies (ICCPCT)*, 1502-1507. https://doi.org/DOI: 10.1109/ICCPCT58313.2023.10244857

Williams, H., Colombi, T., & Keller, T. (2020). The influence of soil management on soil health: An on-farm study in southern Sweden. *Geoderma*, 360, 114010. DOI: 10.1016/j.geoderma.2019.114010

WIPO. (2024). WIPO – Patentscope. https://patentscope.wipo.int/search/es/search.jsf

Wongburi, P., & Park, J. (2023). *Prediction of Wastewater Treatment Plant Effluent Water Quality Using Recurrent Neural Network (RNN)*. Models. Water., DOI: 10.3390/w15193325

Wongburi, P., & Park, J. K. (2022, May). Prediction of sludge volume index in a wastewater treatment plant using recurrent neural network. *Sustainability (Basel)*, 14(10), 6276. DOI: 10.3390/su14106276

Wu, H., He, S., & Zhang, W. (2019). Prediction of industrial wastewater contamination using decision trees. *Environmental Monitoring and Assessment*, 191(11), 679. DOI: 10.1007/s10661-019-7884-4 PMID: 31655913

Wu, Q., & Liu, Y. (2021). Machine learning in water quality monitoring and control: A review. *Journal of Environmental Management*, 297, 113334. DOI: 10.1016/j.jenvman.2021.113334

Wu, S., Han, H., Hou, B., & Diao, K. (2020). Hybrid model for short-term water demand forecasting based on error correction using chaotic time series. *Water (Basel)*, 12(6), 1683. DOI: 10.3390/w12061683

Wu, X., & Li, X. (2021). Application of artificial intelligence in water quality assessment and prediction based on multiple data sources. *Water (Basel)*, 13(21), 3048. DOI: 10.3390/w13213048

Xiang, X., Li, Q., Khan, S., & Khalaf, O. (2021). Urban water resource management for sustainable environment planning using artificial intelligence techniques. *Environmental Impact Assessment Review*, 86, 106515. DOI: 10.1016/j.eiar.2020.106515

Xiao, C., Lu, B., Jing, X., Wu, W., & Zheng, H. (2018). Drinking water distribution system and method based on artificial intelligence [Patent]. https://patentscope.wipo.int/search/es/detail.jsf?docId=CN215328653&_cid=P20-LZCWA8-96371-1

Xing, Y., You, B., & Yang, M. (2017). City Water Demand Forecasting Based on Improved BP Neural Network. *Journal of Residuals Science & Technology*, 14(S1), S111–S117. Advance online publication. DOI: 10.12783/issn.1544-8053/14/S1/15

Xu, T. Y., & Qin, X. S. (2015). A sequential fuzzy model with general-shaped parameters for water supply-demand analysis. *Water Resources Management*, 29(5), 1431–1446. DOI: 10.1007/s11269-014-0884-8

Xu, Y., Zhang, J., Long, Z., & Chen, Y. (2018). A Novel Dual-Scale Deep Belief Network Method for Daily Urban Water Demand Forecasting. *Energies*, 11(5), 1068. DOI: 10.3390/en11051068

Yacelga, A. R. L., Vásquez, R. A. D., Espinoza, J. L. A., & Cabrera, M. A. C. (2023). Implementación de prototipo de medición del consumo de agua potable mediante utilización de tecnologías inteligentes en la parroquia Julio Andrad, Ecuador. *Universidad y Sociedad*, 15(S2), S2.

Yang, L., Feng, Q., Lu, T., Adamowski, J. F., Yin, Z., Hatami, S., Zhu, M., & Wen, X. (2023). The response of agroecosystem water use efficiency to cropland change in northwest China's Hexi Corridor. *Agricultural Water Management*, 276, 108062. DOI: 10.1016/j.agwat.2022.108062

Yaseen, Z., Deo, R., Ebtehaj, I., & Bonakdari, H. (2018). Hybrid Data Intelligent Models and Applications for Water Level Prediction., 121-139. https://doi.org/.DOI: 10.4018/978-1-5225-4766-2.ch006

Yaseen, Z., El-Shafie, A., Jaafar, O., Afan, H., & Sayl, K. (2015). Artificial intelligence based models for stream-flow forecasting: 2000-2015. *Journal of Hydrology (Amsterdam)*, 530, 829–844. DOI: 10.1016/j.jhydrol.2015.10.038

Yetilmezsoy, K., Ozkaya, B., & Cakmakci, M. (2011). ARTIFICIAL INTELLIGENCE-BASED PREDICTION MODELS FOR ENVIRONMENTAL ENGINEERING. *Neural Network World*, 21(3), 193–218. DOI: 10.14311/NNW.2011.21.012

Zaidi, A., Ajibade, S. S., Musa, M., & Bekun, R. V. (2023). New Insights into the Research Landscape on the Application of Artificial Intelligence in Sustainable Smart Cities. *International Journal of Energy Economics and Policy*, 13(4), 287–299. DOI: 10.32479/ijeep.14683

Zamagni, S. (2014). Bienes comunes y economía civil. V32, No 87 (2014): Bien común, justicia y educación. https://erevistas.uca.edu.ar/index.php/CECON/issue/view/140

Zancan, C., Passador, J., & Souza, C. (2023). *Brazilian Intermunicipal Water and Sewage Consortia Management: Harnessing the Power of Artificial Intelligence*. International Journal of Business & Management Studies., DOI: 10.56734/ijbms.v4n8a5

Zapata, G., & Tejeda, I. (2018). Publicidad engañosa en educación superior: Definiciones, regulación y una revisión de la publicidad en prensa escrita. *Calidad en la Educación*, (44), 197–242. DOI: 10.31619/caledu.n44.37

Zhang, J., Huang, Y., Pu, R., Gonzalez-Moreno, P., Yuan, L., Wu, K., & Huang, W. (2019). Monitoring plant diseases and pests through remote sensing technology: A review. *Computers and Electronics in Agriculture*, 165, 104943. DOI: 10.1016/j.compag.2019.104943

Zhang, J., Li, H., & Huang, Z. (2021). Modeling water flow in intelligent irrigation systems using artificial neural networks. *Journal of Water Resources Management*, 35(2), 451–466. DOI: 10.1007/s11269-020-02702-w

Zhang, L., Liu, Y., Ren, L., Teuling, A. J., Zhu, Y., Wei, L., Zhang, L., Jiang, S., Yang, X., Fang, X., & Yin, H. (2022). Analysis of flash droughts in China using machine learning. *Hydrology and Earth System Sciences*, 26(12), 3241–3261. DOI: 10.5194/hess-26-3241-2022

Zhang, L., & Zou, Y. (2019). A survey on artificial intelligence in water quality monitoring and management. *Water Resources Management*, 33(2), 527–544. DOI: 10.1007/s11269-018-2127-5

Zhang, P., Yang, M., Lan, J., Huang, Y., Zhang, J., Huang, S., Yang, Y., & Ru, J. (2023). Water Quality Degradation Due to Heavy Metal Contamination: Health Impacts and Eco-Friendly Approaches for Heavy Metal Remediation. *Toxics*, 11(10), 828. DOI: 10.3390/toxics11100828 PMID: 37888679

Zhang, Q., Zhang, W., & Fang, L. (1998). Application of artificial neural networks in river flow forecasting. *Water Resources Management*, 12(3), 185–198. DOI: 10.1023/A:1007900110595

Zhang, X., Beeson, P., Link, R., Manowitz, D., Izaurralde, R., Sadeghi, A., Thomson, A., Sahajpal, R., Srinivasan, R., & Arnold, J. (2013). Efficient multi-objective calibration of a computationally intensive hydrologic model with parallel computing software in Python. *Environmental Modelling & Software*, 46, 208–218. DOI: 10.1016/j.envsoft.2013.03.013

Zhang, X., Wang, H., & Li, J. (2018). Remote sensing-based detection of water contaminants using convolutional neural networks. *Remote Sensing of Environment*, 213, 382–392. DOI: 10.3390/rs10030382

Zhang, Y., Gao, Y., Liu, Q. S., Zhou, Q., & Jiang, G. (2024). Chemical Contaminants in Blood and Their Implications in Chronic Diseases. *Journal of Hazardous Materials*, 466, 133511. DOI: 10.1016/j.jhazmat.2024.133511 PMID: 38262316

Zhang, Y., & Wang, H. (2022). A review on artificial intelligence in water quality prediction and optimization. *Water Resources and Economics*, 38, 100184. DOI: 10.1016/j.wre.2021.100184

Zhao, L., Dai, T., Qiao, Z., Sun, P., Hao, J., & Yang, Y. (2020). Application of artificial intelligence to wastewater treatment: A bibliometric analysis and systematic review of technology, economy, management, and wastewater reuse. *Process Safety and Environmental Protection*, 133, 169–182. DOI: 10.1016/j.psep.2019.11.014

Zhao, X., Wu, Y. L., & Cui, W. (2019). iForest: Interpreting Random Forests via Visual Analytics. *IEEE Transactions on Visualization and Computer Graphics*, 25(1), 407–416. DOI: 10.1109/TVCG.2018.2864475 PMID: 30188822

Zhong, C. (2017). Domestic smart water dispenser based on artificial intelligence [Patent]. https://patentscope.wipo.int/search/es/detail.jsf?docId=CN207671726&_cid=P20-LZCWA8-96371-1

Zhou, S., McMahon, T. A., Walton, A., & Lewis, J. (2012). Forecasting operational demand for an urban water supply zone. *Water Resources Management*, 26(14), 4147–4163. DOI: 10.1007/s11269-012-0130-8

Zhou, Y., Guo, W., & Li, Z. (2020). Bayesian networks for risk assessment in potable water supply systems. *Journal of Risk Research*, 23(5), 655–672. DOI: 10.1080/13669877.2019.1626002

Zhou, Z., & Huang, G. (2021). Machine learning for water quality prediction: Challenges, strategies, and applications. *Science China. Technological Sciences*, 64(6), 1044–1060. DOI: 10.1007/s11431-020-1685-7

Zhu, S., Zhao, N., & Sha, J. (2019). Predicting battery life with early cyclic data by machine learning. *Energy Storage*, 1(6), e98. Advance online publication. DOI: 10.1002/est2.98

Zubaidi, S. L., Ortega-Martorell, S., Al-Bugharbee, H., Olier, I., Hashim, K. S., Gharghan, S. K., Kot, P., & Al-Khaddar, R. (2020c). Urban water demand prediction for a city that suffers from climate change and population growth: Gauteng province case study. *Water (Basel)*, 12(7), 1885. DOI: 10.3390/w12071885

Zyl, J., Savić, D., & Walters, G. (2004). Operational Optimization of Water Distribution Systems using a Hybrid Genetic Algorithm. *Journal of Water Resources Planning and Management*, 130(2), 160–170. DOI: 10.1061/(ASCE)0733-9496(2004)130:2(160)

About the Contributors

Jorge Ruiz-Vanoye is a Professor and Researcher at Universidad Politécnica de Pachuca (UPP), a National Researcher under the CONAHCYT program, and an Adjunct Researcher at the National Laboratory of Autonomous Vehicles and Exoskeletons (LANAVEX). Additionally, he holds the title of Lecturer-level Researcher for the Agency for the Quality of the University System of Catalonia (Generalitat de Catalunya). His research interests include Smart Cities, Smart Water, Smart Education, Smart Government, Smart Healthcare, Smart Farming, Smart Energy, Smart Sports, Smart Food, Smart Tourism, Algorithmic Finance, Algorithmic Portfolio Management, Applications and Theory of Algorithms, Bio-inspired Algorithms, Combinatorial Optimization Problems, Compilers, Computational Intelligence, Computational Complexity Theory, Computational Financial Intelligence, Complexity of Algorithms, Complexity of Instances, Computational Statistics, Computer Networks, Computer Science Security, Cybercrimes, Data Mining, Education, Evolutionary Computation, Heuristic Optimization Techniques in Bioinformatics, Hybrid Evolutionary Algorithms, Hybrid Optimization Algorithms, Machine Learning, Meta-heuristics, Operations Research, Operations Management, Parallel & Distributed Computing, Production Planning and Logistics Optimization, Project Scheduling Problems, Software Engineering, and Transgenic Algorithms.

Octolan Diaz-Parra is a Professor and Researcher at Universidad Politécnica de Pachuca (UPP), a National Researcher under the CONAHCYT program, and an Adjunct Researcher at the National Laboratory of Autonomous Vehicles and Exoskeletons (LANAVEX). His research interests include Smart Cities, Smart Water, Smart Education, Smart Government, Smart Healthcare, Smart Farming, Smart Energy, Smart Sports, Smart Food, Smart Tourism, Algorithmic Finance, Algorithmic Portfolio Management, Applications and Theory of Algorithms, Bio-inspired Algorithms, Combinatorial Optimization Problems, Compilers,

Computational Intelligence, Computational Complexity Theory, Computational Financial Intelligence, Complexity of Algorithms, Complexity of Instances, Computational Statistics, Computer Networks, Computer Science Security, Cybercrimes, Data Mining, Education, Evolutionary Computation, Heuristic Optimization Techniques in Bioinformatics, Hybrid Evolutionary Algorithms, Hybrid Optimization Algorithms, Machine Learning, Meta-heuristics, Operations Research, Operations Management, Parallel & Distributed Computing, Production Planning and Logistics Optimization, Project Scheduling Problems, Software Engineering, and Transgenic Algorithms.

***

Jaime Aguilar-Ortiz has a diverse background in mathematics and artificial intelligence, holding a Ph.D. in Mathematical Sciences from the Centro de Investigación y Estudios Avanzados del Instituto Politécnico Nacional de México and multiple masters from Tecnológico de Monterrey. His career includes significant contributions to educational software development, authoring numerous educational books and developing over 60 educational films. He has also been a part of various innovative projects receiving recognition and funding from multiple Mexican governmental bodies. Currently, he serves as the General Director at CICAINTE and TEDNOVA GROUP, and is a tenured research professor at Universidad Politécnica de Pachuca. His ongoing professional development includes participation in several international academic programs, most recently at Harvard University.

Shalom Akhai has established himself as a prominent figure in the field of engineering, leveraging over 17 years of academic, research, and technical experience to make significant contributions. With a robust portfolio of over 50 publications in prestigious journals and conferences, Dr. Akhai demonstrates a commitment to advancing knowledge through scholarly work. Additionally, he has authored/co-authored several technical textbooks, further solidifying his expertise and dedication to education in the field. Beyond academia, Dr. Akhai's impact extends to real-world applications through the establishment of IGNIS Technical Solutions, a consultancy firm to translate his knowledge into real-world applications. His skills have resulted in the successful development of over 50 patents+, underscoring his innovative approach and influence in shaping the engineering landscape. Moreover, Dr. Akhai's involvement in professional societies, notably as a member of the Indian Society of Heating Ventilation and Air Conditioning Engineers (ISHRAE) Chandigarh Chapter, reflects his commitment to fostering collaboration and knowledge exchange within the society. As an influential figure, Dr. Akhai continues to drive progress and innovation, leaving an indelible mark on the field of engineering landscape.

Ricardo Barrera-Cámara is a PhD in Computer Systems from Universidad del Sur. Professor of Computer Science at the Universidad Autónoma del Carmen. Research Professor at the Universidad Autónoma del Carmen in Mexico. Member of the National System of Researchers Level 1. Leader of the academic body of scientific and technological informatics. Leader of the Academy of Management and Information Processing. His research areas are Scientific Computing, Data Science, Optimization, Quality, Sustainable Development Goals.

Sumanta Bhattacharya from Salkia, Howrah, West Bengal is an Expert commissioner member IUCN Environmental, economics and social policy, Individual Policy Analyst (ADB) . He is among the top 2% world scientist (AD Index), Special civilian member Ministry of Defence, (USI) Member at Ministry of Culture (AS) . Member of Ministry of Defence { [Defence and strategic analysis] IDSA }, Member of World Economic Forum, littérateur at National Economic Research Bureau, Member of Ministry of Personnel, Pension and Public grievance (IIPA) He completed his B.Tech, M.Tech in Textile Technology along with that he has MA in Development studies, M.Sc in Environmental Science, LLB, MA in security and defence law from NALSAR, MA in Economics, MBA, Post Graduate Diploma in Environment and Sustainable Development, DIA&D, DS&DS, MPI(Oxford University) . Being a research scholar he has 216 research papers published in International and Scopus index journal, Wiley, Springer, Hindawi in various sectors like law, economic, Development studies, Textiles, Climate Change, SDGs, Public policy, Nanotechnology, Biotechnology, cancer, environment studies, 57 Book chapters published, Edited 2 books, wrote 1 book, 80 patents International and National, 4 copyrights, attended 93 International and National conference and presented papers and won 25 awards for his excellent in education, policy making and innovation which includes Best Water Management Solution Award 2022, Young Researcher Award 2020, Bravo International Book of Records, Mahatma Gandhi Ratna Award 2020, Sustainable Technology Award 2022, Global Change Maker Award 2020, Best Water Management Solution Award at the Rural & Urban Development Summit & Awards 2022, New Delhi, The Progress Global Awards 2023 for Excellence in Research, GIS -Person of the year – Water Conservation, Green India award, 2023, Global Legend Award 2023 by GSIL Educational Society . He is also a member of various International and National organization which includes IEI, IICHE, Asiatic Society, IIPA, IDSA, YOUNGO, IUCN, USI and many more .

Abhijeet Das is currently working as a Research Scholar in the Department of Civil Engineering at C.V. Raman Global University (CGU), Bhubaneswar, Odisha, India. His area of specialization is Water Resource Engineering. His research interests focus on watershed management, Water quality analysis, hydrological forecasting, Machine learning assessment, hydrologic modelling, and developing sustainable means of managing the environment. During his academic journey, I gained practical experience as an intern in a water-based company, working on projects related to MCDM tools, AI, ML, and GIS. However, he has been awarded with the “Young Research Scholar Award” at Society for Ecological Sustainability in association of Department of Environmental Science PAH Solapur University, Maharashtra, India.

Rafael Alfonso Figueroa-Díaz received PhD degree in Mechanical Engineering in National Center for Research and Technological Development, Cuernavaca, México, in 2012. His made a master science in mechatronics in 2007. His current research interests in modal balancing in rotating systems and fault diagnosis using vibration.

Alejandro Fuentes Penna is a distinguished Mexican academic and researcher specializing in strategic planning, artificial intelligence, and educational technology. He earned his Ph.D. in Strategic Planning and Technology Management from the Universidad Popular Autónoma del Estado de Puebla (UPAEP) in 2012. Prior to that, he completed a Master's degree in Computer Science at the Instituto Tecnológico y de Estudios Superiores de Monterrey (ITESM), where he published two articles in international conferences and secured first place in the Technology category of the Third Banamex Award for Internet Evolution. Dr. Fuentes Penna has held various academic positions, including serving as a Full-Time Research Professor at the Escuela Superior de Tlahuelilpan of the Universidad Autónoma del Estado de Hidalgo (UAEH). His research interests encompass educational management, educational technology, complexity in social sciences, and bio-inspired intelligent scientific computing. He has contributed to multiple publications, including articles indexed in JCR and Scopus, as well as books in his areas of expertise. Throughout his career, Dr. Fuentes Penna has received numerous national and international recognitions for academic and research excellence. He has been awarded the PRODEP Profile (formerly PROMEP) three times, is a member of the State System of Researchers in Morelos, and was recently included in the National System of Researchers at the Candidate level. Additionally, he has collaborated with various universities and research institutes, including the Universidad Nacional Autónoma de México (UNAM) and the Mexican Institute of Transportation. Dr. Fuentes Penna has extensive experience in teaching at both undergraduate and graduate levels, in both face-to-face and online modalities. He has supervised theses, coordinated educational programs, and organized significant academic events such as the International Space Week and the National Science and Technology Week.

Raúl Gómez Cárdenas is a Mexican academic affiliated with El Colegio de Morelos. He holds a degree in Economics, a Master's in Administration, and a Ph.D. in Higher Education. With extensive experience in municipal and federal public sectors, as well as in higher education teaching, he currently serves as a full-time professor at El Colegio de Morelos, located in Cuernavaca, Morelos. His research areas include economics, political and social sciences, administration, and education. In October 2022, Dr. Gómez Cárdenas was appointed President of the Ethics Committee at El Colegio de Morelos. In this role, he has promoted training initiatives and the dissemination of a culture of integrity within the institution. In July 2024, he was recognized for his dedication and talent by being incorporated as a candidate into the National System of Researchers and Investigators of the National Council of Humanities, Sciences, and Technologies (CONAHCYT). Dr. Gómez Cárdenas has also contributed to academic literature. In May 2024, he published an article titled "Higher Education and Artificial Intelligence: Transitioning from Teaching to Tutoring" in the International Journal of Studies on Educational Systems (RIESED). This work discusses the current challenges in contemporary teaching, emphasizing the impact of new information technologies and artificial intelligence on education.

Sandeep Gupta is presently working as an Associate Professor at Graphic Era (Deemed to be University), Dehradun (India). He has a total of 15 years of experience. His areas of interest in research are the application of artificial intelligence to power system control design, FACTS devices, renewable energy, power electronics, and stability of power systems with machine learning. He is a member of numerous professional organizations, including IE, IRED, IDES, IAENG, IAER, and the Society of Professional Engineers. He has been the author or co-author of more than 90 papers published in international journals and presented at different national and international conferences. He is currently guiding the research scholars under Machine Learning, Control & Power Systems, Wireless Technology, and the Internet of Things.

Azmirul Hoque is a research scholar at the Department of Agricultural Engineering, Triguna Sen School of Technology, Assam University, Silchar, India. His research areas are artificial intelligence, machine learning, and refractance window drying and food processing.

Kundan Kumar is a researcher at the Department of Agricultural Engineering, Triguna Sen School of Technology, Assam University, Silchar, India. His research areas are artificial intelligence, machine learning, automation, and agriculture.

T.C. Manjunath is a distinguished Professor in the Department of Computer Science and Engineering at Rajarajeswari College of Engineering, Bangalore. He specializes in IoT Cybersecurity and Blockchain Technology and serves as the Dean of Research & Development. With a strong background in advanced technologies, Dr. Manjunath has contributed significantly to the academic and research communities. His work continues to bridge the gap between theory and practical application, particularly in the realms of cybersecurity and emerging technologies. Located in Kumbalgodu, Bangalore, he is actively involved in promoting cutting-edge research and fostering innovation within his field.

Francisco Marroquín-Gutiérrez is a distinguished academic and researcher affiliated with the Universidad Politécnica de Pachuca. He holds extensive expertise in advanced technologies, particularly in fields such as data analytics, artificial intelligence, and computational systems. Dr Marroquín-Gutiérrez has contributed significantly to the development of innovative methodologies aimed at enhancing smart city frameworks and sustainable technological solutions. His research spans interdisciplinary domains, combining theoretical foundations with practical applications to address complex societal challenges. Dr Marroquín-Gutiérrez is also an advocate for academic collaboration, regularly participating in international conferences and contributing to high-impact scientific publications. His commitment to education is reflected in his role as a mentor, guiding postgraduate students in their pursuit of cutting-edge research. At the Universidad Politécnica de Pachuca, he continues to spearhead initiatives that bridge the gap between academia and industry, fostering technological advancement and societal progress.

P. Selvakumar is currently an Associate Professor in the Department of Science and Humanities at Nehru Institute of Technology, Coimbatore, Tamil Nadu, India. Since 2023, he has served as the Department IQAC Coordinator at the institute. Prior to joining Nehru Institute of Technology, Dr. Selvakumar held positions as an Assistant Professor in the Department of Science and Humanities at Info Institute of Engineering and Dr. N.G.P Institute of Technology, both located in Coimbatore, Tamil Nadu, India, spanning from 2011 to 2023. Dr. Selvakumar's academic and research interests lie primarily in the field of phytochemistry. He has made significant contributions to his field, with an impressive record that includes 30 journal publications, 12 patents, and 20 conference papers. Additionally, he has organized 5 seminars and conferences, further demonstrating his commitment to academic and scientific exchange. Throughout his career, Dr. Selvakumar has been recognized for his dedication and excellence in teaching, having been honored with several awards, including the Best Teacher Awards. He holds a prominent position in the academic community, contributing actively to research.

Mrutyunjay Padhiary is an Assistant Professor in the Department of Agricultural Engineering at Assam University, Silchar, India, with a deep academic background. His previous positions included significant teaching and research duties at the Faculty of Technology, UBKV in West Bengal, and Tribhuvan University in Nepal, as well as research experience at IIT Kharagpur. He has been awarded multiple awards, including an International NUS Research Fellowship and an excellent score on the 2001 GATE Exam, demonstrating his proficiency in precision agriculture. He has made substantial contributions to biofuel production, artificial intelligence vision, aquaculture, precision agriculture, agricultural automation, and artificial intelligence. He is currently working on several research projects, which range from autonomous surface vessels to precision agriculture all-terrain vehicles.

Gajendra Prasad is currently employed as an Assistant Professor in the Department of Agricultural Engineering at the Triguna Sen School of Technology, Assam University, Silchar, India. He has earned a Master of Technology (M.Tech.) in Dairy and Food Engineering from IIT Kharagpur, India. His areas of interest include the Internet of Things, Food Processing, and modeling. He possess a comprehensive understanding of topics such as food packaging, refrigeration, preservation techniques, and milk processing. His work encompass academic teachings, research contributions, or industry expertise in food safety, automation, and emerging technologies in dairy and food engineering. His work concentrate on the development of advanced food processing techniques, automation, smart sensors, and IoT applications for the purpose of real-time monitoring, quality control, and supply chain optimization.

Julio Cesar Ramos-Fernåndez obtained his PhD degree in the international programme ECOS-NORD/ANUIES--SEP-CONAHCYT (France-Mexico). His scientific research interests are in the modelling and control of dynamic systems, with application to precision agriculture. Modelling in disease prediction based on machine learning with neuro-fuzzy networks.

R. Renugadevi completed her PhD in 2022 from Anna University, Chennai. .She is currently working as Associate professor in Saveetha Engineering College. She has more than 16 years of teaching experience in engineering colleges.She published more than 30 papers in international journals, conferences and holds 8 patents. Her research interest includes Machine learning, Internet of Things, Wireless Sensor Networks and Cloud computing. She is a life member of ISTE, CSI and IAENG.

Daniel Robles-Camarillo is a Professor and Researcher at the Polytechnic University of Pachuca (UPP), Mexico. He earned his PhD in Communications and Electronics from the National Polytechnic Institute (IPN) in 2011. At UPP, he is affiliated with the Master's programme in Information and Communication Technologies and the PhD programme in Advanced Sciences and Technologies. His research interests encompass embedded systems applications, electronic instrumentation and communication, wireless sensor networks, and cyber-physical systems, complemented by machine learning algorithms. Dr Robles-Camarillo has contributed to various academic publications, including: "The Effect of Transcorneal Electrical Stimulation in Visual Acuity: Retinitis Pigmentosa" published in the Journal of Biomedical Science and Engineering. "Adaptive Electrical Stimulation to Improve In-Vitro Cell Growth" featured in Procedia Technology by Elsevier. He has also been involved in international conferences, presenting works such as "Telecontrol de Procesos Operado a través de una Aplicación Web" at the Ibero-American Multidisciplinary Symposium on Science and Engineering (SIMCI) 2013. In recognition of his contributions, Dr Robles-Camarillo received the Hidalgo Award for Science, Technology, and Innovation in 2014, in the category of Technological Development, for his work on a transcorneal electrical stimulator aimed at treating visual impairments.

Julio C. Salgado-Ramírez is a highly respected academic and researcher at the Universidad Politécnica de Pachuca. With a strong background in information and communication technologies, his work focuses on innovative solutions for real-world challenges, particularly in areas such as advanced algorithms, cybersecurity, and the optimisation of computational systems. Dr Salgado-Ramírez has earned recognition for his contributions to applied research and his dedication to advancing knowledge in cutting-edge technological domains. In addition to his research, Dr Salgado-Ramírez is deeply committed to fostering academic excellence. He actively collaborates with industry leaders and academic peers to promote technological innovation and sustainable development. As an educator, he inspires students to engage in high-impact research, equipping them with the skills and insights necessary to address emerging challenges in the field. Through his leadership and expertise, Dr Salgado-Ramírez continues to drive meaningful progress within the academic and technological communities.

Anayeli Silva Aguilar is a distinguished Mexican academic affiliated with El Colegio de Morelos. She has held significant roles within the institution, notably serving as the Coordinator of Research and Secretary of the Academic Council. In the 2024-2 semester, Dr. Silva Aguilar is scheduled to lead several seminars, including: Digital Tools for Research Design and Development of Educational Resources with Generative Artificial Intelligence Learning, Interaction, and Evaluation in Face-to-Face, Virtual, and Hybrid Environments Beyond her contributions to El Colegio de Morelos, Dr. Silva Aguilar has been associated with the Universidad Autónoma del Estado de Morelos (UAEM). She has contributed to the journal "Derecho y Globalización," published by UAEM's Faculty of Law and Social Sciences. Dr. Silva Aguilar's research interests encompass educational methodologies and digital learning environments. She has explored topics such as the integration of instructional design elements in e-learning courses, emphasizing the importance of teacher presence in virtual settings. Her academic endeavors reflect a commitment to advancing educational practices through the incorporation of technology and innovative teaching strategies.

Eric Simancas Acevedo (Phd. in Electro-communications), graduated from the Postgraduate Studies and Research Section (SEPI-ESIME Culhuacan) of the National Polytechnic Institute (IPN) in 2015; he had a postgraduate study stays in Tokyo, Japan 2001-2003; He is currently a Senior Research Professor and Posgraduate teacher at the Polytechnic University of Pachuca in Engineering and Graduate degrees. His research interests are: Signal and image processing, artificial intelligence, neuroengineering, adaptive filters, neural networks, fuzzy logic, stochastic models, embedded systems, machine learning, security and control systems based on biometric features and embedded systems.

Ratan Rajan Srivastava is a distinguished academician currently serving as an Assistant Professor in the Department of Computer Science & Engineering at Shri Ramswaroop Memorial College of Engineering & Management, Lucknow . With a wealth of experience spanning over 16 years in academia, he has established himself as a respected figure in the field of computer science. His expertise encompasses a range of subjects, with a primary focus on Machine Learning, Deep Learning, Big Data, Analysis of Algorithms, Computer Networks, and Cyber Security. His commitment to advancing knowledge in these areas is evident through his extensive academic contributions and research pursuits.His work has been recognized in both national and international journals, & patents showcasing the depth of his understanding and the significance of his research contributions.

Tanu Taneja, a dedicated professional, holds a B.Tech in Civil Engineering (2012) and an M.Tech in Environment Science & Engineering (2015). Currently pursuing a Ph.D. at RIMT University, Tanu has over 11 years of teaching experience along with 9 years of proven leadership as Head of Civil Engineering at Baba Farid College of Engineering & Technology, Bathinda. Her expertise extends to consultancy, water testing, concrete testing, and structural design. Beyond her technical knowledge, Dr. Taneja possesses strong leadership and administrative skills and has actively participated in various administrative roles, including serving as the Dean of Admissions. She has been instrumental in organizing conferences, workshops, and faculty development programs. Her commitment to academia is evident through her publications, patents, and awards, showcasing her unwavering dedication to professional growth. As a valued member of the Indian Society for Technical Education and the Council of Engineers & Valuers organization, Tanu remains actively engaged in the technical community, further enriching her impressive career.

Francisco Trejo-Macotela is a Professor and Researcher at Polytechnic University of Pachuca (UPP), a National Researcher under the CONAHCYT program, and an Adjunct Researcher at the National Laboratory of Autonomous Vehicles and Exoskeletons (LANAVEX). Dr. Trejo-Macotela received the PhD. Degree in Electronics from the National Institute of Astrophysics, Optics and Electronics (INAOE), Puebla, Mexico. His main research interests are Integrated Circuit Design, Analog and Digital Design, Embedded Systems, RF Electronics, Telecommunications Systems, Precision Agriculture, Internet of Things, among others.

Juan Xicotencatl-Perez received a Ph.D. degree from the National Institute of Astrophysics of Optics and Electronics in the area of electronics, interested in embedded systems, computer vision, and FPGAs. Currently working in the mechatronics department at the Polytechnic University of Pachuca

Index

A

B

C

D

E

F

H

I

L

M

N

P

R

S

U

W

www.ingramcontent.com/pod-product-compliance
Lightning Source LLC
LaVergne TN
LVHW080610300926
849162LV00005B/1103